AF752156

This
latest dat
another
be returned to th
charged if it is kep

14 JUN 1988
30 SEP 1988
31 JAN 1990
22 FEB 1990
26 APR 1990
16 MAY 1990
30 JAN 1991
CANCELLED

VOLUME 41

IN THE SOCIETY OF GENERAL PHYSIOLOGISTS SERIES

PROTEINS OF EXCITABLE MEMBRANES

PROTEINS OF EXCITABLE MEMBRANES

Editors:

BERTIL HILLE
Department of Physiology and Biophysics
University of Washington
Seattle, Washington

DOUGLAS M. FAMBROUGH
Department of Biology
The John Hopkins University
Baltimore, Maryland

Society of General Physiologists and

WILEY-INTERSCIENCE

New York • Chichester • Brisbane • Toronto • Singapore

574.875 P

UNIVERSITY COLLEGE
LIBRARY
CARDIFF

Copyright © 1987 by John Wiley & Sons, Inc.

All rights reserved. Published simultaneously in Canada.

Reproduction or translation of any part of this work beyond that permitted by Section 107 or 108 of the 1976 United States Copyright Act without the permission of the copyright owner is unlawful. Requests for permission or further information should be addressed to the Permissions Department, John Wiley & Sons, Inc.

Library of Congress Cataloging in Publication Data:

Proteins of excitable membranes.
(Society of General Physiologists series ; v. 41)
Based on a symposium held in 1985.
Includes index.
1. Membrane proteins—Congresses. 2. Acetycholine—Receptors—Congresses. 3. Adenosine triphosphatase—Congresses. I. Hille, Bertil, 1940-
II. Fambrough, Douglas M. III. Society of General Physiologists. IV. Series. [DNLM: 1. Membrane Proteins—physiology—congresses. W1 S0872G v.41 / QU 55 P96955 1985]
QP552.M44P76 1987 591.87′5 86-18914
ISBN 0-471-01150-9

42

Printed in the United States of America

10 9 8 7 6 5 4 3 2 1

CONTRIBUTORS

Robert L. Barchi Mahoney Institute of Neurological Sciences and Departments of Neurology and of Biochemistry and Biophysics, University of Pennsylvania School of Medicine, Philadelphia, Pennsylvania

Christopher J. Brandl Banting and Best Department of Medical Research, C. H. Best Institute, University of Toronto, Toronto, Ontario, Canada

A. Buonanno Department of Pharmacology and Division of Biomedical Sciences, Washington University Medical School, St. Louis, Missouri

B. Carlin Department of Pharmacology, Washington University Medical School, St. Louis, Missouri

Jan M. Casadei Mahoney Institute of Neurological Sciences and Department of Biochemistry and Biophysics, University of Pennsylvania School of Medicine, Philadelphia, Pennsylvania

William A. Catterall Department of Pharmacology, University of Washington, Seattle, Washington

J. Covault Department of Anatomy and Neurobiology, Washington University Medical School, St. Louis, Missouri

C. M. Crowder Department of Pharmacology and Division of Biomedical Sciences, Washington University Medical School, St. Louis, Missouri

Peter Csermely Department of Biochemistry, State University of New York, Upstate Medical Center, Syracuse, New York

Benson M. Curtis Department of Pharmacology, University of Washington, Seattle, Washington

T. Dietz Department of Pharmacology and Division of Biomedical Sciences, Washington University Medical School, St. Louis, Missouri

Mario Dipaola Departments of Biochemistry and Molecular Biophysics, and Neurology, College of Physicians and Surgeons, Columbia University, New York, New York

Laszlo Dux Institute of Biochemistry, University Medical School, Szeged, Hungary

Aaron P. Fox Department of Physiology, Yale University School of Medicine, New Haven, Connecticut

D. Frail Department of Pharmacology, Washington University Medical School, St. Louis, Missouri

Robert D. Gordon Mahoney Institute of Neurological Sciences, University of Pennsylvania School of Medicine, Philadelphia, Pennsylvania

N. Michael Green National Institute for Medical Research, The Ridgeway, Mill Hill, London, United Kingdom

Robert Hartshorne Departments of Biology and Physics, University of California at San Diego, La Jolla, California

Peter Hess Department of Physiology, Yale University School of Medicine, New Haven, Connecticut

Bertil Hille Department of Physiology and Biophysics, University of Washington School of Medicine, Seattle, Washington

George Holan Commonwealth Scientific and Industrial Research Organization, Melbourne, Australia

Giuseppe Inesi Department of Biological Chemistry, University of Maryland School of Medicine, Baltimore, Maryland

K. Isenberg Departments of Pharmacology and Psychiatry, Washington University Medical School, St. Louis, Missouri

Istvan Jona Department of Biochemistry, State University of New York, Upstate Medical Center, Syracuse, New York

Peter N. Kao Departments of Biochemistry and Molecular Biophysics, and Neurology, College of Physicians and Surgeons, Columbia University, New York, New York

Arthur Karlin Departments of Biochemistry and Molecular Biophysics, and Neurology, College of Physicians and Surgeons, Columbia University, New York, New York

B. Keller* Departments of Biology and Physics, University of California at San Diego, La Jolla, California

**Present address:* Max Planck Institüt für Biophysikalische Chemie, Göttingen, West Germany.

H. Gobind Khorana Departments of Biology and Chemistry, Massachusetts Institute of Technology, Cambridge, Massachusetts

S. Kokubun Department of Pharmacology, University of Berne, Berne, Switzerland

Bozena Korczak Banting and Best Department of Medical Research, C. H. Best Institute, University of Toronto, Toronto, Ontario, Canada

J. Lawrence Department of Pharmacology, Washington University Medical School, St. Louis, Missouri

Mark D. Leibowitz Department of Physiology and Biophysics, University of Washington School of Medicine, Seattle, Washington

Jerry B. Lingrel Department of Microbiology and Molecular Genetics, University of Cincinnati, College of Medicine, Cincinnati, Ohio

Peter Lobel Departments of Biochemistry and Molecular Biophysics, and Neurology, College of Physicians and Surgeons, Columbia University, New York, New York

David H. MacLennan Banting and Best Department of Medical Research, C. H. Best Institute, University of Toronto, Toronto, Ontario, Canada

Edwin W. McCleskey Department of Physiology, Yale University School of Medicine, New Haven, Connecticut

Anthony Martonosi Department of Biochemistry, State University of New York, Upstate Medical Center, Syracuse, New York

John P. Merlie Department of Pharmacology, Washington University Medical School, St. Louis, Missouri

Mauricio Montal* Departments of Biology and Physics, University of California at San Diego, La Jolla, California

J. Mudd Department of Pharmacology, Washington University Medical School, St. Louis, Missouri

Nandor Müllner Department of Biochemistry, State University of New York, Upstate Medical Center, Syracuse, New York

Mark T. Nelson Department of Pharmacology, University of Miami School of Medicine, Miami, Florida

B. Nilius Department of Physiology, Yale University School of Medicine, New Haven, Connecticut

**Present address:* Department of Neurosciences, Roche Institute of Molecular Biology, Nutley, New Jersey.

Martha C. Nowycky Department of Physiology, Yale University School of Medicine, New Haven, Connecticut

Sandor Papp Department of Biochemistry, State University of New York, Upstate Medical Center, Syracuse, New York

H. Porzig Department of Pharmacology, University of Berne, Berne, Switzerland

B. Prod'hom Department of Pharmacology, University of Berne, Berne, Switzerland

Harald Reuter Department of Pharmacology, University of Berne, Berne, Switzerland

Richard H. Roberts Mahoney Institute of Neurological Sciences and Department of Biochemistry and Biophysics, University of Pennsylvania School of Medicine, Philadelphia, Pennsylvania

R. L. Rosenberg Department of Physiology, Yale University School of Medicine, New Haven, Connecticut

Joshua Sanes Department of Anatomy and Neurobiology, Washington University Medical School, St. Louis, Missouri

Jürgen R. Schwarz Institute of Physiology, University of Hamburg, Hamburg, West Germany

V. Shah Department of Pharmacology, Washington University Medical School, St. Louis, Missouri

Gary E. Shull Department of Microbiology and Molecular Genetics, University of Cincinnati, College of Medicine, Cincinnati, Ohio

S. M. Sine Department of Physiology and Biophysics, Yale University School of Medicine, New Haven, Connecticut

M. M. Smith Department of Pharmacology, Washington University Medical School, St. Louis, Missouri

Joe Henry Steinbach Department of Anesthesiology, Washington University School of Medicine, St. Louis, Missouri

Charles F. Stevens Molecular Neurobiology, Yale Medical School, New Haven, Connecticut

Robert M. Stroud Department of Biochemistry and Biophysics, University of California, San Francisco, San Francisco, California

Jeffrey B. Sutro Department of Physiology and Biophysics, University of California at Irvine, Irvine, California

Kenneth A. Taylor Department of Anatomy, Duke University Medical Center, Durham, North Carolina

L. L. Thio Department of Pharmacology and Division of Biomedical Sciences, Washington University Medical School, St. Louis, Missouri

H. P. Ting-Beall Department of Anatomy, Duke University Medical Center, Durham, North Carolina

Richard W. Tsien Department of Physiology, Yale University School of Medicine, New Haven, Connecticut

Sandor Varga Department of Biochemistry, State University of New York, Upstate Medical Center, Syracuse, New York

PREFACE

For nearly 150 years general physiologists have sought to describe processes general to all life—absorption, reproduction, irritability, motility, and so forth—in terms of laws of physics and chemistry. As physiologists, they concentrated on function, deducing mechanisms from careful measurement of rates and flows as the environment was varied. In their reductionist view they moved from organs, to cells, and ultimately toward molecules.

During the last year, the new molecular genetic techniques have succeeded in identifying and sequencing nucleic acid messages coding for several of the most important of the functional proteins of excitable membranes. For the first time, membrane physiologists have seen complete amino acid sequences of the proteins underlying the functions they have studied for so long. We have entered a period requiring intense collaborations and interdisciplinary discussions to understand these molecular machines. The symposium on which this volume is based was planned to bring physiologists together with molecular biologists and protein chemists just at the time when the first sequences were completed. The resulting papers show the beginnings of the union between molecular and physiological approaches for membrane proteins.

We focused on five proteins that underlie propagation of action potentials in nerve and muscle, regulation of calcium fluxes, transmission at the neuromuscular junction, and restoration of ionic gradients after excitation. These are the voltage-gated sodium and calcium channels, the nicotinic acetylcholine receptor, and the Na^+–K^+ and Ca^{2+} ATPases. In addition, the visual pigment rhodopsin was described. The functions of these proteins have been studied in exceptional physiological detail, and in all but one case, encoding DNAs have now been cloned and sequenced. Thus, thc chapters represent the best studied and most promising systems for structure–function analysis. They illustrate the present state of knowledge and successful strategies for the study of membrane proteins.

For much practical advice in the organization of the symposium we would like particularly to thank Paul De Weer. The Society of General Physiologists gratefully acknowledges financial assistance from the National Institutes of Health, the Muscular Dystrophy Association, the U.S. Army Research Office, and the Office of Naval Research.

BERTIL HILLE
DOUGLAS M. FAMBROUGH

Seattle, Washington
Baltimore, Maryland
November 1986

CONTENTS

PROTEINS OF EXCITABLE MEMBRANES

CHAPTER 1

RECENT WORK ON BACTERIORHODOPSIN

H. GOBIND KHORANA
Departments of Biology and Chemistry
Massachusetts Institute of Technology
Cambridge, Massachusetts

The purpose of this chapter is to review the work of the present author's group on an integral membrane protein, bacteriorhodopsin, that was discovered by Stoeckenius and his colleagues in the early 1970s (19). Bacteriorhodopsin is the only protein in the differentiated purple membrane of the extremely halophilic bacterium, *Halobacterium halobium*. It is synthesized and inserted into the cytoplasmic membrane when growth conditions become relatively anaerobic. In aerobic conditions the organism can carry out energy production through electron transport in the usual way. However, in 4 *M* salt, in which the organism grows, dissolved oxygen is low and convection may also be low, and thus the conditions tend to be relatively anaerobic. In sunlight, the organism carries out the synthesis of the purple membrane that forms patches containing highly organized arrays of two-dimensional crystals of bacteriorhodopsin. Bacteriorhodopsin in the purple membrane carries out a light-dependent proton translocation from inside to the outside of the cell as in the chemiosmotic hypothesis of Peter Mitchell. The electrochemical gradient thus generated is used by the cell for ATP synthesis. Two other retinal-containing pigments are also synthesized by the same organism; one is involved in pumping chloride ions, whereas the second carries out a phototactic response. The review presented in this chapter is concerned only with bacteriorhodopsin.

Halobacterium halobium can be grown easily, and the isolation of relatively large amounts (several hundred milligrams) of purple membrane can be carried out simply by an equilibrium density gradient step after lysis that occurs when the salt concentration is reduced. The purple membrane fragments survive this shock and remain intact during the osmotic shock.

Before the work of the present author's group is reviewed here, two important discoveries concerning bacteriorhodopsin in the 1970s are described. First, the chromophore in bacteriorhodopsin is *all trans* retinal, linked to the ε-amino group of a lysine residue as a Schiff base. The discovery of retinal in bacteriorhodopsin was highly intriguing because until that time only the visual pigments, rods, and cone photoreceptors in both vertebrates and invertebrates were known to contain the retinal (vitamin A aldehyde) chromophore. Following the chromophore analogy, bacteriorhodopsin, like the visual pigments, was shown to have a photochemical cycle, although differing in detail from those of the visual pigments. By working within certain ranges of temperatures, one can demonstrate a succession of intermediates with discrete spectroscopic properties. An approximate representation of the bacteriorhodopsin photocycle is shown in Fig. 1. The first step (indicated by the wiggly arrow) to the K intermediate is the only light-dependent step in the cycle. All subsequent steps can proceed in the dark. The *cis–trans* isomerization in retinal molecule is the basis of light transduction in bacteriorhodopsin and all the known visual pigments. The ultimate aim of work in bacteriorhodopsin is to understand the photochemical cycle in chemical terms— in other words, the photochemical steps in relation to the steps in proton translocation.

The second important discovery of the mid-1970s was the electron diffraction work of Henderson and Unwin (7) on bacteriorhodopsin. Because the purple membrane contains highly organized two-dimensional crystalline arrays of the

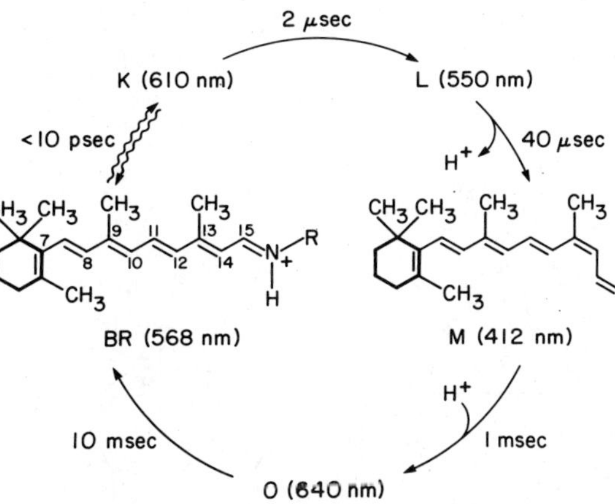

Figure 1. A current model showing the intermediates detected in the photochemical cycle of the purple membrane by low temperature and flash spectroscopy. One proton is released and taken up again at the positions shown. The protonation states of the chromophore, also shown, are derived by resonance Raman spectroscopy. A similar series of intermediates is well known in the visual pigment, rhodopsin, which also contains the chromophore retinal in a 1:1 complex with the protein.

protein, it lends itself to electron diffraction study by the method developed by Aaron Klug at the Medical Research Council Laboratory in Cambridge, England.

Electron diffraction patterns show that the protein is present as trimers that form a hexagonal lattice. Protein monomers are believed to contain seven helical segments embedded in the membrane. Therefore, the protein that contains a single polypeptide chain presumably traverses the membrane seven times. Unfortunately, proof of this interpretation has not been forthcoming since the original work was carried out 10 years ago. For the present we will assume that there are seven helices per protein molecule although the possible presence of some β-sheet structure cannot be completely ruled out at this stage.

Thus bacteriorhodopsin is a single relatively small integral membrane protein present in the purple membrane. It performs an important chemical reaction, proton translocation by means of light energy. The presence of retinal chromophore suggests similarities in chemistry between bacteriorhodopsin, rhodopsin, and related pigments. From its overall properties and well-defined function, bacteriorhodopsin undoubtedly provides a highly attractive system for the study of an important biological function.

1. STUDIES AT THE PROTEIN LEVEL AND QUESTIONS REGARDING FUNCTION

Figure 2 shows three phenomena. First, in the orientation of bacteriorhodopsin in the membrane, the carboxyl end is at the cytoplasmic side with about 20 amino acids

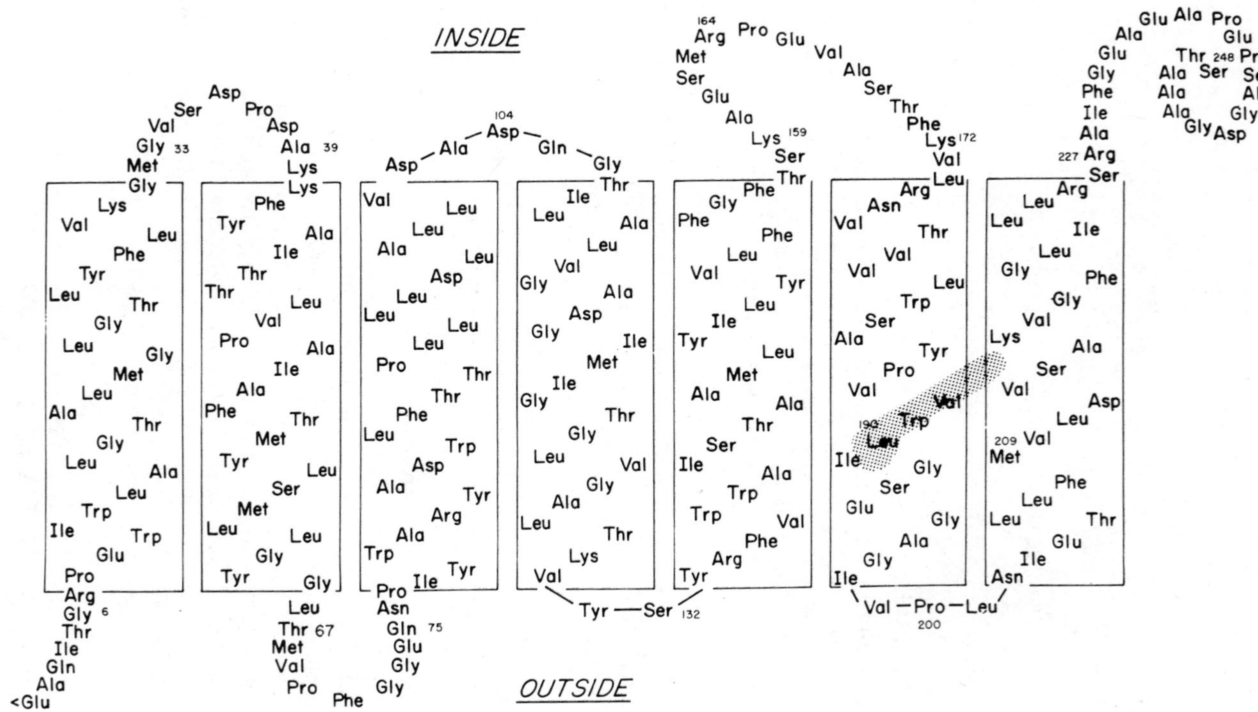

Figure 2. A secondary structure model for bacteriorhodopsin as present in the purple membrane. The shaded area represents the location and orientation of retinal in the protein. For details and further discussion of this model, see Huang et al. (10).

protruding out of the membrane (5). The amino end is concluded to be protruding to the exterior of the membrane. Second, we know the complete amino acid sequence. This was determined first by amino acid sequencing (12) and, subsequently, by sequencing of the gene (3). The sequences obtained in these two ways were in complete agreement. As is evident from the amino acid sequence, the protein is very hydrophobic. (About 70% of the total amino acids are hydrophobic.) Since the protein dissolves in only 70% formic acid, manipulations and techniques required for degradation and sequencing of the protein had to be very different from the usual techniques applied to soluble proteins. Third, the secondary structure shown in Fig. 2 is an attempt to fit the amino acid sequence to the electron diffraction pattern determined by Henderson and Unwin (7). This model was originally proposed by Don Engelman and co-workers (4). It was modified to accommodate the photoaffinity crosslinking data of K.-S. Huang et al. (10). There are large uncertainties in this model. For example, all helices are shown to be approximately equal, but this probably will not turn out to be the case. Further, the sizes of the loops that are out of the membrane and connect the successive helices are unknown. This is also a very important unknown at this time, not only for this protein, but also probably for all the integral polytopic membrane proteins. The probable manner of the folding of the linear secondary structure in Fig. 2 to form the tertiary structure is not known. The folded structure must have strong interhelical interactions. Probably there are no phospholipids intervening between the helices. Rather, the lipid molecules are present only at the periphery of the helical bundle, which seems to work in a cooperative fashion.

Two further aspects of our studies with this protein will now be mentioned. In working with membrane proteins, it is highly desirable and advantageous to be able to remove endogenous lipids and reconstitute the proteins under defined conditions. For bacteriorhodopsin, this has been accomplished very efficiently by using a combination of detergents and phospholipids. Solubilized bacteriorhodopsin can be kept in a fully functional state in deoxycholate, and one can subsequently reconstitute this into mixed micelles or vesicles that have very high proton pumping activity (8). We usually obtain 50–60 protons pumped per bacteriorhodopsin molecule, and the initial rate of pumping is about one proton per second (9).

The second remarkable property of this protein is its unique ability to refold correctly after denaturation (9). One can treat bacteriorhodopsin with sodium dodeocyl sulfate (SDS) detergent at 100° C or dissolve the protein in strong acids such as trifluoroacetic acid. Invariably, following these treatments, the protein folds quantitatively to regenerate the typical chromophore when appropriate mixtures of detergents, phospholipids, and retinal are provided. Indeed, proteolytic fragments of the protein prepared by selective proteolytic cuts will also reassociate to form functional bacteriorhodopsin-like structures (14).

This ability of the denatured protein to refold correctly and regenerate bacteriorhodopsin-like chromophore is proving extremely useful in our current work with the bacteriorhodopsin mutants. Indeed, regeneration of the chromophore provides a sensitive assay for the phenotypes of the mutants.

Summarizing bacteriorhodopsin as a protein (19), we know the primary structure of the protein and have rough ideas about its secondary and tertiary structure (7); the protein is very robust, and the fact that it renatures so easily and quantitatively is a highly useful feature.

Finally, we can formulate some of the main questions in regard to bacteriorhodopsin.

1. What is the nature of protein–retinal interactions? Or, what causes the large red shift in the retinal spectrum? Retinal and retinylamine absorb at 380–400 nm, while bacteriorhodopsin, rhodopsin, and cone pigments absorb at 500–600 nm.
2. What is the role of electrostatic interactions in the folded structure within the membrane?
3. What is the mechanism of proton translocation? Does it involve proton conductance or a light-activated channel?
4. Do photochemical intermediates represent steps in proton translocation?
5. Does the mechanism of proton translocation in bacteriorhodopsin apply to other proton translocating systems such as ATPases?

It is very likely that the electrostatic interactions do exist. In particular, it would appear that stabilization of the positive Schiff base charge at Lys-216 by interaction with a negative charge such as aspartic acid at 212 (Fig. 3) may be crucial. A scale model (Fig. 3) indeed suggests that such an interaction is very likely (1). Then there is the postulate of an external point charge as a basis for the red ''opsin'' shift described by Nakanishi et al. (17). It would be significant if the removal of the negative charge from a glutamic acid, such as glutamic acid 194, would influence the spectral properties.

Finally, an important and attractive proposal in regard to the mechanism of proton translocation is whether the proton is conducted across the membrane by an extended hydrogen-bonded system (wire). The idea comes from the higher conductance of protons in ice relative to water and has been proposed in the past by a number of eminent physical chemists (Shedlovsky, Griffith, Onsager, Eigen, and others).

2. CLONING AND EXPRESSION OF THE BACTERIORHODOPSIN GENE IN *E. COLI*

The procedure used for the isolation and characterization of the bacteriorhodopsin gene has been described elsewhere (3). The first requirement is the identification of this gene in *H. halobium* DNA and the cloning of the gene in a form suitable for integration into expression vectors.

The nucleotide sequence of the gene and the flanking regions is shown in Fig. 4. The sequence of the gene showed the presence of an unusual precursor sequence of 13

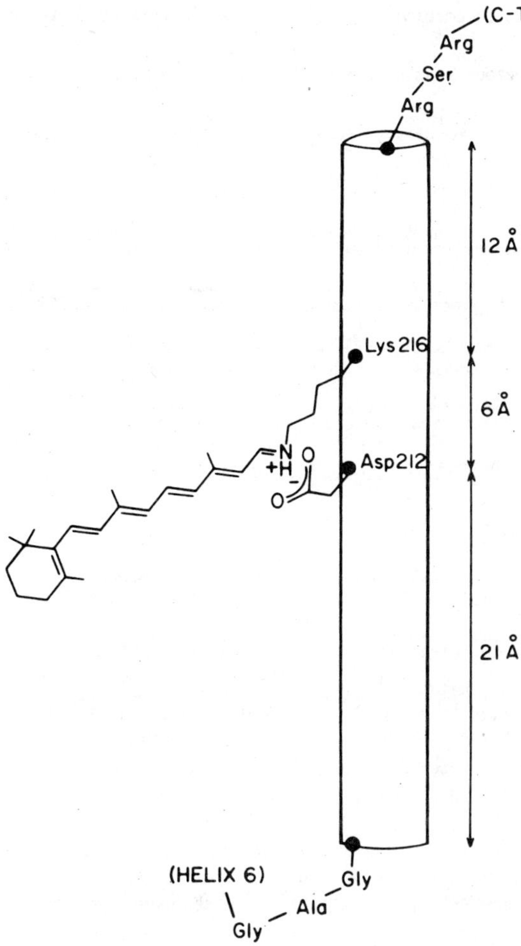

Figure 3. A scale model of the *C*-terminal helix (helix 7) of bacterioopsin showing the formation of an electrostatic bond between Lys-216, Schiff base, and Asp-212.

amino acids and one additional amino acid, aspartic acid, at the carboxyl end. The precursor sequence is probably cleaved in a cotranslational fashion to give the *N*-terminal pyroglutamic acid. No information is available on the processing of the *C*-terminal aspartic acid residue.

2.1 Expression Vectors Based on *E. coli* Lipoprotein Gene.

First, a plasmid pEN-22 (Fig. 5) containing bacteriorhodopsin gene without the *N*-terminal precursor sequence was prepared. Nakamura et al. (18) have constructed a series of expression vectors using the gene for the *E. coli* outer membrane lipoprotein

```
GGGTGCAACCGTGAAGTCCGCCACGACCGCGTCACGACAGGAGCCGACCAGCGACACCCAGAAGGTGCGAACGGTTGAGTGCCGCAACGA
-360                                                                                   -271

TCACGAGTTTTCGTGCGCTTCGAGTGGTAACACGCGTGCACGCATCGACTTCACCGCGGGTGTTTCGACGCCAGCCGGCCGTTGAACCA
-270                                                                                   -181

GCAGGCAGCGGGCATTTACAGCCGCTGTGGCCCAAATGGTGGGGTGCGCTATTTTGGTATGGTTTGGAATCCGCGTGTCGGCTCCGTGTC
-180                                                                                    -91

TGACGGTTCATCGGTTCTAAATTCCGTCACGAGCGTACCATACTGATTGGGTCGTAGAGTTACACACATATCCTCGTTAGGTACTGTTGC
-90                                                                                      -1
                                          ↓
Met Leu Glu Leu Leu Pro Thr Ala Val Glu Gly Val Ser Gln Ala Gln Ile Thr Gly Arg Pro Glu Trp Ile Trp Leu Ala Leu Gly Thr
ATGTTGGAGTTATTGCCAACAGCAGTGGAGGGGGTATCGCAGGCCCAGATCACCGGACGTCCGGAGTGGATCTGGCTAGCGCTCGGTACG
1                                                                                        90

Ala Leu Met Gly Leu Gly Thr Leu Tyr Phe Leu Val Lys Gly Met Gly Val Ser Asp Pro Asp Ala Lys Lys Phe Tyr Ala Ile Thr Thr
GCGCTAATGGGACTCGGGACGCTCTATTTCCTCGTGAAAGGGATGGGCGTCTCGGACCCAGATGCAAAGAAATTCTACGCCATCACGACG
91                                                                                      180

Leu Val Pro Ala Ile Ala Phe Thr Met Tyr Leu Ser Met Leu Leu Gly Tyr Gly Leu Thr Met Val Pro Phe Gly Gly Glu Gln Asn Pro
CTCGTCCCAGCCATCGCGTTCACGATGTACCTCTCGATGCTGCTGGGGTATGGCCTCACAATGGTACCGTTCGGTGGGGAGCAGAACCCC
181                                                                                     270

Ile Tyr Trp Ala Arg Tyr Ala Asp Trp Leu Phe Thr Thr Pro Leu Leu Leu Leu Asp Leu Ala Leu Leu Val Asp Ala Asp Gln Gly Thr
ATCTACTGGGCGCGGTACGCTGACTGGCTGTTCACCACGCCGCTGTTGTTGTTAGACCTCGCGTTGCTCGTTGACGCGGATCAGGGAACG
271                                                                                     360

Ile Leu Ala Leu Val Gly Ala Asp Gly Ile Met Ile Gly Thr Gly Leu Val Gly Ala Leu Thr Lys Val Tyr Ser Tyr Arg Phe Val Trp
ATCCTTGCGCTCGTCGGTGCCGACGGCATCATGATCGGGACCGGCCTGGTCGGCGCACTGACGAAGGTCTACTCGTACCGCTTCGTGTGG
361                                                                                     450

Trp Ala Ile Ser Thr Ala Ala Met Leu Tyr Ile Leu Tyr Val Leu Phe Phe Gly Phe Thr Ser Lys Ala Glu Ser Met Arg Pro Glu Val
TGGGCGATCAGCACCGCAGCGATGCTGTACATCCTGTACGTGCTGTTCTTCGGGTTCACCTCGAAGGCCGAAAGCATGCGCCCCGAGGTC
451                                                                                     540

Ala Ser Thr Phe Lys Val Leu Arg Asn Val Thr Val Val Leu Trp Ser Ala Tyr Pro Val Val Trp Leu Ile Gly Ser Glu Gly Ala Gly
GCATCCACGTTCAAAGTACTGCGTAACGTTACCGTTGTGTTGTGGTCCGCGTATCCCGTCGTGTGGCTGATCGGCAGCGAAGGTGCGGGA
541                                                                                     630

Ile Val Pro Leu Asn Ile Glu Thr Leu Leu Phe Met Val Leu Asp Val Ser Ala Lys Val Gly Phe Gly Leu Ile Leu Leu Arg Ser Arg
ATCGTGCCGCTGAACATCGAGACGCTGCTGTTCATGGTGCTTGACGTGAGCGCGAAGGTCGGCTTCGGGCTCATCCTCCTGCGCAGTCGT
631                                                                                     720
                                                                    ↓
Ala Ile Phe Gly Glu Ala Glu Ala Pro Glu Pro Ser Ala Gly Asp Gly Ala Ala Ala Thr Ser Asp ***
GCGATCTTCGGCGAAGCCGAAGCGCCGGAGCCGTCCGCCGGCGACGGCGCGGCCGCGACCAGCGACTGATCGCACACGCAGGACAGCCCC
721                                                                                     810

ACAACCGGCGCGGCTGTGTTCAACGACACACGATGAGTCCCCCACTCGGTCTTGTACTC
811                                                     869
```

Figure 4. The DNA sequence of the bacteriorhodopsin gene and adjacent regions. The amino acid sequence of the bacteriorhodopsin precursor is given; the *N*-terminal methionine codon is at nucleotide residue 1. Arrows indicate the amino and carboxyl termini of the mature protein.

(Lpp). In the pIN-II series of vectors (Fig. 5*b*), the *lac*UV5 promoter–operator fragment has been inserted immediately after the constitutive Lpp promoter. Thus the system is inducible and is under the control of both the Lpp and the lac promoters. The desirability of the use of an inducible transcription system was strikingly evident in our initial work on the expression of the very hydrophobic bacteriorhodopsin. Additional features of pIN-II vectors are illustrated in Fig. 5*b*. Thus the ribosome binding site (Shine-Delgarno sequence) and the initiator codon are those of the Lpp gene. Three cloning sites have been created within the coding sequence of the *pro*-lipoprotein. In pIN-IIA, a nucleotide sequence of 22 base pairs that contains EcoRI, HindII, and BamHI restriction sites has been inserted at the position of the third amino acid of the *pro*-lipoprotein (in three reading frames). Similar cloning sites have been inserted before the twentieth amino acid downstream from the *N* terminus (pIN-IIB) and after the eighth amino acid into the structural lipoprotein sequence (pIN-IIC).

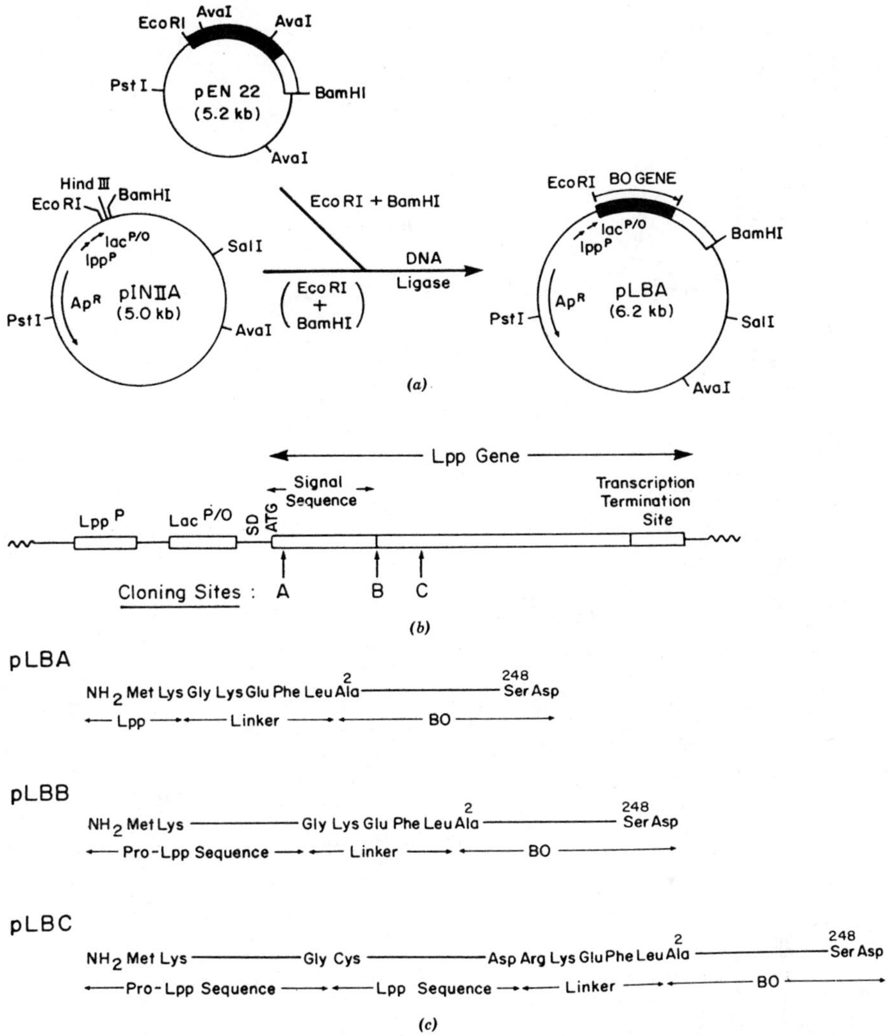

Figure 5. Expression vectors for the bacteriorhodopsin gene and the expected *N*-terminal amino acid sequences expected in proteins expressed in *E. coli*. The plasmid pEN-22 containing the BR gene without the precursor sequence and an EcoRI restriction site at the amino terminus was constructed as described elsewhere (R. Dunn et al., ref. 2, and unpublished observations). The expression vector pIN-IIA and the related vectors, pIN-IIB and pIN-IIC, were obtained from Professor M. Inouye. The bacteriorhodopsin gene from pEN-22 was inserted as shown in part *a* to form the plasmid pLBA. Plasmids pLBB and pLBC were prepared similarly by using the cloning sites *B* and *C* in Fig. 5*b*. Part *c* shows the additional amino acids expected following expression of the BR gene in the three plasmids (see text).

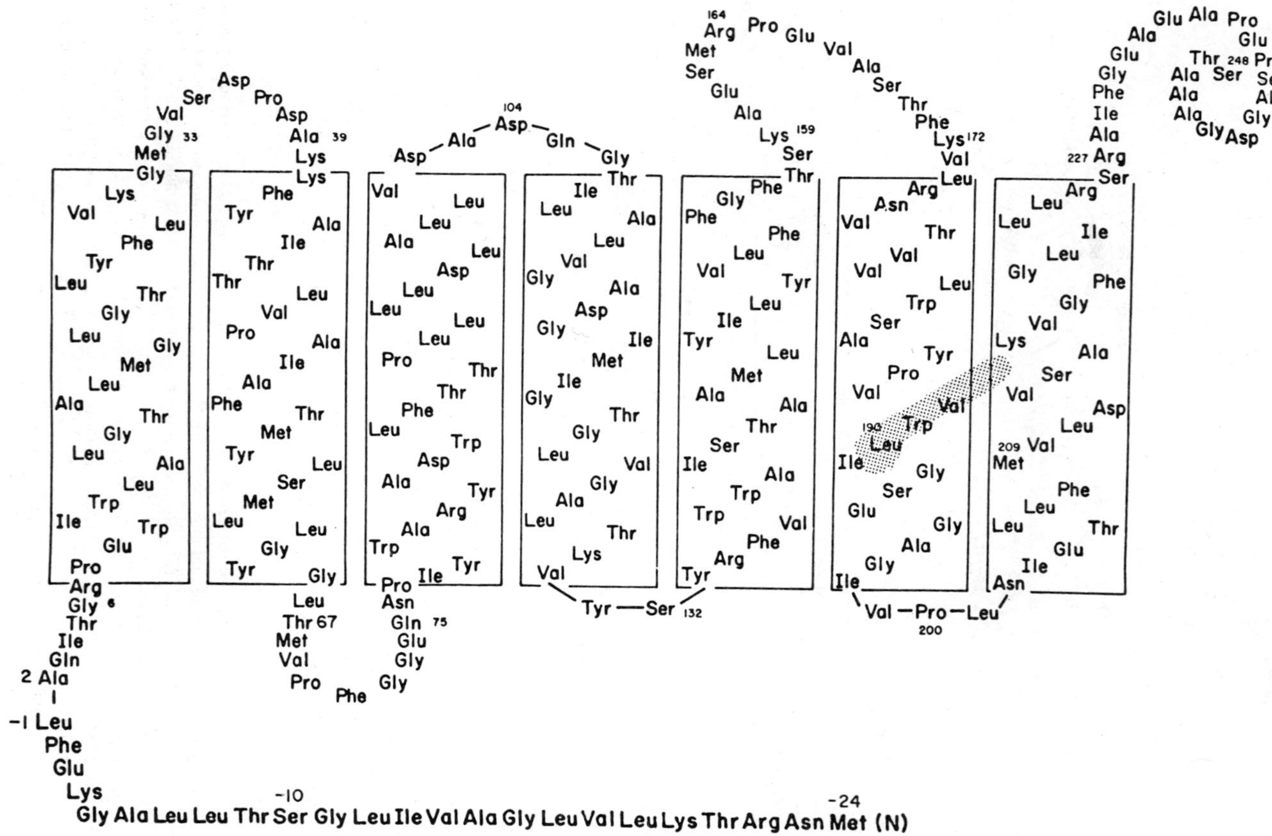

Figure 6. The fusion protein with the extension at the amino terminus as expected from vector pLBB.

Bacteriorhodopsin expression plasmids were prepared by insertion of EcoRI–BamHI fragment of pEN-22 containing the bacteriorhodopsin gene into pIN-IIA, pIN-IIB, and pIN-IIC (Fig. 5*a*) (2). The resulting plasmids, pLBA, pLBB, and pLBC, encode the opsin derivatives with extensions at the *N* terminus shown in Fig. 5*c*. Thus all three plasmids encode a Leu residue in place of the *N* terminal pyroglutamic residue in natural bacterioopsin (BO). They also have the additional aspartic residue at the *C* terminus. The protein from the plasmid pLBA encodes six additional *N*-terminal amino acids, Met–Lys from the *pro*-lipoprotein amino terminus, and the next four amino acids from the linker sequence. The plasmid pLBB encodes a total of 24 additional amino acids at the amino terminus of bacterioopsin. The *C* site plasmid (pLBC) includes the complete lipoprotein signal sequence and eight amino acids from the *N* terminus of the mature lipoprotein.

Work so far has been carried out mainly with the fusion protein derived from the pLBB plasmid (Fig. 6). Figure 7 shows the experiment on chromophore regeneration from the fusion proteins and BO from *H. halobium*. Two differences may be noted. Bacterioopsin from *E. coli* regenerates to the extent of about 50%, and the rate of regeneration (not shown) is slower than that of the native BO. Furthermore, the yields of the opsin with the preceding expression vectors were very low (about 0.02–0.05%). Therefore, in the last few years, we have devoted a great deal of effort

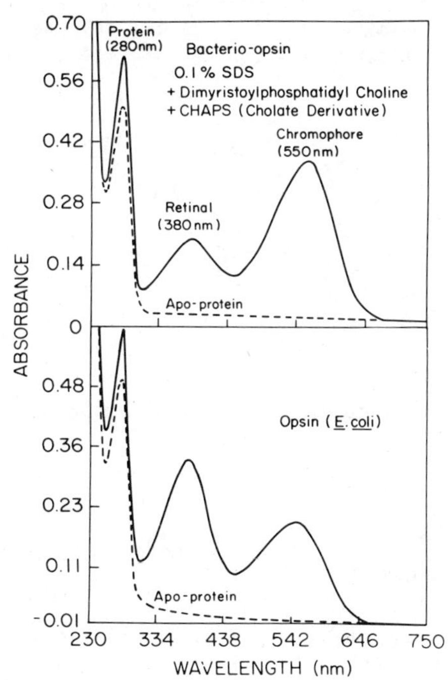

Figure 7. Regeneration of the BR-like chromophore from bacterioopsin (BO) and from the fusion protein shown in Fig. 6.

in trying to improve the expression. Further, the sequence at the amino terminus of BO to be expressed in *E. coli* was changed so that it resembled as closely as possible the *N*-terminal sequence in the native BO.

A variety of vectors were constructed (11). We show only one of the inducible vectors that we now use (Fig. 8). Noteworthy are the nucleotide changes made in the first six codons after ATG; these were in the third letter and involved changing GC to AT wherever possible. The expression of the BO protein from this plasmid has been much more satisfactory (about 1% of *E. coli* proteins). The only change from the native *N*-terminal sequence is Met instead of pyroglutamic acid. Rate of regeneration of the chromophore from this expression product is more like that of the native BO, although for reason(s) not understood, the extent of regeneration is still around 50% of theoretical.

Brief comments on our efforts to purify BO expressed in *E. coli* should be made. Especially in view of the very low yields of the protein, a number of methods were investigated, including selective absorption of detergent-solubilized BO on a monoclonal antibody column (13). This single step purified the protein to about 80%, and further purification is achieved by preparative gel electrophoresis. Second, it was found that BO can be extracted from *E. coli* cells or membranes by using an organic

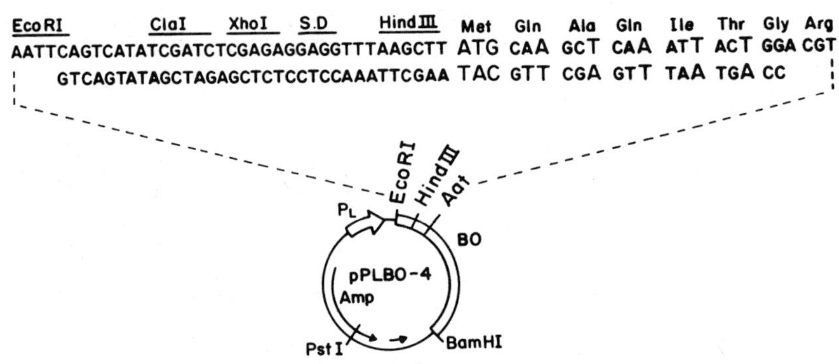

Alterations of sequences following the ribosome binding site

Plasmid clone	Sequence	Restriction sites	Intent
BOTC-I	met gln ala gln ile thr gly arg AGCTT ATG CAG GCC CAG ATC ACC GGA CGT A TAC GTC CGG GTG TAG TGG CC	HindIII-AatII	Wild Type
BOTC-4	agctt atg caA gcT caA atT acT gga cgt a tac gtT cgA gtT taA tgA cc	HindIII-AatII	Third position of the codons changed

Figure 8. An inducible vector for expression of BO with codon changes introduced in the amino acids at the amino-terminus.

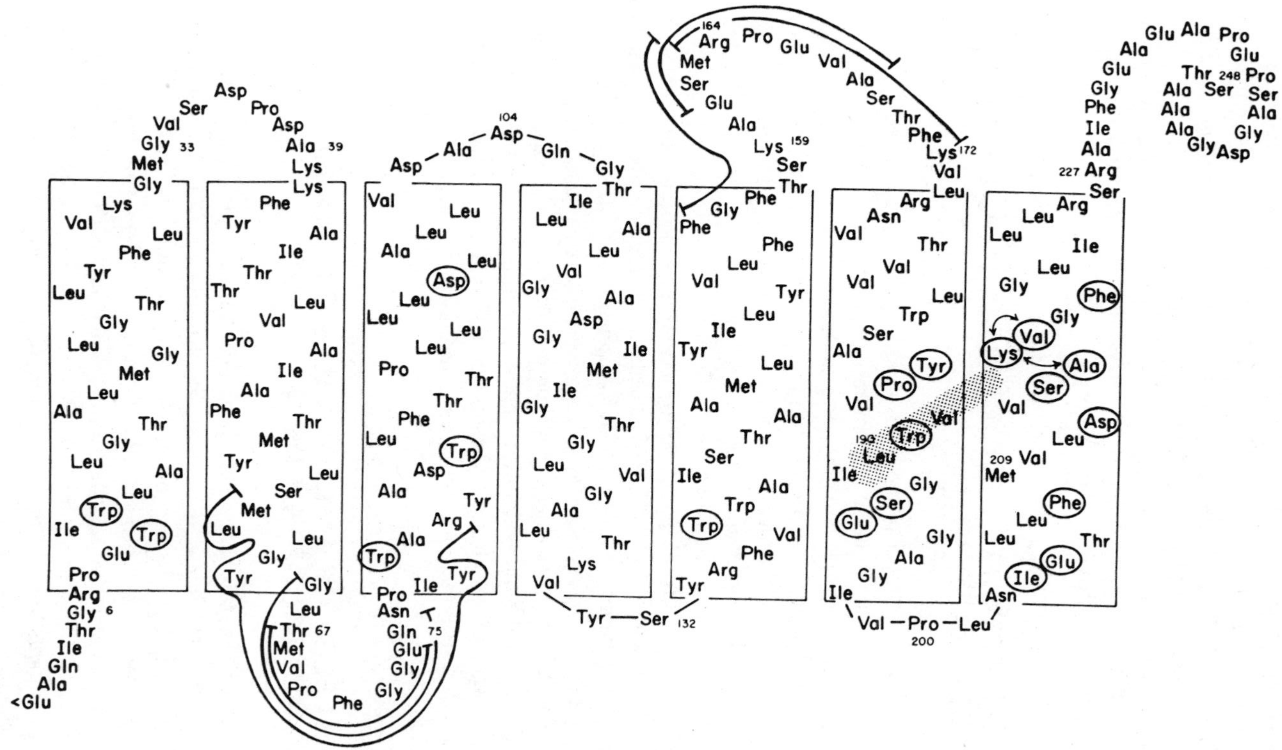

Figure 9. Summary of mutations in BO: the mutations (e.g., Trp-12, 80, 86, 137) to ambers were made by the use of Mut D strain (16). Mismatched oligonucleotide-directed mutagenesis was used for introduction of XhoI restriction site at amino acid 203 (15). Mutations in helices 6 and 7 were introduced mainly by excision of fragments and their substitution by synthetic counterparts (15). The deletion mutations were performed by a procedure described elsewhere (6).

solvent mixture (chloroform–methanol–triethylamine). This rapid method gives BO with purity in the range of 50%. While other methods (HPLC, anion-exchange columns, hydroxyapatite) are being tried for further purification, small-scale purification can be accomplished by gel electrophoresis.

In summary, the expression of BO in *E. coli* and its purification is now reasonably satisfactory, and we can now proceed to carry out structure–function studies by preparing mutants and studying their phenotypes.

3. SITE-SPECIFIC MUTAGENESIS IN AMINO ACID REPLACEMENTS

We have investigated a number of methods for site-specific mutagenesis (Fig. 9). In brief, we used (1) introduction of amber mutations by the use of an *E. coli* mutator strain (*Mut* D) (16); (2) mismatched oligonucleotide-directed point mutagenesis (15); (3) replacement of a suitable restriction fragment by a totally synthetic counterpart containing desired codon changes (15); and (4) deletions of a few to some amino acids, so far mainly in the presumed loops. This was done by opening the circular DNA at the desired place, limited digestion with nucleases and circularizing the DNA to the covalently closed form. Figure 9 shows a catalog of the mutations that we have introduced so far. Because we will clearly need a very large number of mutants, we have concluded that the most facile method for preparation of these mutants is by the synthetic approach. We have, therefore, designed and carried out a total synthesis of the BO gene. By design, the synthetic gene contains only unique restriction sites, and these are placed at regular intervals. Mutagenesis, then, involves replacement of any excised short fragment by altered synthetic counterparts. The principles for the synthesis of DNA fragments containing altered codons have been described previously (15).

4. EFFECTS OF AMINO ACID SUBSTITUTIONS ON BACTERIORHODOPSIN

The results described in this section are very recent and are far from complete. The main assay so far has been to test for the folding of BO and chromophore regeneration. We still have to test for the effects on proton translocation. Figure 10 shows a summary of the results obtained to date. Thus we have observed three types of phenotypes. In the first group, such as amino acid residues 193 and 204, the protein folds and regenerates the chromophore that is identical to that of the native BO. In the second class of mutants, the chromophore is regenerated, but that there is a shift in the adsorption spectrum of the reconstituted molecule. Such mutations with spectral shifts would be of great interest in studying the nature of interactions between the amino acid side chains and retinal. In the third group of single amino acid changes, no chromophore regeneration is observed. This could be due to the loss of ability to fold

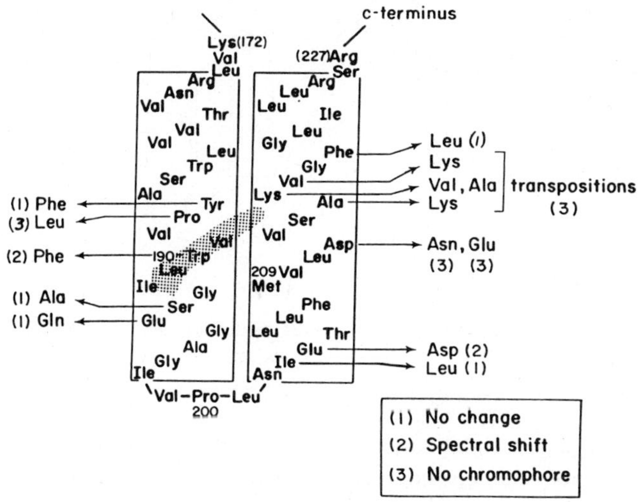

Figure 10. Summary of the phenotypes of bacteriorhodopsin mutants obtained by amino acid replacements.

or bind retinal or to form a stable Schiff base. Thus, on changing aspartic acid 212 to asparagine, no chromophore regeneration is observed. It seems possible that in this case the protein does fold and forms the retinal binding pocket, but that the Schiff base is not stabilized because of the absence of an electrostatic interaction with the negative charge of aspartic acid. These possibilities are being tested.

In addition to studying different types of mutations by physical methods such as Fourier transform infrared (FTIR), laser Raman spectroscopy, and photokinetics, our immediate objective is to screen for those mutations that have defects in proton pumping. This screening would first be done with those mutants that form unaltered chromophore on reconstitution. Hopefully, we can begin to see possibilities for the path that the proton takes in being translocated. These studies should be complemented by studies of the photochemical cycle in each of these mutants.

5. CONCLUDING REMARKS

Bacteriorhodopsin represents the simplest and most attractive system for transduction of light energy to chemical energy. In addition, we are optimistic that insights into the mechanism of proton translocation by bacteriorhodopsin will clarify the process of proton translocation in general.

The author takes this opportunity to compare bacteriorhodopsin with rhodopsin. Cartoons of the current secondary structure models for these proteins are shown in Fig. 11. The following points of similarities and dissimilarities may be noted: (1) both are integral membrane proteins; (2) both use retinal as the chromophore; (3) in both

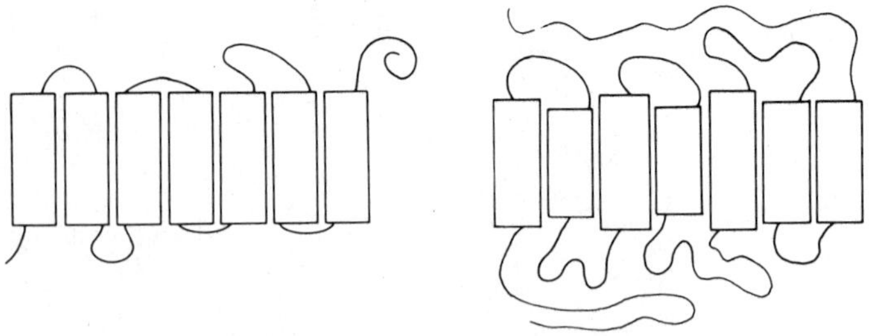

Figure 11. Cartoons of current secondary structure models of bacteriorhodopsin and rhodopsin.

cases, the first event after photon capture is cis–trans isomerization; and (4) bacteriorhodopsin is a self-contained system for proton translocation; whereas (5) rhodopsin, on photoexcitation, interacts with a number of proteins that are involved in transmission of the sensory signal and regulation of a number of functions, such as light and dark adaptation.

In the years ahead, we hope to be able to study rhodopsin in the way we have studied bacteriorhodopsin and study the nature of its interactions with the regulatory proteins in the rod cell.

Finally, we note the rapid progress that is now being made in studies with integral membrane proteins. Although this class of proteins, such as rhodopsin, ionic channels, mitochondrial proteins, ATPases and others, have remained intractable until recently, a great deal will be learned about their structure and function in the near future.

ACKNOWLEDGMENTS

The work described has been supported by the National Institutes of Health, Grants GM28289 and AI11479; the National Science Foundation, Grant PCM78-13713; and the Office of Naval Research, Arlington, VA, contract number N00014-82-K-0668.

I am indebted to a large number of able associates whose efforts, theoretical and practical, made this work possible. Work on bacteriorhodopsin began some 10 years ago. I wish to acknowledge particularly my recent and current colleagues: Mark Braiman, Betty Chao, Rob Dunn, Marie Gilles-Gonzalez, Neil Hackett, Sadu Karnik, Simon Jones, K. Anne Kronis, Dae-Sil Lee, L. Bernard Lerch, Kin-Ming Lo, John McCoy, Michael Nassal, Shigeru Noguchi, and Larry Stern.

REFERENCES

1. Bayley, H., Radhakrishnan, R., Huang, K.-S., and Khorana, H. G. (1981). *J. Biol. Chem.*, **256**, 3797–3801.

2. Dunn, R. J., Hackett, N. R., Huang, K.-S., Jones, S. S., Khorana, H. G., Lee, D.-S., Liao, M.-J., Lo, K.-M., McCoy, J., Noguchi, S., Radhakrishnan, R., and RajBhandary, U. L. (1983). *Cold Spring Harbor Symp. Quant. Biol.*, **48**.
3. Dunn, R., McCoy, J., Simsek, M. Majumdar, A., Chang, S. H., RajBhandary, U. L., and Khorana, H. G. (1981). *Proc. Natl. Acad. Sci.* (USA), **78**, 6744–6748.
4. Engelman, D. M., Henderson, R., McLachlan, A. D., and Wallace, B. A. (1980). *Proc. Natl. Acad. Sci.* (USA), **77**, 2023–2027.
5. Gerber, G. E., Gray, C. P., Wildenauer, D., and Khorana, H. G. (1977). *Proc. Natl. Acad. Sci.* (USA) **74**, 5426–5430.
6. Gilles-Gonzalez, M. A., Hackett, N. R., Jones, S. S., Khorana, H. G., Lee, D.-S., Lo, K.-M., and McCoy, J. M. (in press). *Methods Enzymol.*
7. Henderson, R. and Unwin, P. N. T. (1975). *Nature*, **257**, 28–32.
8. Huang, K.-S., Bayley, H., and Khorana, H. G. (1980). *Proc. Natl. Acad. Sci.* (USA), **77**, 323–327.
9. Huang, K.-S., Bayley, H., Liao, M.-J., London, E., and Khorana, H. G. (1981). *J. Biol. Chem.*, **256**, 3802–3809.
10. Huang, K.-S., Radhakrishnan, R., Bayley, H., and Khorana, H. G. (1982). *J. Biol. Chem.*, **257**, 13616–13623.
11. Karnik, S., Braiman, M. S., Chao, B. H., Dunn, R., Gilles-Gonzalez, M. A., Hackett, N. R., Jay, E., Kronis, K. A., Nassal, M., and Khorana, H. G. (in press). *Protein Eng.*
12. Khorana, H. G., Gerber, G. E., Herlihy, W. C., Gray, C. P., Anderegg, R. J., Nihei, K., and Biemann, K. (1979). *Proc. Natl. Acad. Sci.* (USA), **76**, 5046–5050.
13. Kimura, K., Mason, T. L., and Khorana, H. G. (1981). *J. Biol. Chem,.*, **257**, 2859–2867.
14. Liao, M.-J., London, E., and Khorana, H. G. (1983). *J. Biol. Chem.*, **258**, 9949–9955.
15. Lo, K.-M., Jones, S. S., Hackett, N. R., and Khorana, H. G. (1984). *Proc. Natl. Acad. Sci.* (USA), **81**, 2285–2289.
16. McCoy, J. M., and Khorana, H. G. (1983). *J. Biol. Chem.*, **258**, 8456–8461.
17. Nakanishi, K., Balogh-Nair, V., Arnaboldi, M. Tsujimoto, K., and Honig, B. J. (1980). *Am. Chem. Soc.*, **102**, 7945–7947.
18. Nakamura, K., Masui, Y., and Inouye, M. (1982). *J. Molec. Appl. Genet.*, **1**, 289.
19. Stoeckenius, W., Lozier, R. H., and Bogomolni, R. A. (1979). *Biochim. Biophys. Acta*, **505**, 215-278.

CHAPTER 2

FUNCTION OF NICOTINIC ACETYLCHOLINE RECEPTORS

J. H. STEINBACH

Department of Anesthesiology
Washington University School of Medicine
St. Louis, Missouri

S. M. SINE

Department of Physiology and Biophysics
Yale University School of Medicine
New Haven, Connecticut

This research has been supported by grants NS-13719 and NS-22356 (NINCDS).

Synaptic transmission at vertebrate neuromuscular junctions is a rapid and effective process. Acetylcholine (ACh) released from the nerve terminal diffuses across a narrow cleft and binds to nicotinic acetylcholine receptors (AChRs) present at a high density in the muscle postjunctional membrane. A cation-selective channel formed by the receptor proteins then opens, resulting in a net inward flow of cations and depolarization of the muscle membrane.

The topic considered here is how the AChR functions—how many states with closed or open channels the receptor has and what the rates for transitions between those states are. In broad outline, the answer to this question has been known for some time. The receptor has at least one state with an open ion channel. There must be more than one state with a closed ion channel, however. Most receptors with open channels have two ACh molecules bound. This means that there are three forms of the receptor with a closed channel in various degrees of ligation. It is also known that the receptor can adopt long-lived desensitized closed channel states, so there are at least four receptor states with closed channels.

How are these states distinguished? Physiologically, they are first characterized by the conductance of the ion channel of the receptor. The states are then divided by kinetic properties, on the basis of how long they last. Finally, the durations at different conductance levels are determined following experimental manipulations, such as variations of the concentration of ligand, to examine the nature of the transitions involved in entering and leaving a given state.

The use of single-channel recordings has resulted in a large amount of new information about the function of the nicotinic AChR in particular; hence previous ideas of AChR function have had to be reevaluated. However, both single-channel records and macroscopic current records result from transitions of channels between the same conductance levels, at the same rates, and thus should contain the same information. Macroscopic records provide a picture of the average functional status of a relatively large number of channels, with the caveats that the initial and final conditions reflect the average population distribution, and the amplitudes and time constants of kinetic components reflect the relaxation of the entire population among several states.

Single-channel records contain a more detailed and precise description of the timing and nature of transitions. One reason for this is simply improved resolution of both channel conductance and the duration of states; another is that single-channel records give different and often less ambiguous information about channel function. For example, the conductance of a channel can be readily separated from the probability that the channel is open, an operation that is considerably more difficult with macroscopic records. It is also possible to define initial conditions to obtain a particular set of data, such as the distribution of closed times starting at the time when a channel closed. These sets of data contain combinations of transition rates that differ from the relaxation rates seen in macroscopic records and are often more readily interpretable. Finally, the temporal association between single-channel events often can give information about how various states of the receptor are connected. The major drawback to single-channel records is time, since statistically reliable data sets must be built up by examination of a large number of individual transitions.

1. STATES WITH OPEN CHANNELS

Although it is not yet known exactly how many states with open channels the AChR has, it clearly is more than one. It has been known for some years that there are two types of receptor, initially described as innervated junctional and denervated nonjunctional in adult muscle, that differ in their open channel states (29). Innervated junctional channels have a larger mean single-channel conductance than do denervated nonjunctional channels in amphibian and mammalian muscle fibers, as well as a shorter apparent mean open time. Since the denervated nonjunctional receptors are synthesized after denervation (12), it seemed likely that the two functional classes resulted from the existence of two physically distinct types of receptor. However, studies of muscle development led to the idea that the mean open time for a population of receptors can change without the complete replacement of all parts of each receptor (28). Also, single-channel records have indicated that the same AChR can show both single-channel conductance levels (refs. 10, 36, and 40 and Steinbach, unpublished observations). This indication is based on rare observations of multilevel openings of a single channel, showing transitions between the conductance levels of the two types of receptor. These transitions have been observed so rarely with AChRs that they have not been quantitatively examined, but other channels activated by neurotransmitters show several conductance levels with far more frequent transitions among them (16).

A second open channel state, or group of states, of the AChR has been termed a *subconductance level*. This state has been reported most often in studies of AChRs on cultured muscle cells (3,4,17,35,50,51). The conductance and duration of subconductance states vary in different preparations; the conductance is 10–30% of the main level, and the mean duration is usually shorter than that of the main level. Subconductance levels are rarer for AChRs at the adult neuromuscular junction (10,11,25) and for receptors on some cultured cells (42,45).

Finally, Sigworth (41) has studied the properties of the open channel current noise during a main conductance dwell for AChRs on cultured rat myotubes. He favors the idea that the main conductance level actually is composed of several closely spaced levels with dwell times of a few milliseconds at each level.

The surfeit of information on open channel conductance for the AChR channel leads to two complications. The first is semantic. The nomenclature is clearly inadequate; it probably arose because, on the basis of macroscopic measurements and the initial single-channel records, AChRs were expected to show only one open channel conductance—the "main" level. Hence the lower conductance dwells must be subconductance states. It still seems adequate to call the (group of closely spaced) larger conductance states the "main conductance level" and the group of states with conductances less than 50% of the main level "subconductance levels."

The second complication is more fundamental. How do we account for the various open channel states in our understanding of receptor function, particularly in attempts to quantitatively analyze function? In physiological terms, relatively little current flows during subconductance dwells, but the existence of these dwells must be accounted for when the data are interpreted. Two studies have begun the task of characterizing and analyzing subconductance states (3,4,50). Most of the analysis of receptor activation, however, has been done with the use of systems where subconductance levels are very rarely seen in the data (10,11,25,42,44,45). Because subconductance levels are not resolved in these records, it has been assumed that they do not contribute significantly to the data. This approach is clearly ad hoc and, as Sachs (35) has pointed out, makes the skeptical observer a little suspicious of any interpretation of the data. However, the assumption does make it possible to extract estimates for rate constants of considerable physiological interest. If the estimates are obtained under several conditions and are internally consistent, the simplifying assumption may be justified.

The postulated existence of several interconnected main conductance open states, with slightly different channel conductances, means that even the full openings should show complicated kinetic behavior. At the present level of understanding, however, full openings are simply lumped into one state. This is reasonable, so long as the rates for moving between adjacent open states are higher than those for closing.

2. KINETICALLY DISTINCT STATES

How many states of the AChR can be distinguished on the basis of their mean durations? This is the topic for the rest of this chapter. Most of the discussion will involve two measurements: "closed times" and "open times" (Fig. 1). A closed time is the length of time spent at baseline, starting with the time at which a channel closed. An open time is the length of time a single channel stays open, excluding multiple openings. Neither open times nor closed times have a directly comparable measure in macroscopic measurements. A "burst duration" can also be measured in single-channel records as the time between the first opening in a closely spaced series of openings to the last closing of the series. The mean burst duration is the same as the macroscopic mean open time, at low agonist concentrations.

The analysis of single-channel data has been done in terms of kinetic states on the basis of absolute rate theory in chemical kinetics. Each state of the receptor is assumed to be a relatively stable conformation of the receptor, which the receptor adopts for a greater or lesser period of time. The length of time a state persists depends on how readily the receptor can move to another state. In the analysis it is assumed that the time spent in transition to another state is considerably shorter than that spent in any distinguishable state. For example, the channel closing rate does not reflect the amount of time required to swing a gate closed; instead, it reflects the average length of time required for a receptor with an open channel to gain enough thermal energy to

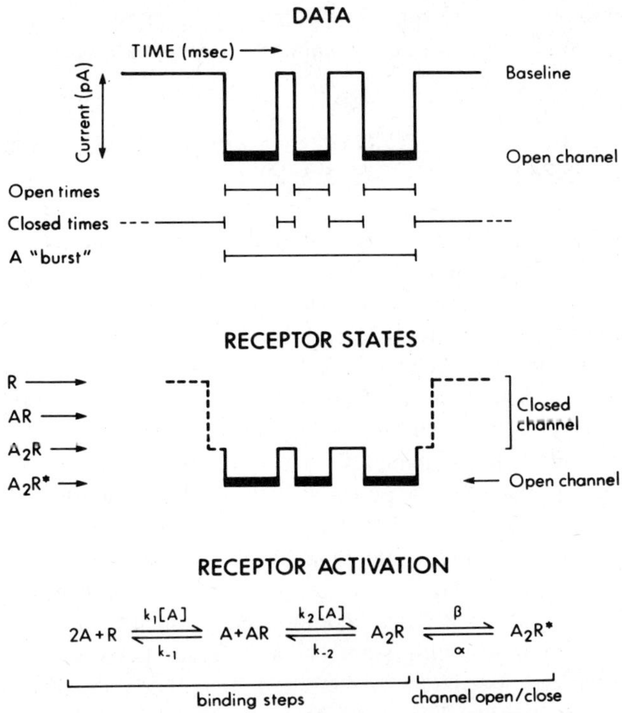

Figure 1. This figure illustrates the data obtained and the framework for analysis. The top portion shows an idealized data segment and illustrates the definitions of closed time, open time, and burst duration. The middle portion shows transitions between several receptor states. (The receptor activation scheme is shown at the bottom of the figure.) Three receptor states with closed channels are shown (R, AR, and A_2R), in which R represents a receptor with a closed ion channel and A an agonist molecule. A single state with a conducting ion channel exists (A_2R^*) with two bound agonist molecules. Before a channel opening occurs, the receptor can make multiple transitions between states with closed channels (dashed lines). During a burst at low concentration, the receptor oscillates between doubly liganded states with open and closed channels (A_2R^* and A_2R). At the end of a burst, the receptor begins another series of transitions between states with closed ion channels. The bottom section of this figure illustrates the simplest kinetic scheme for receptor activation (see text). The rate constants for transitions in the scheme are shown; ligand-binding and channel open–close steps are indicated.

leave the open channel state. The actual process of closing the gate is thought to take a very short period of time. The analysis then continues with the idea that a given state does not change with time: the rate for leaving a given state is constant no matter how long the occupancy of the state has lasted and no matter how the state was reached. The application of this approach to channel kinetics has been described by Stevens (49), Colquhoun and Hawkes (6–8), and Fredkin et al. (15). Some of the assumptions underlying the approach have recently been questioned (14,18).

With these assumptions, the distributions of open and closed times is described by the sums of a number of exponential components. The numbers of open or closed states are greater than or equal to the number of components seen in the data.

3. ACETYLCHOLINE RECEPTORS ON CLONAL MAMMALIAN BC3H-1 CELLS

Mammalian BC3H-1 cells are a clone of cells derived from an intracranial tumor in a C3H mouse, which grow and differentiate as mononucleated cells in culture (38). The progenitor cell type for this tumor is not known; it may have been vascular smooth muscle. Differentiated BC3H-1 cells synthesize the subunits for the skeletal muscle nicotinic AChR and incorporate functional receptors into their surface membrane. There is now a great deal of information on the receptor subunits in BC3H-1 cells, including their primary sequences (5,24), synthesis, processing, and degradation (Chapter 5, this volume). Further, the basic features of ligand binding and receptor state changes have been characterized by studies of ion flux and ligand binding (46,47). Recently we have been studying receptor activation on these cells, using single-channel records. Our basic goal has been to obtain estimates for some of the rates governing receptor activation. To do this, we have had to sort out the different components seen in open times and closed times and associate them with steps in a receptor activation scheme. The experimental variables we have manipulated have been the activating ligand (agonist) used to elicit channel openings and the concentration of agonist applied. The experimental methods are described in the original papers, which should be consulted for details (42–45).

These cells are rather nice for biophysical studies of the AChR, as their membrane contains few other channel types. Also, in the cells we have used, the AChRs behave in a simple way. Subconductance levels are very rare, and the receptor population appears to be homogeneous. The cells are induced to differentiate by reducing the concentration of serum added to the growth medium. Receptor function is then studied 10–20 days later. At longer times after inducing differentiation, the AChR population is more diverse, showing two main conductance levels, subconductance states, and several kinetic classes. We have not analyzed the data from these older cells as yet and do not know the mechanism by which a homogeneous population of AChRs on young cells is replaced by the more complicated situation seen in older cells. All the data presented and discussed pertain to the AChRs on young BC3H-1 cells. All data were obtained from on-cell patches at 10–11°C. Currents were recorded with high K^+ solutions in the pipette.

Only one main conductance level is seen. Single-channel events of this level are not seen in the absence of agonist or after cells are treated with α-neurotoxin (42,44). At low agonist concentrations, channel openings are often seen in "bursts" (Fig. 2). Several openings occur with short gaps between them, and then a long closed period occurs, separating the burst from the next opening. Three

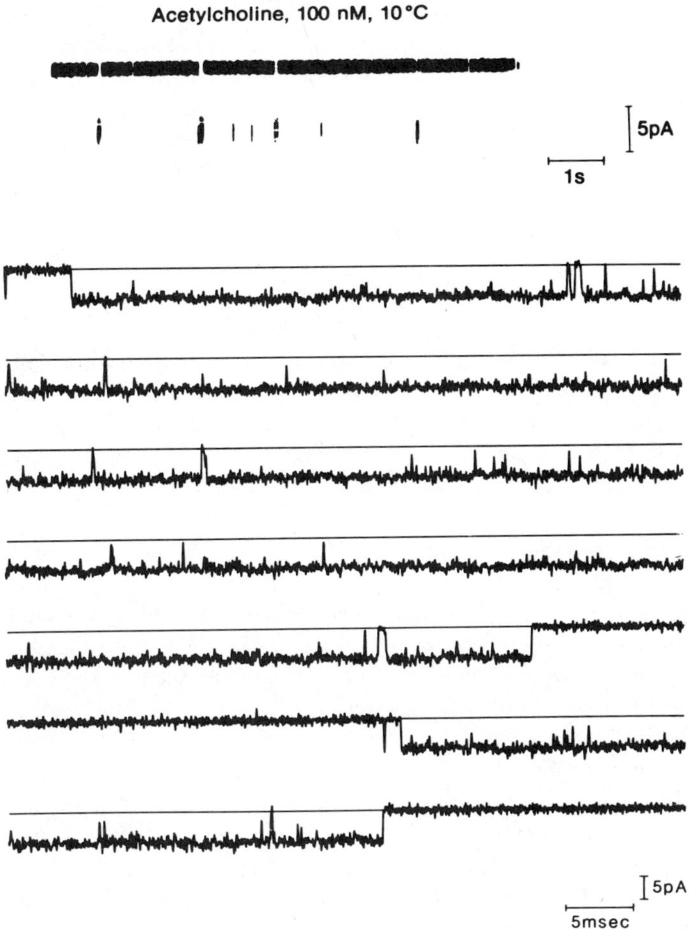

Figure 2. Some data traces obtained with a low concentration of ACh are shown. The top portion shows a section of a record at a slow sweep speed. Note that channel openings appear separated by long closed periods; these closed intervals constitute the longest component in the closed duration distribution (mean duration >1 sec). The lower traces show data at a faster sweep speed. Openings shown at slow sweep speed in the upper portion actually consist of bursts of openings with short-duration closures separating openings. The first five traces in the lower portion are sequential and show a relatively long burst (about 200 msec long). The bottom two traces show a second burst (about 50 msec long). The closures within bursts make up the brief duration components of the closed time distribution (Fig. 4) (−160 mV; 10°C; 100 n*M* ACh; upper trace filtered at 1000 Hz, lower at 4000 Hz).

components are seen in the closed time histogram at low concentrations, indicating that there are at least three closed states. The long closed periods reflect the time intervals between independent activation events. During one, none of the receptors in a patch opens. The brief closed periods result from transient closings of the channel of a receptor with bound agonist molecules (6).

At high concentration the records show long closed periods (seconds long) separating "clusters" of activity—a series of openings separated by gaps lasting between tens of microseconds and a few hundred milliseconds (Fig. 3). Within a cluster the openings occur in "groups," a series of openings separated by closings of less than 10 msec, with groups separated by the longer intracluster closed periods. The longest, intercluster, closed periods are thought to reflect a long-lived desensitized state (37). During such a gap, all the receptors are refractory to activation, even though they have bound agonist molecules for much of the time. During a cluster a single receptor leaves this long-lived closed state and repeatedly opens and closes. The longest of the intracluster closed periods is also considered to reflect a shorter-lived desensitized state (37).

How are we to interpret these data? The open times contain information on the number of receptor states with open channels and the rates for leaving open states. The closed times contain information on the number of receptor states with closed channels and the rates for leaving closed states. Finally, the temporal sequence or association of dwells can give information on how states are connected.

3.1 Open States

There are at least two kinetically distinct open states for the AChR channel, as there are two exponential components in the open-duration histogram (42,44,45). The mean duration of the brief openings is about 100 μsec and that of the long-duration openings, 20–30 msec. Brief openings make up about 30% of the total number of openings seen in the records.

A possible interpretation would be that there are two completely separate types of channel, say, fast and slow ones. This is quite unlikely, though, because the brief-duration and long-duration openings do not occur independently of each other. Instead, brief openings occur close to long openings significantly more often than expected for a chance association (42). It seems, therefore, that the same AChR can make both brief and long openings. It has been suggested that brief openings might arise from receptors with only a single ACh molecule bound, whereas long ones would arise from doubly liganded receptors (9,50). If this were the case, the brief openings should constitute a progressively smaller proportion of all the openings seen as the agonist concentration is raised (since fewer receptors would have only one agonist molecule bound). With the AChRs on BC3H-1 cells, this prediction is clearly incorrect. We have used four agonists, and in every case there is little or no change in the relative numbers of brief and long openings over a wide concentration range (42,44).

How are brief- and long-duration open states connected? We don't know the answer yet, but both do not arise from the same closed state (i.e., they do not occur in a linear sequence, or a simple Y branch.) We reach this conclusion because there are long closed periods both before and after most brief openings. If brief and long openings arose from the same closed state, the distribution of closed periods around them should be the same. Since they are not the same, brief

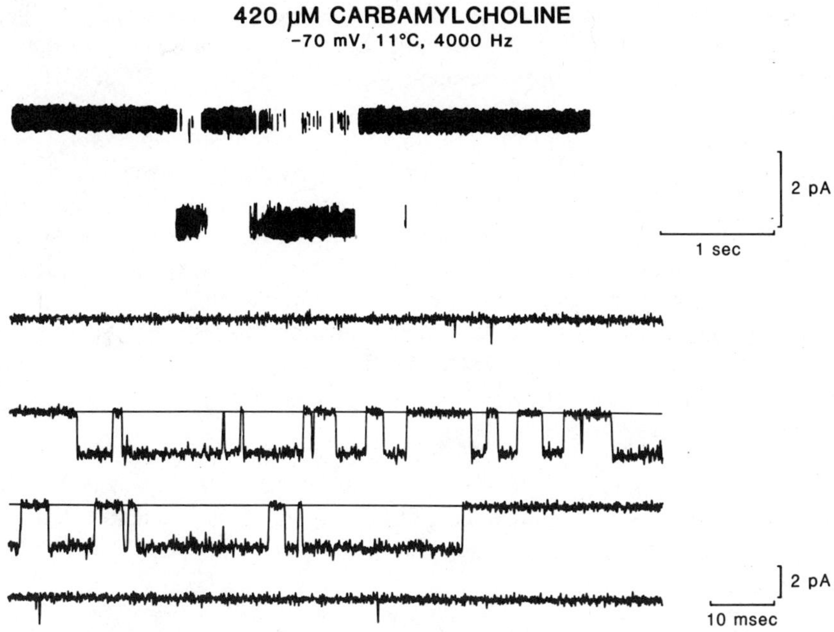

Figure 3. Data traces obtained at 420 μ*M* Carb illustrate the features of the data obtained at high agonist concentrations. The upper trace shows a single cluster of openings, with long-duration closed periods preceding and following the cluster. This cluster is composed of two groups of openings, separated by a closure of about 400 msec. The closed periods illustrated here make up the two longest duration components of the closed time histogram at high agonist concentrations (mean durations of several seconds and about 300 msec). The bottom traces show data at a higher sweep speed. The top and bottom traces show isolated, brief-duration openings. The middle two traces show a continuous record of a single group of openings. Each group consists of a number of openings separated by brief closed periods. The closed periods within groups constitute the brief-duration components seen in the closed time histogram at high agonist concentrations (Fig. 5) (−70 mV; 11°C; 420 μ*M* Carb; all traces filtered at 4000 Hz).

openings and long openings must open from and close to different closed states. (Note that according to the mono- and diliganded hypothesis, the open states are connected to different closed states.)

The overall picture, then, in terms of open states is that AChRs on BC3H-1 cells have at least two kinetically distinct open states of the same conductance, brief duration and long duration. The same receptor can produce either brief or long openings, and both brief and long openings arise from receptors with the same number of bound ligands. Brief and long openings do not arise from the same closed state.

Brief openings by themselves are probably not of any great physiological relevance, as relatively little current flows during a brief opening. They also are less common than long openings. However, the fact that receptors appear to enter a very inefficient mode of operation may be significant. A mode characterized by brief openings and long closed periods could result in significant changes in postsynaptic responsiveness if a reasonable proportion of the receptor population entered it. A class of brief-duration openings like those seen in BC3H-1 cells has been seen in records from AChRs at frog neuromuscular junctions (see ref. 10 and and section 4).

We have analyzed the kinetic behavior of the long duration openings, the major type of opening seen. Since most brief openings occur in isolation, we have concentrated on studying the behavior of bursts or groups of long-duration openings separated by short closed periods. This approach assumes that it is legitimate to separate the processes resulting in brief- and long-duration openings, which appears reasonable as a first approximation simply because most brief openings do occur as isolated events.

3.2 Closed States

Previous studies of agonist-elicited fluxes in BC3H-1 cells have provided evidence that most receptors with open channels have two bound agonist molecules (1,47). The simplest scheme for receptor activation, therefore, is Scheme 1 (described in Fig. 1):

$$2A + R \underset{k_{-1}}{\overset{k_1[A]}{\rightleftharpoons}} A + R \underset{k_{-2}}{\overset{k_2[A]}{\rightleftharpoons}} A_2R \underset{\alpha}{\overset{\beta}{\rightleftharpoons}} A_2R^*$$

Scheme 1

At a very low concentration of agonist the receptor will spend most of the time with no bound agonist. Occasionally two agonist molecules will be bound and the channel may open, producing a signal that can be seen (Fig. 1). A burst of successive openings will occur when an open channel closes but reopens before either agonist molecule dissociates, producing short-duration closings or gaps reflecting dwells in the doubly liganded closed state, A_2R. Sometimes an agonist molecule will dissociate, resulting in a monoliganded receptor. Occasionally (very rarely, at low agonist concentrations), a second ligand will bind to the monoliganded receptor and the channel may reopen. The dwell in the compound state, $AR + A_2R$, will result in a second and longer-duration class of closed times. Usually, at low concentration both bound agonist molecules will dissociate and the receptor will end up with no bound agonist molecules, R. This progression results in a long-duration gap reflecting the time it takes for the receptor to lose both bound agonists, then get them back and finally open (the compound state $R + AR + A_2R$). Even at low concentration, there are three components in the distribution of closed times, but the middle component has a vanishingly small amplitude and so only two components would be experimentally resolvable.

As the agonist concentration is increased, all three components may become apparent. At very high concentrations the forward binding rate becomes much larger than the unbinding rate, and so dwells in unliganded or monoliganded states become uncommon and of very short duration. At high concentrations, therefore, only one component will be resolvable in the closed time distribution, which will reflect only the time spent in A_2R before an opening—the channel opening rate.

Strictly speaking, this discussion applies only to an experimental situation in which the behavior of a single channel is monitored. Practically, though, we can apply the same ideas to a patch with many channels, particularly if the probability that any channel is open is very low (19,45). At low agonist concentrations most receptors are unliganded with closed channels and at high concentrations most receptors are desensitized with closed channels. Data are most difficult for us to obtain at intermediate concentrations—in our experiments concentrations of ACh between about 1 and 10 μM and of carbamylcholine between 30 and 300 μM—at which concentrations channel activity is high and overlapping openings common.

In terms of Scheme 1, then, we would expect that at low agonist concentrations the distribution of closed times would show two components, at high concentrations only one, and at intermediate concentrations two or three components. The relative numbers and mean durations of these components would change with agonist concentration. Most likely the proportions and durations would also depend on the agonist used, reflecting differences among agonists in binding and channel opening rates. Long-duration closed time components cannot be interpreted immediately, as they depend on the number of receptors in the patch. Hence our attention is focused on relatively short duration closed periods, those within "bursts" at low concentrations and "groups" at high concentrations.

3.3 Closed Times Within Bursts at Low Agonist Concentrations

There are two short-duration closed time components at low agonist concentrations, rather than the single one expected (Fig. 4). Two possibilities immediately come to mind—either the concentration is not really low or Scheme 1 does not describe the behavior of the AChR on BC3H-1 cells. The agonist concentrations used are low with respect to the dissociation constant, insofar as we can determine: (1) they are orders of magnitude lower than apparent K_D values determined from binding and flux experiments (46,47), and (2) they are also low compared to the K_D extracted from single-channel records at high agonist concentrations (see the following paragraphs). Therefore, it is likely that Scheme 1 does not describe receptor function and must be modified. The identification of an appropriate scheme rests on two approaches: (1) the use of several agonists to examine the agonist specificity of the two components at low concentration and (2) experimentation at higher concentrations, to obtain independent estimates for rate constants and dissociation constants.

We used four agonists, three effective agonists (ACh; carbamylcholine [Carb]; and suberyldicholine [Sub]) and one ineffective agonist (+ −dimethyltubocurarine [DMT]). The compound DMT was chosen because it blocks agonist-elicited sodium fluxes in BC3H-1 cells (47) at low concentrations (ED_{50} of 200 nM) and hence is likely to have a low channel opening rate. All four agonists produce closed time

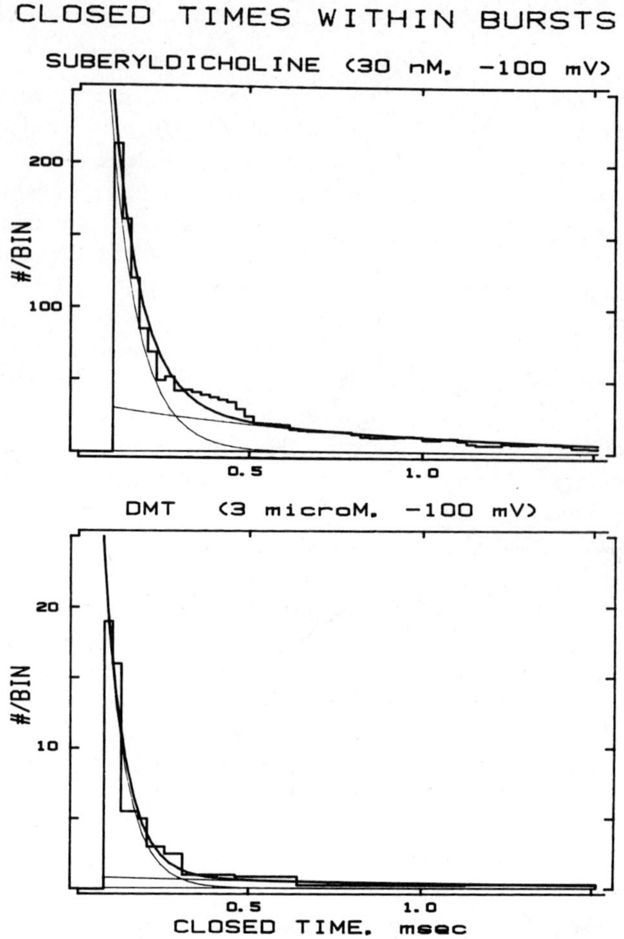

Figure 4. This figure shows histograms of the brief gaps seen within bursts of openings at low concentrations of two agonists. The long-duration closed periods are not shown in these histograms. For both agonists (Sub upper panel, DMT lower panel) the total closed time histogram was fit with the sum of three exponential components. The exponentials are shown as thin lines, and the sum of the three exponentials is shown as a thick line over the data histogram. (Note that the longest-duration component on this time scale is shown as the flat thin line.) The data are shown in 25-μsec bins. Closed periods lasting less than 100 μsec could not be fully resolved, so they were ignored in data reduction and are not shown in the histograms. The exponentials fitted to the data had the following mean durations and fractions of the total closed periods (corrected for the 100-μsec minimum resolution).

	Slow	Intermediate	Brief
Sub	2.5 sec	500 μsec	50 μsec
(n = 3167)	(0.33)	(0.33)	(0.33)
DMT	1.1 sec	800 μsec	60 μsec
(n = 1300)	(0.85)	(0.03)	(0.15)

With respect to brief closed periods within bursts (the closed periods shown in this figure), Sub and DMT produce similar brief gaps. Intermediate gaps are less common with DMT.

TABLE 1. COMPARISON OF BRIEF CLOSURES[a]

Agonist (Concentration)	Duration (μsec)	Rate of Occurrence (sec^{-1})	β (sec^{-1})	k_{-2} (sec^{-1})	β/(β + α)
ACh (100 n*M*)	48 ± 8	48 ± 24	14,000	6,500	0.99
Carb (2 μ*M*)	50 ± 9	59 ± 30	12,000	8,600	0.99
Sub (50 n*M*)	61 ± 15	49 ± 24	12,000	5,000	0.99
DMT (3 μ*M*)	59 ± 7	70 ± 29	5,500	11,500	0.96

[a] This table summarizes the properties of brief closures seen within bursts of openings at low agonist concentration. The mean duration and rate of occurrence (per second of open time) of brief closures are very similar for the four agonists used. The three columns to the right give the channel opening rate β, agonist dissociation rate k_{-2}, and the probability of being open β/(β + α) calculated under the assumption that brief closures reflect transitions $A_2R^* \rightarrow A_2R \rightarrow A_2R^*$.

distributions showing the brief gap component (Fig. 4). Indeed, the properties of brief gaps (their rate of occurrence per second of open time and their mean duration) are essentially identical for all four agonists (Table 1). If these gaps are interpreted as resulting from transitions $A_2R^* \rightarrow A_2R \rightarrow A_2R^*$, all agonists have extremely large channel opening rates β, very high probabilities of being open [β/(β + α)], and high dissociation rates k_{-2}. It had been expected that Carb would be less effective than ACh or Sub (1,48) and that DMT would have a very low channel opening rate. This comparison, therefore, indicates that the brief gaps are not likely to reflect transitions in the activation pathway between R and A_2R^*. Instead, the brief gaps result from "flickers" of the open channel to a nonconducting state (A_2R', Scheme 2). (Parenthetically, these flickers could occur to a closed state between A_2R and A_2R^*. This is unlikely, since interpretation of our data in terms of this idea results in negative rate constants for some transitions.)

$$2A + R \rightleftharpoons A + AR \rightleftharpoons A_2R \rightleftharpoons A_2R^* \rightleftharpoons A_2R'$$

Scheme 2

These observations indicate that the intermediate closed time component is more likely to reflect activation processes. The same data were then analyzed in terms of this idea, resulting in considerably lower values of β and k_{-2} (Table 2). These estimates are in line with earlier observations, as Carb is a less effective agonist in terms of β and of β/(β + α). We should point out, however, that data obtained with DMT also show a small but discernible intermediate-duration component. Analysis of this component shows DMT to be a relatively ineffective agonist, but more effective than expected. We do not have a satisfactory resolution for this observation, although channel block by DMT could play a role (44,45).

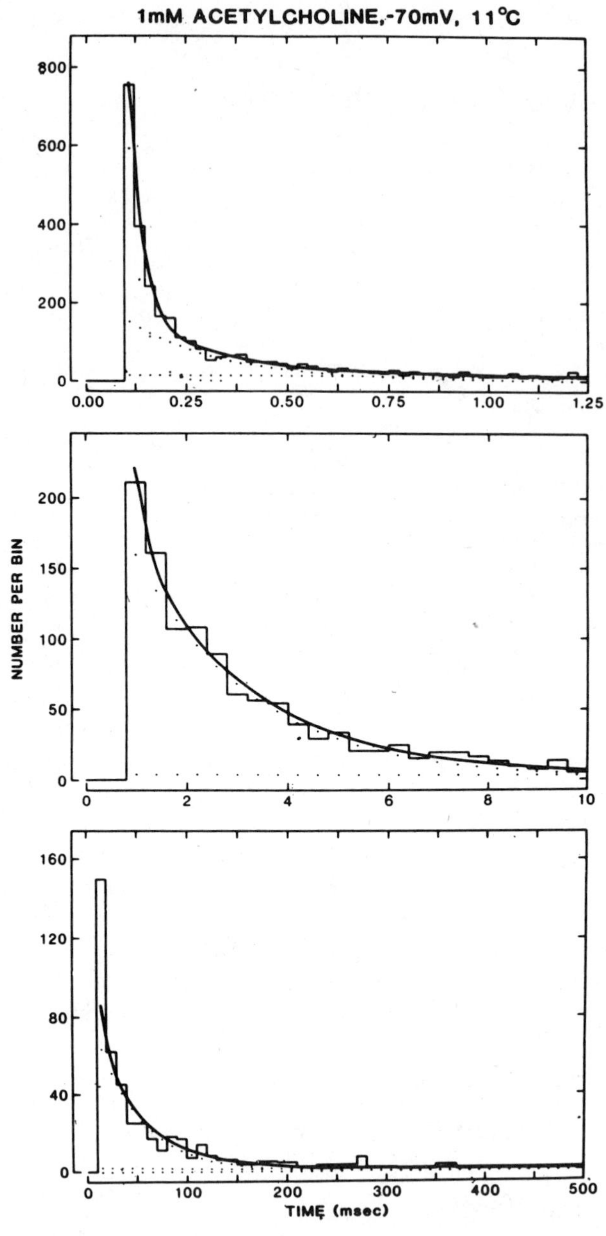

Figure 5. This figure shows the closed periods within groups elicited with 1 m*M* ACh. The total closed duration histogram was fitted with the sum of six exponential components; this figure shows the four short-duration closed times within groups. The data are shown in three histograms to illustrate the components more clearly. In all, 4487 closed periods were collected. The two longest-duration components (mean durations 500 msec and 5.6 sec) are not shown in this figure. (The fitted exponentials can be seen in the two dotted lines at the bottom of the lower panel.) In each panel the continuous line through the data shows the sum of the six exponential components fitted, and dotted lines show individual components. The top panel shows the two briefest components (bin size 23 μsec, maximum duration shown 1.25 msec).

TABLE 2. RATE CONSTANTS FOR ACTIVATION BY DIFFERENT AGONISTS[a]

$$\xleftarrow[k_{-2}]{} A_2R \underset{\alpha}{\overset{\beta}{\rightleftharpoons}} A_2R^*$$

Agonist	$\beta(\text{sec}^{-1})$	$k_{-2}(\text{sec}^{-1})$	$\alpha(\text{sec}^{-1})$	$\beta/(\beta + \alpha)$
ACh	480	1000	23	0.93
Carb	190	1000	51	0.77
Sub	1310	840	45	0.97
DMT	30	920	163	0.15

[a] This table gives estimates for three rate constants for three transitions in receptor activation (shown at the top of the table). The estimates were obtained by analyzing intermediate gaps as transitions $A_2R^* \rightarrow A_2R \rightarrow A_2R^*$.

Scheme 2 is a minimal modification of the simplest scheme for receptor activation, adding the idea that an open AChR channel can briefly close to a state other than A_2R.

3.4 Closed Times Within Groups at High Agonist Concentrations

We have used only ACh and Carb in these experiments, as Sub and DMT are too effective as channel blockers (43,44). In the simplest case, we would see only one component in the closed time histogram at a saturating agonist concentration, reflecting the transition $A_2R \rightarrow A_2R^*$. This simple situation cannot apply, as we also expect two long-duration components reflecting desensitization (37), and brief components reflecting flickers (see section 3.4) and channel block by agonists (32,43).

Even at 1 m*M* ACh, a concentration that is likely to be saturating, we see six exponential components in the closed time histogram. The mean durations range from about 50 μsec to several seconds, separated by factors of sixfold or more. Even after discounting the two longest-duration closed time classes as resulting from desensitization, we are left with four closed times within groups (Fig. 5). The first question is whether any one is likely to reflect activation processes.

There is one closed time component with a mean duration of about 2.3 msec, close to the inverse of the channel opening rate β estimated at low concentration. This closed time occurs at a rate of 32 sec^{-1} of channel open time, close to the channel closing rate α estimated at low concentration. These observations suggest that this component in the closed time histogram at high concentration might well reflect the

The exponentials fitted have mean durations of 30 and 400 μsec. The middle panel (bin size 400 μsec; the first 800 μsec have been blanked for clarity) is dominated by a single component. The mean duration fitted is 2.3 msec (a small amount of the 400-μsec component is seen in the first bin from 800 to 1200 μsec). The bottom panel (bin size 10 msec, first 10 msec blanked) is dominated by a component fitted with a mean duration of 45 msec.

transition $A_2R \rightarrow A_2R^*$. We can now examine another experimental parameter, the probability of being open in a group of openings P_O. At a saturating concentration of agonist, P_O should equal the value of $\beta/(\beta + \alpha)$ from low-concentration measurements. To estimate P_O, the "intergroup" gap duration was defined to be three times the mean duration of the apparent activation-related closed time, and the proportion of time spent open in a group of openings separated by shorter closings was determined. The value of P_O is 0.91, close to that predicted from low-concentration data (0.93).

What about the other three classes of closed times within groups? The briefest one (about 50 μsec) probably is a combination of flickering and channel block. The other two (300 μsec and 30 msec) are considered further below. These observations made it reasonable to examine the closed time histograms at various ACh concentrations. A component reflecting activation should become briefer as the concentration is raised, saturating at the inverse of the channel opening rate, and P_O should increase, saturating at $\beta/(\beta + \alpha)$.

These predictions are reasonably well fulfilled for ACh (Fig. 6). The relationship makes it possible to estimate the dissociation constant for ACh binding. Although there is a fair degree of uncertainty, there appears to be some positive cooperativity in

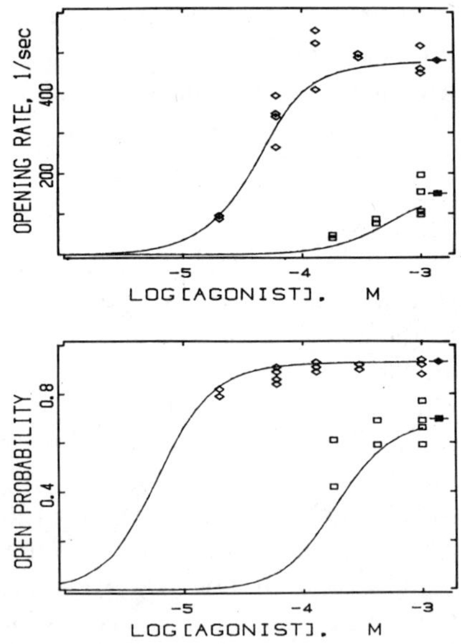

Figure 6. This figure shows plots of two parameters as a function of agonist concentration. The upper panel shows the inverse of the mean duration of a major component in the closed time distribution, which we have identified as reflecting the effective opening rate (see text). The lower panel shows the probability of being open within a group P_O. In each panel, diamonds show data obtained by using ACh and squares show data with Carb. The filled symbols at the right margin show saturated values predicted from low-concentration data [upper panel, β; lower panel, $\beta/(\beta + \alpha)$]. The continuous curves show theoretical predictions calculated with the kinetic constants given in Table 3.

TABLE 3. ACTIVATION "CORE"[a]

$$2A + R \underset{k_{-1}}{\overset{k_1[A]}{\rightleftharpoons}} A + AR \underset{k_{-2}}{\overset{k_2[A]}{\rightleftharpoons}} A_2R \underset{\alpha}{\overset{\beta}{\rightleftharpoons}} A_2R^*$$

	ACh	Carb	Sub
$\beta(\text{sec}^{-1})$	480	150	740
k_{-2}	1000	1000	800
α	35	65	56
$K_{D_1}(\mu M)$	50	1000	—
K_{D_2}	5	100	—
$k_2(M^{-1}\ \text{sec}^{-1})$	1×10^8	1×10^7	—
k_1	1×10^7	3×10^5	—
$k_{-1}(\text{sec}^{-1})$	500	300	—

[a] This table gives estimates for rate constants in a four-state linear core activation scheme for AChRs on BC3H-1 cells. The activation scheme is shown at the top of the table; the rate constants given were estimated at 11°C and a membrane potential of −70 mV. The estimate for β was made at both low and high agonist concentrations; k_{-2} and α were estimated at low concentrations. The macroscopic dissociation constants for the two binding steps were estimated from the concentration dependences of the opening rate and P_O (Fig. 6). The value for k_2 was calculated from k_{-2} and K_{D_2}; k_1 and k_{-1} were calculated from k_{D_1}, with k_{-1} constrained to describe adequately closed time distributions at intermediate agonist concentrations. All the values given are macroscopic values and have not been corrected for statistical factors (e.g., k_{-2} reflects the total rate of loss of either bound agonist molecule).

binding; the macroscopic dissociation constant for the first binding step K_{D_1} is about 50 μM and K_{D_2} is about 5 μM.

We have less extensive data with Carb, and the data are more difficult to interpret. There are two reasons for the difficulty in interpretation: (1) we have been unable to obtain clear saturation with Carb, even at 1 mM; and (2) the predicted activation-related closed times with Carb are within threefold of some of the additional closed times seen at high ACh concentrations. This means that we are not able to reliably separate out activation components when Carb is used at high concentrations, assuming that these additional components are also present. The data with Carb are in qualitative agreement with predictions from low-concentration data, however (Fig. 6). Our current estimates for dissociation constants for Carb are K_{D_1} = 1000 μM and K_{D_2} = 100 μM.

To summarize, the data at high agonist concentrations show some major features that are in agreement with our interpretation of the data at low concentrations in terms of Scheme 2. However, the high-concentration data also show that several additional states of the receptor with closed channels must exist. In view of the number of states the receptor is able to adopt at various agonist concentrations, it is almost surprising that any consistent pattern emerges over the concentration range studied. The major argument in favor of our interpretation of the data is quite simply the observation that the data obtained with the use of several agonists, over a wide concentration range, all are consistent with a simple scheme for the main aspects of AChR activation on BC3H-1 cells.

3.5 Activation of ACh Receptors on BC3H-1 Cells

Overall, then, studies of different agonists over a wide concentration range give a consistent picture of a "core" of receptor activation (Scheme 1; see also Table 3).

With estimates for β, k_{-2}, and the dissociation constants for agonist binding, we can now analyze the data at low concentration to estimate the number of channels required to produce the observed frequency of bursts of long-duration openings (45). We obtain values of about 50–100 receptors per patch (47 with 100 n*M* ACh, 89 with 2 μ*M* Carb, and 58 with 50 n*M* Sub, guessing that Sub binds twice as tightly as ACh). These numbers are rather close to the expected 50–200 AChRs/μm^2 based on studies of α-neurotoxin binding (33), since our patches probably contain 1–2 μm^2 of cell membrane.

The agonist dissociation rate estimated at low concentration k_{-2} shows essentially no dependence on membrane potential, in line with the idea that binding occurs at a site outside the membrane potential gradient. Both channel opening β and closing α rates depend on potential. The channel opening rate β increases with hyperpolarization *e*-fold per 70 mV, whereas α decreases *e*-fold per 150 mV. A burst elicited by a low concentration of ACh at -100 mV contains about 1.4 openings, on average, so the "mean burst duration" underestimates the true channel closing rate. Since β increases with hyperpolarization, the frequency of bursts at low agonist concentration should also increase. This is seen (34,45), and the observed voltage dependence is close to the predicted value (45).

Two points remain to be considered: how do these observations compare to those made on other nicotinic AChRs, and what might be the nature of the states not included in the "core?"

4. COMPARISON WITH OTHER MUSCLE AChRS

What type of nicotinic AChRs are expressed on BC3H-1 cells? The mean burst duration is longer than the time constant for the decay of miniature endplate currents at adult mouse junctions (26). Also, the values for β and k_{-2} are almost certainly too small to account for the rise of miniature endplate currents (26). The single-channel currents resemble most closely events recorded from neonatal mouse muscle fibers (J. Steele, personal communication), showing a longer burst duration and a lower main conductance level than adult junctional receptors, but far fewer sublevels than adult nonjunctional receptors. Therefore, it seems likely that the homologous receptor on skeletal muscle is expressed early in development and is probably not seen in the adult.

Most studies of both junctional and nonjunctional receptors have reported two kinetic classes of open channels with the same conductance (3,9,10,20,21,50). One exception is the receptors at snake junctions, which usually show only a single component in the open time distribution (11,25). When the point has been examined, the data have indicated that brief and long openings do not arise from the same closed state (10,20,22,42,45). With BC3H-1 receptors the relative frequency of brief openings does not change much with concentration (42,44), indicating that the same number of agonist molecules are bound to receptors producing brief and long openings. For frog junctional receptors, on the other hand, the proportion of openings that are brief decreases markedly at higher agonist concentrations, although even at high concentrations about 10% of the total openings are brief (10). These observations suggest that some of the brief openings seen at frog junctions with a low agonist concentration result from monoliganded receptors; some other mechanism is required to account for the brief openings seen at high concentrations of agonist at frog junctions or on BC3H-1 cells.

Few reports have appeared on the distribution of closed times at high agonist concentrations. Two long-duration closed times have been seen (31,37,39) that presumably reflect fast and slow desensitization processes (2,13,30). There are at least three brief gap components within groups of openings recorded from frog junctions with 100 μM ACh, the only data that have been published (31).

All AChRs studied have shown at least one short duration closed time component at low agonist concentrations, when adequate temporal resolution is used. Receptors at snake junctions show a single major brief gap component (11,25), although the method of data analysis used did not separate exponential components. Receptors on BC3H-1 cells on primary chick myotubes (3) and at frog junctions (9,10) all show more than one brief closed time component, usually described as brief- and intermediate-duration closings.

Auerbach and Sachs (3,4) have studied the brief closings of chick ACh receptors and have shown that many of them are not true closings but flickers to a subconductance level. This observation certainly suggests the possibility that incompletely resolved brief closings in other systems reflect similar processes.

Data from snake and frog receptors have been analyzed to estimate rates for channel transitions. The channel opening rate β and agonist dissociation rate k_{-2} for ACh estimated at snake junctional receptors are rather similar to the estimates from BC3H-1 receptors (11,25), when corrected for temperature. The estimated voltage dependence for β is roughly comparable, but k_{-2} also shows a steep dependence on voltage at snake junctions (25). Carbamylcholine gave so few brief gaps at snake junctions that rates could not be estimated; presumably β is less than for ACh and k_{-2} is greater.

A more complete study has been made of frog junctional receptors (9,10). The closed time distributions at low agonist concentrations are qualitatively identical for receptors on BC3H-1 cells and at frog junctions, showing brief, intermediate, and long closures. However, the brief gap components behave differently in the two preparations. Colquhoun and Sakmann (10) have used four agonists (ACh, Carb, Sub, and a longer-chain subanalogue—DecCh). They concluded that brief gaps with Carb probably reflect channel block by agonist. With the other three agonists, however, block cannot account for the brief gap component. Both the mean brief gap duration and the rate of occurrence are more dependent on the nature of the agonist than in BC3H-1 receptors. Also, the mean duration and rate of occurrence are less dependent on the membrane potential. For these reasons, it is possible that, regardless of any similarities, the "brief gaps" in data from frog junctional receptors and BC3H-1 receptors reflect different processes. Colquhoun and Sakmann have analyzed the brief gaps seen with ACh, Sub, and DecCh in terms of transitions $A_2R^* \rightarrow A_2R \rightarrow A_2R^*$. The rate constants they derive for β and k_{-2} are at least tenfold higher than those obtained from intermediate gaps in records from BC3H-1 cells. For ACh (at 10°C and -130 mV), β is about 30000 sec^{-1} and k_{-2} 8000 sec^{-1}; for Sub, the values are 18,000 sec^{-1} and 2400 sec^{-1}. Neither β nor k_{-2} showed any significant dependence on membrane potential. Ogden and Colquhoun (31) have also determined the probability that a channel is open, P_O, at higher concentrations of ACh and Sub and found that the values exceed 0.9. This is an important point, for if Colquhoun and Sakmann assumed that the intermediate-gap component seen at low concentrations of ACh reflected activation, the predicted maximal value of P_o was less than 0.4. It is also the case that the intermediate-gap component is too long in duration to account for the rise of frog miniature endplate currents. Indeed, an analysis of the time course of miniature endplate currents at lizard endplates has estimated that β is about 20,000 sec^{-1} at room temperature for these receptors (23).

At this point, therefore, it cannot be said that either interpretation of short-duration closed time components is incorrect. The overall consistency of results obtained with BC3H-1 receptors is appealing. However, it seems unlikely that the intermediate-gap component seen at frog neuromuscular junctions could reflect transitions $A_2R^* \rightarrow A_2R \rightarrow A_2R^*$. One problem in interpretation is the assumption that there is only one source for any of the components seen in distributions of durations. For example, there appear to be at least two sources for "brief-duration openings" at frog neuromuscular junctions, with different dependences on agonist concentration. If either brief- or intermediate-duration closings at low concentration had several sources (e.g., both "flickers" and $A_2R^* \rightarrow A_2R \rightarrow A_2R^*$ transitions), the estimates for rate constants calculated assuming that only a single source existed would be inaccurate.

4.1 Physiological Implications

The qualitative conclusion from these experiments is that the AChR on BC3H-1 cells has the major functional properties expected of a receptor that responds quickly to a large, rapid ACh concentration transient. Acetylcholine binds rapidly and in a positively cooperative fashion, so that production of doubly liganded receptors is both rapid and a steep function of ACh concentration. The dissociation rate is also fairly large, resulting in a large dissociation constant, but it can be argued that the rapid loss of bound agonist is appropriate for a receptor that gives a short-lived response to transmitter. The channel opening rate is much larger than the closing rate, so a doubly liganded receptor has a high probability of having an open channel. However, BC3H-1 receptors are physiologically inefficient in two ways: (1) the

TABLE 4. ADDITIONAL STATES OF THE ACh RECEPTOR[a]

State	Characteristics
1. Agonist blocked channel	Mean duration <20 μsec, K_B = 5 m*M*
2. Brief closures ("flickers")	Mean duration 50 μsec, rate of occurrence 60 sec^{-1} of open time; seen at all concentrations
3. A closed state	Lasting about 300 μsec; seen at concentrations >60 μ*M*
4. A closed state	Lasting about 30 msec; seen at concentrations >60 μ*M*
5. Short-lived desensitized state	Mean duration 200–300 msec
6. Long-lived desensitized state	Mean duration >100 sec
7. Brief duration open state	Mean duration 100 μsec, seen at all concentrations
8. One closed state (or more)	Associated with the brief open state

[a] This table lists states of the AChR on BC3H-1 cells that are not shown in Table 3. The parameters given are appropriate for ACh as agonist at 11°C.

channel opening and closing rates are both relatively low, too low to generate the rapid rise and fall of the endplate current; and (2) the opening rate is relatively small compared to the agonist dissociation rate. This means that a doubly liganded receptor with a closed channel has only a one-in-three chance of opening its channel as opposed to losing a bound agonist molecule. The estimated rates for receptors at the frog junction, on the other hand, give the chances as about two out of three. The frog junctional receptors also respond much more rapidly to ACh, as they have higher apparent opening, closing, and agonist dissociation rates.

5. ADDITIONAL STATES OF THE ACh RECEPTOR

Table 3 presented a summary of our current picture of a core scheme for activation of AChRs on BC3H-1 cells. Clearly, this table does not contain all the states that the AChR shows. The additional states that are apparent in the data are summarized in Table 4. Most of these states have been inferred or observed by others.

1. Agonist block of open AChR channels has been reported with BC3H-1 receptors (43) and frog junctional receptors (32).
2. A brief closed state at low agonist concentration (channel flicker) has been reported for receptors on chick primary myotubes by Auerbach and Sachs (3,4).
3. The closed time component lasting about 300 μsec, seen at high ACh concentrations with BC3H-1 receptors, has not been reported for other receptors. The mean duration and rate of occurrence of this component do

not change much with agonist concentration. We do not know what process this component reflects.

4. Magleby and Pallotta (27) have proposed that a "predesensitized" closed channel state exists for frog junctional receptors, lasting less than 30 msec. This may be analogous to the 30-msec closed time component seen with BC3H-1 receptors at high ACh concentrations.

5,6. Two desensitized states have been inferred in many studies of AChRs from several sources (31,37,39).

7. Brief-duration openings have been seen with receptors from several sources. For frog junctional receptors, some appear to result from the opening of channels of monoliganded receptors. Others, however, apparently result from an additional open state of the receptor (see section 4).

8. Brief duration openings generally occur in isolation, implying that they do not have the same adjacent closed state as long-duration openings.

Overall, the states listed in Table 4 are postulated because of the data obtained in studies of AChRs on BC3H-1 cells, but almost all of them have been seen in other studies as well.

6. SUMMARY AND CONCLUSIONS

In summary, the AChRs on BC3H-1 cells behave in a fashion consistent with the major properties expected of muscle nicotinic AChRs. The results we have discussed here provide a quantitative description of some of the steps in receptor activation, and give a general picture of the functional states the AChR adopts.

The results provide a framework for biochemical and ultrastructural studies of the AChR as well. One clear problem is that the receptor with an open channel is a very low probability state at equilibrium, at any concentration. A second problem is the postulated existence of closed channel states associated with brief-duration openings. It is not known what the probability is that a receptor is in such a state, so even "resting" receptors at low agonist concentrations may be in several functional states.

At high agonist concentrations the receptor population goes through several transient states before reaching the final equilibrium state. At 11°C, with a high concentration of ACh (> 0.3 m*M*) receptors with open channels appear rapidly (rate $\sim 500\ sec^{-1}$). Initially, there is a high probability that a receptor is both doubly liganded and has an open channel (0.9, Fig. 6). However, the predesensitized state develops rapidly ($\sim 30\ sec^{-1}$), and eventually the probability reaches about 0.4 that doubly liganded receptors are predesensitized. The short-lived desensitized state develops at a rate of about 3 sec^{-1}. After a second or so, the distribution is about 0.3:0.2:0.5 for AChRs between receptors with open channels, predesensitized receptors, and short-lived desensitized receptors. Long-lived desensitization develops at a rate of about 0.5 sec^{-1}, but at equilibrium the probability that a receptor is in this state is above 0.99.

For AChRs on BC3H-1 cells, and most likely all AChRs, biochemical and structural studies need to be made in a narrow time window to catch an appreciable fraction of the receptors in any of the transient functional states described. Only time-resolved biochemical studies will provide the structural information necessary to work out the relationship between structure and function for the ACh receptor and give some substance to the ghostly kinetic states necessary to describe channel function.

REFERENCES

1. Adams, P. R. (1981). *J. Membrane Biol.*, **58**, 161–174.
2. Anwyl, R. and Narahashi, T. (1980). *Br. J. Pharmacol.*, **69**, 91–98.
3. Auerbach, A. and Sachs, F. (1983). *Biophys. J.*, **42**, 1–11.
4. Auerbach, A. and Sachs, F. (1984). *Biophys. J.*, **45**, 187–198.
5. Boulter, J., Luyten, W., Evans, K., Mason, P., Ballivet, M., Goldman, D., Stengelin, S., Martin, G., Heinemann, S. and Patrick, J. (1985). *J. Neurosci.*, **5**, 2545–2552.
6. Colquhoun, D. and Hawkes, A. G. (1981). *Proc. Roy. Soc. B,* **211**, 205–235.
7. Colquhoun, D. and Hawkes, A. G. (1982). *Phil. Trans. Roy. Soc. Lond. B,* **300**, 1–59.
8. Colquhoun, D. and Hawkes, A. G. (1983). In Sakmann, B. and Neher, E., Eds., *Single Channel Recording*, Plenum, New York, pp. 135–176.
9. Colquhoun, D. and Sakmann, B. (1981). *Nature*, **294**, 464–466.
10. Colquhoun, D. and Sakmann, B. (1985). *J. Physiol.*, **369**, 501–557.
11. Dionne, V. E. and Leibowitz, M. D. (1982) *Biophys. J.*, **39**, 253–261.
12. Fambrough, D. (1970). *Science*, **168**, 372–373.
13. Feltz, A. and Trautmann, A. (1982). *J. Physiol.*, **322**, 257–272.
14. Frauenfelder, H. and Wolynes, P. G. (1985). *Science*, **229**, 337–345.
15. Fredkin, D. R., Montal, M., and Rice, J. A. (1986). In *Proceedings of the Berkeley Conference in Honor of Jerzy Neyman and Jack Kiefer*, Wadsworth, Belmont, CA.
16. Hamill, O. P., Bormann, J., and Sakmann, B. (1983). *Nature*, **305**, 805–808.
17. Hamill, O. P., and Sakmann, B. (1981). *Nature*, **294**, 462–464.
18. Hille, B. (1984). *Ionic Channels of Excitable Membranes*, Sinauer Associates, Sunderland, MA, pp. 329–353.
19. Jackson, M. (1985). *Biophys. J.*, **47**, 129–137.
20. Jackson, M. B., Wong, B. S., Morris, C. E., Lecar, H., and Christian, C. N. (1983). *Biophys. J.*, **42**, 109–114.
21. Labarca, P., Lindstrom, J., and Montal, M. (1984). *J. Neurosci.*, **4**, 502–507.
22. Labarca, P., Rice, J. A., Fredkin, D. R., and Montal, M. (1985). *Biophys. J.*, **47**, 469–478.
23. Land, B. R., Harris, W. V., Salpeter, E. E., and Salpeter, M. M. (1984). *Proc. Natl. Acad. Sci.* (USA), **81**, 1594–1598.
24. LaPolla, R. J., Mayne, K. M., and Davidson, N. (1984). *Proc. Natl. Acad. Sci.* (USA), **81**, 7970–7974.
25. Leibowitz, M. D. and Dionne, V. E. (1984). *Biophys. J.*, **45**, 153–164.
26. Linder, T. M., Pennefather, P., and Quastel, D. M. J. (1984). *J. Gen. Physiol.* **83**, 435–468.
27. Magleby, K. S. and Pallotta, B. S. (1981). *J. Physiol.*, **316**, 225–250.
28. Michler, A. and Sakmann, B. (1980). *Dev. Biol.*, **80**, 1–17.

29. Neher, E. and Sakmann, B. (1976). *J. Physiol.*, **258**, 705–730.
30. Neubig, R. R. and Cohen, J. B. (1980). *Biochemistry*, **19**, 2770–2779.
31. Ogden, D. C. and Colquhoun, D. (1983). *Pflügers Arch. Eur. J. Physiol.*, **399**, 246–248.
32. Ogden, D. C. and Colquhoun, D. (1985). *Proc. Roy. Soc. Lond. B*, **225**, 329–355.
33. Patrick, J., McMillan, J., Wolfson, H., and O'Brien, J. C. (1977). *J. Biol. Chem.*, **252**, 2143–2153.
34. Redmann, G. A., Adams, P. R., and Clark, R. B. (1982). *Soc. Neurosci. Abstr.*, **8**, 728.
35. Sachs, F. (1983). In Sakmann, B. and Neher, E., Eds., *Single Channel Recording*, Plenum, New York, pp. 365–376.
36. Sakmann, B., Bormann, J., and Hamill, O. P. (1983). *Cold Spring Harbor Symp. Quant. Biol.*, **48**, 247–257.
37. Sakmann, B., Patlak, J., and Neher, E. (1980). *Nature*, **286**, 71–73.
38. Schubert, D., Harris, A. J., Devine, C. E., and Heinemann, S. (1974). *J. Cell Biol.*, **61**, 398–413.
39. Siegelbaum, S. A., Trautmann, A., and Koenig, J. (1984). *Dev. Biol.*, **104**, 366–379.
40. Sigworth, F. J. (1983). In Sakmann, B. and Neher, E., Eds., *Single Channel Recording*, Plenum, New York, pp. 301–321.
41. Sigworth, F. J. (1985). *Biophys. J.*, **47**, 709–720.
42. Sine, S. M. and Steinbach, J. H. (1984). *Biophys. J.* **45**, 175–185.
43. Sine, S. M. and Steinbach, J. H. (1984). *Biophys. J.* **46**, 277–284.
44. Sine, S. M. and Steinbach, J. H. (1986). *J. Physiol.* **370**, 357–379.
45. Sine, S. M. and Steinbach, J. H. (1986). *J. Physiol.* **373**, 129–162.
46. Sine, S. M. and Taylor, P. (1980). *J. Biol. Chem.*, **255**, 10144–10156.
47. Sine, S. M. and Taylor, P. (1981). *J. Biol. Chem.*, **256**, 6692–6699.
48. Steinbach, J. H. (1980). In Cotman, C., Poste, G., and Nicolson, G., Eds., *Cell Surface Reviews*, Vol. 6, Elsevier, New York.
49. Stevens, C. F. (1972). *Biophys. J.*, **12**, 1028–1047.
50. Takeda, K. and Trautmann, A. (1984). *J. Physiol.*, **349**, 353–374.
51. Trautmann, A. (1982). *Nature*, **298**, 272–275.

CHAPTER 3

FUNCTIONAL SITES AND TRANSIENT STATES OF THE NICOTINIC ACETYLCHOLINE RECEPTOR

ARTHUR KARLIN
MARIO DIPAOLA
PETER N. KAO
PETER LOBEL

Departments of Biochemistry and Molecular Biophysics, and Neurology
College of Physicians and Surgeons
Columbia University
New York, New York

The goal of our research is the molecular mechanism for signal transduction by the nicotinic acetylcholine receptor (AChR) (for reviews, see refs: 6, 18, and 50). The issues regarding the AChR are applicable to other membrane receptors and transport proteins, as well: how binding of specific ligands initiates or drives the functional cycles of proteins, how information is transferred through and along cell membranes, what the structures of specific ion channels are, what the mechanisms of permeation through these channels are, and how they are gated. The nicotinic AChR is a favorable protein for investigating these general problems because its functional properties are well characterized, it is readily isolated in large quantities for chemical and physical characterization, and it can be functionally incorporated into artificial membranes. Additionally, the availability of the genes for the AChR subunits and of heterologous cells that will express these genes, and altered versions of them, opens up another powerful approach to structure–function relationships (9,63,64).

Our approach to the molecular mechanism has been to try to locate the functional sites of the AChR. This has meant, (1) associating a function with a particular subunit of the receptor, (2) locating in the sequence of that subunit amino acid residues contributing to the functional site, and (3) locating the subunit and site in the three-dimensional structure of the receptor. We are applying this approach to the acetylcholine (ACh) binding sites and to noncompetitive inhibitor (NCI) sites. This program is far along with the ACh binding sites and at an early stage with the NCI binding sites.

1. STRUCTURE OF THE RECEPTOR

It is now over 10 years since we found that the AChR from *Torpedo californica* electric tissue contains four types of subunits and that only one of these, α, of apparent molecular weight 40,000 contains the ACh binding site (91). Sedimentation equilibrium analysis of the intact AChR yielded a molecular weight of 250,000 (81). This value was combined with the number of ACh binding sites [determined by curarimimetic toxin (CTX) binding] per mass of protein to yield the stoichiometry of two ACh binding sites per receptor and, thus, two α chains per receptor. Given the apparent molecular weights of the α (40,000), β (48,000), γ (58,000), and δ (64,000) chains, we inferred that the subunit composition of the *Torpedo* AChR is $\alpha_2\beta\gamma\delta$ (81), which was confirmed for *Torpedo* AChR (59,80) and found to apply also to the AChR from *Electrophorus electricus* electric tissue (22). This complex of five subunits is both necessary, as shown by expression of combinations of the subunit mRNAs in *Xenopus* oocytes (63), and sufficient, as shown by reconstitution of purified AChR into liposomes (44,58,94), for the function of the receptor.

We now know that both the molecular weight of the intact receptor derived from analytical centrifugation and the apparent molecular weights of the subunits (especially α) derived from sodium dodecyl sulfate–polyacrylamide gel electrophoresis (SDS–PAGE) were low. The amino acid sequences of the four types of subunits have been inferred from their corresponding cDNAs (21,26,72,73,86) and the molecular weights of these sequences are about 50,000 (α), 54,000 (β), 56,000 (γ), and 58,000(δ). In addition, these chains are glycosylated to the extent of 4–7% (59,89). The actual molecular weight of the Torpedo receptor, therefore, is about 280,000 (see ref. 79 for a review of physical properties).

What are the shape and the dimensions of the receptor? This has been investigated by electron microscopy, x-ray scattering, and neutron scattering (51,79). The current view is that the AChR is like a pipe of irregular diameter passing through the membrane (Fig. 1). The length perpendicular to the membrane is 110 Å (55) to 140 Å (16), the extracellular domain is 50–70 Å long, the domain within the lipid bilayer is

Figure 1. A structural model of the AChR. The distribution of mass in the extracellular (ca. 50%), membrane (ca. 30%), and cytoplasmic (ca. 20%) domains correspond to the folding schemes in which four hydrophobic segments span the membrane per subunit (see text), with additional residues added to the membrane domain to clear the polar head groups of the phospholipids (see ref. 51 for details). The dimensions approximate those derived from electron microscopic and x-ray scattering data. Taking into account the neutron scattering densities of the individual amino acid residues in each domain, the radius of gyration of neutron scattering density of the model is equal to that determined for the receptor (4.6 nm; ref. 93).

30–40 Å long, and the cytoplasmic domain is 20–40 Å long. The extracellular domain has a diameter of 80–90 Å and contains a central pit about 30 Å in diameter. The receptor narrows within the bilayer to 50–60 Å in diameter. The extracellular pit is probably continuous with a cation-conducting pore (channel) through the region of the lipid bilayer. This channel, when open, is at least 6 Å in diameter (27,43).

Models of the folding of the AChR in the membrane, based on the sequences of the subunits, have been consistent with the preceding structural model. In each subunit, there are four segments of mostly hydrophobic amino acids that are likely to span the bilayer as α helices (21,26,72,73). In addition, each subunit contains a subsequence that may form an amphipathic, membrane-spanning helix (29,34); thus there may be five membrane-spanning helices per subunit (95). [Anderson and Blobel (5) have shown that each subunit spans the membrane at least once.] The similarity of the secondary structures predicted from the sequences of the subunits and the approximately 40% residue identity between pairs of optimally aligned subunit sequences support the notion that the subunits may be arranged pseudosymmetrically around the axis through the channel (73). One electron microscopic study of two-dimensional, ordered arrays of AChR in membrane supports the notion that the subunits are arranged with pentagonal pseudosymmetry around the central axis of the receptor (16); other structural studies, however, have indicated an asymmetric arrangement of subunits (reviewed in ref. 79).

What is the order of the subunits around the central channel? Given the assumption that the subunits are arranged like barrel staves around the central channel, there are 12 possible orders of the five subunits, the six shown in Fig. 2 and their mirror images. Decision regarding these possibilities requires the identification in images of the receptor of the individual subunits. We identified four of the five subunits in electron microscopic images of negatively stained receptor (42,52). The α subunits were tagged with a complex of avidin and monobiotinylated principal CTX of *Naja naja siamensis* venom, and the 68,000-dalton avidins attached to the AChR were easily visualized. The δ subunit was located in images of native receptor dimer. In *Torpedo*, receptor is largely in the form of disulfide crosslinked dimers, and the disulfide bond is between δ chains. Thus we assumed that the point of closest contact between the monomers in a dimer located the δ subunits in the monomers. The equivalent assumption is that there is dyad symmetry around the δ–δ disulfide bond, which is supported by some observations (15,42,97) but not by others (12,55). Finally, we could locate the β subunits in artificially generated disulfide-crosslinked dimers in which the disulfide bonds were specifically between β subunits (35,52).

First, we could estimate the angular separation between the CTX binding sites on the two α subunits in a monomeric unit by examining the images of monomers and dimers. The avidins crosslinked to the AChR by the biotinyl–CTX were on the average 110° apart, too far for the α chains to be adjacent (42). [Zingsheim et al. (96) and Kistler et al. (55), using approaches different from ours, reached the same conclusion.] Thus we needed to determine which subunit separates the two α subunits. Examination of δ–δ dimers labeled with biotinyl–CTX and avidin showed that the α chains were about 60 and 180° from the δ–δ crosslink; thus the δ subunit

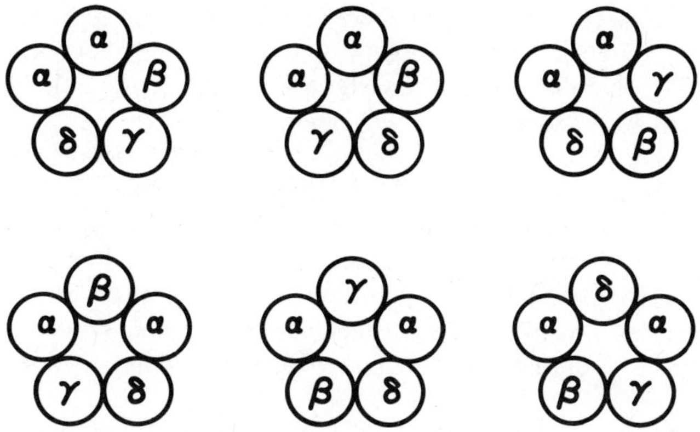

Figure 2. The possible arrangements of the receptor subunits around the central axis, as projected onto the plane of the membrane (from ref. 52). There are 12 possible arrangements, the six shown and their mirror images.

could not lie between the α subunits. Examination of the β–β dimer showed that the α subunits were also about 60 and 180° from the crosslink; thus the β subunit could not lie between the α subunits. By elimination of the other possibilities, we inferred that γ most likely lies between the α subunits, in the order α γ α β δ (Fig. 2, bottom row middle). We still need to locate the γ subunit relative to the α subunits directly to corroborate this conclusion.

We can also draw some conclusions about the location of the CTX binding sites on the α subunits in the intact AChR. The observation that the avidins crosslinked to AChR through monobiotinyl CTX appeared at the periphery of the AChR suggested that the CTX site was relatively far from the central channel and close to the outer edge of the AChR. In the course of the avidin–biotin study, we prepared the six possible *N*-linked monobiotinyl derivatives of *Naja naja siamensis* CTX (60) and tested their binding to avidin, their binding to AChR, and their ability to crosslink avidin to AChR. We demonstrated that two of the five lysines (Lys-12 and Lys-69) are on the side of CTX opposite to the side that contacts the AChR in the AChR–CTX complex; that is, biotin attached to either of these two lysines is unobstructed in the complex and is free to bind avidin.

We also prepared fluorescent derivatives of CTX in which the fluophors were attached to either of these two lysines (Lobel and Karlin, in preparation). The 3-thiopropionyl derivative of CTX at either Lys-12 or Lys-69 was reacted with either 5 - [2-(iodoacetyl)aminoethyl]aminonaphthalene-1-sulfonic acid or 5-iodoacetamidofluorescein to form 12- or 69-ANS–CTX and 12- or 69-AAF–CTX, respectively. The ANS–CTX and the AAF–CTX complexes are a donor–acceptor pair for nonradiative singlet–singlet energy transfer. The efficiency of such transfer depends on the inverse sixth power of the ratio of the distance between the fluorophors and a characteristic distance R_0 at which the efficiency of transfer between a donor–acceptor pair would be 50% (30). The R_0 value for these pairs is 44

Å. We observed that the efficiency of energy transfer between 12- or 69-ANS–CTX and 12- or 69-AAF–CTX, bound to AChR (both monomer and dimer) in detergent, was less than 5%. In dilute solution, transfer occurs only between fluorescent CTXs on the same AChR, and the small efficiency of transfer implies that the two fluorophors, and hence the two CTX binding sites, on AChR monomer are at least 75 Å apart. Given the diameter of the AChR monomer (80–90 Å), we conclude that the CTXs are bound close to the periphery of the AChR with the fluorophores oriented outward (also see ref. 46). [In AChR dimer, the two ACh binding sites on different monomers closest to the disulfide crosslink are about 60° from the crosslink (42) and, hence, are about one AChR monomer diameter apart; therefore, there is even less energy transfer between fluorophors bound to them than to the two sites on the same monomer.]

A peripheral location of the CTX binding sites is consistent with our observations of energy transfer in membrane, in which the surface density of AChR is of the order of $10^4/\mu m^2$. In this case, we found much greater energy transfer than in solution due to the interaction of fluorescent CTXs on separate but nearby AChRs (Lobel and Karlin, in preparation). The average distances (arising from a distribution of separations and rotational orientations of the AChRs) for 12-ANS–CTX to 12-AAF–CTX and for 69-ANS–CTX to 69-AAF–CTX were both about 50 Å. Since there is a distribution of distances and since the fluorophors are attached to the CTX by a long and flexible arm, we cannot use these measurements to locate the CTX binding sites (and ACh binding sites) exactly. They do appear, however, to be close to the periphery of the AChR (in agreement with refs. 28 and 46).

2. LOCATION OF THE ACh BINDING SITE IN THE SEQUENCE OF THE α SUBUNIT

On the basis of the effects of chemical modification, particularly affinity labeling on the AChR-mediated responses of intact single cells from the electric tissue of *Electrophorus electricus*, we inferred 15 years ago that there is a readily reduced disulfide bond about 10 Å from the negative subsite of the ACh binding site (Fig. 3) (49,50). (The site of interaction of the quaternary ammonium group that is present on all potent activators and competitive inhibitors of the AChR is presumed to contain one or more negatively charged side chains.) Evidence for a disulfide bond close to the ACh binding site of nicotinic AChR has since been obtained in a variety of species and phyla (annelids, mollusks, and vertebrates) and in various types of cells (striated muscle, electrocytes, sympathetic neurons, central neurons, PC12) (48,50).

Using an affinity label, 4-(*N*-maleimidobenzyl)tri-[^{3}H]methylammonium ([^{3}H]MBTA), that is highly specific for the reduced disulfide of the ACh binding site of *Torpedo californica* AChR, we have located in the sequence of the α chain the half-cystinyl residues that form the binding site disulfide bond (48;48a). The procedure was to reduce the AChR with dithiothreitol and to affinity alkylate the binding site with [^{3}H]MBTA. The labeled α chain was isolated by polyacrylamide gel

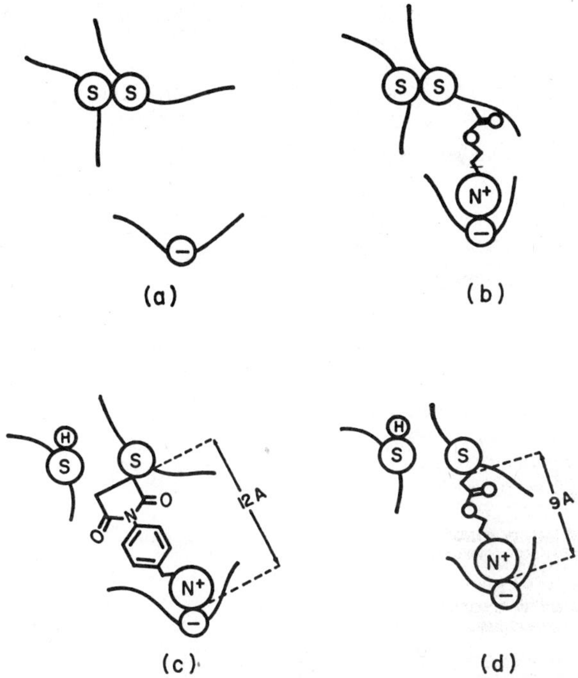

Figure 3. A schematic model of the ACh binding site based on the effects of covalent modification of the receptor in intact cells (from ref. 49). It shows a negative subsite and a readily reducible disulfide bond. It was inferred that the region next to the disulfide is a hydrophobic subsite. (*a*) Native site in the resting state; (*b*) native site binding ACh, in the active state; (*c*) site with disulfide reduced and alkylated by 4-(*N*-maleimido)benzyltrimethylammonium, in an inactive state; (*d*) site with disulfide reduced and alkylated by bromoacetylcholine, in an active state. The maximally extended lengths of the attached moieties are shown.

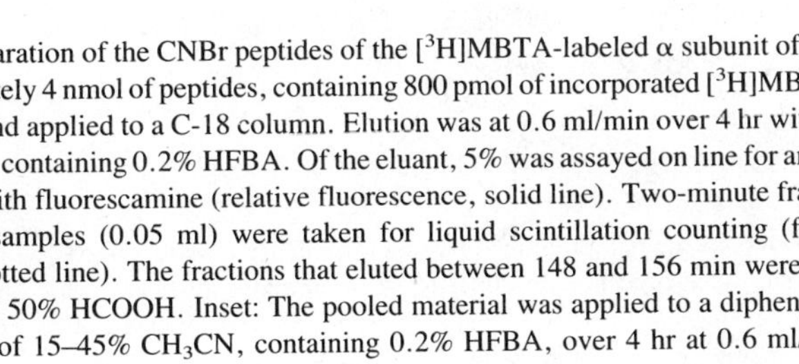

Figure 4. Separation of the CNBr peptides of the [^{3}H]MBTA-labeled α subunit of the AChR (from ref. 48). Approximately 4 nmol of peptides, containing 800 pmol of incorporated [^{3}H]MBTA, was dissolved in 90% HCOOH and applied to a C-18 column. Elution was at 0.6 ml/min over 4 hr with a linear gradient of 0–60% CH_3CN, containing 0.2% HFBA. Of the eluant, 5% was assayed on line for amines by post column derivatization with fluorescamine (relative fluorescence, solid line). Two-minute fractions (1.2 ml) were collected, and samples (0.05 ml) were taken for liquid scintillation counting (fraction of recovered radioactivity, dotted line). The fractions that eluted between 148 and 156 min were pooled, lyophilized, and dissolved in 50% HCOOH. Inset: The pooled material was applied to a diphenyl column and eluted with a gradient of 15–45% CH_3CN, containing 0.2% HFBA, over 4 hr at 0.6 ml/min.

electrophoresis, the α chain was cleaved with CNBr, and the peptide fragments were separated by high-pressure liquid chromatography (HPLC) (Fig. 4). A uniquely labeled CNBr fragment was isolated and sequenced by gas-phase, automated Edman degradation (48).

We identified the first five residues, starting at the amino terminus, as Lys–Asp–Tyr–Arg–Gly (KDYRG). By comparison with the sequence of α (Fig. 5) (72), we unambiguously identified the labeled peptide as the CNBr fragment starting at Lys-179 and ending at Met-207. This fragment contains adjacent Cys residues at 192 and 193. To sequence through to these residues, we cleaved the CNBr fragment at Trp and isolated the shortest labeled fragment. This corresponded to a fragment starting at Trp-187, which could be sequenced through the Cys-192 and Cys-193 and up to Tyr-198. Radioactivity, due to covalently attached [^{3}H]MBTA, was first released in the sequencing cycle corresponding to Cys-192. In the next cycle, corresponding to Cys-193, two-thirds as much radioactivity was released, more than would be expected from carry-over from the previous cycle.

These results demonstrated that Cys-192 contributes to a binding-site disulfide and that perhaps Cys-193 does also. The currently accepted folding scheme for α in the membrane places four of the seven Cys residues of α, Cys-128, Cys-142, Cys-192, and Cys-193, on the extracellular side of the membrane. Since the ACh binding site is

```
                   CNBR FRAGMENTS OF ALPHA CHAIN
 1. SEHETRLVANLLENYNKVIRPVEHHTHFVDITVGLQLIQLISVDEVNQIVETNVRL
    RQQWIDVRLRWNPADYGGIKKIRLPSDDVWLPDLVLYNNADGDFAIVHM 1-1Ø5
 2. TKLLLDYTGKIM 1Ø6-117
 3. WTPPAIFKSYCEIIVTHFPFDQQNCTM 118-144
 4. KLGIWTYDGTKVSISPESDRPDLSTFM 145-171
 5. ESGEWVM 172-178
 6. KDYRGWKHWVYYTCCPDTPYLDITYHFIM 179-2Ø7
 7. QRIPLYFVVNVIIPCLLFSFLTGLVFYLPTDSGEKM 2Ø8-243
                    M1                   M2
 8. TLSISVLLSLTVFLLVIVELIPSTSSAVPLIGKYM 244-278
            M2                       M3
 9. LFTM 279-282
     M3
1Ø. IFVISSIIITVVVINTHHRSPSTHTM 283-3Ø8
            M3
11. PQWVRKIFIDTIPNVM 3Ø9-324
12. FFSTM 325-329
13. KRASKEKQENKIFADDIDISDISGKQVTGEVIFQTPLIKNPDVKSAIEGVKYIAEHM 33Ø-386
                                                 MA
14. KSDEESSNAAEEWKYVAM 387-4Ø4
15. VIDHILLCVFM 4Ø5-415
        M4
16. LICIIGTVSVFAGRLIELSQEG 416-437
       M4
```

Figure 5. The predicted cyanogen bromide fragments of the α subunit of receptor from *Torpedo californica* electric tissue. Cleavage is *C*-terminal to the methionine residues, which are converted to homoserine residues. The four hydrophobic sequences and one amphipathic sequence (364–386) that are postulated to span the membrane are underlined.

extracellular, our results confirm that Cys-192 and Cys-193 are extracellular. In addition, Cys-142 is next to the one Asn in α that is potentially N-glycosylated. Since extracellular Cys are almost always disulfide crosslinked (87), it is likely that all four extracellular Cys amino acids form disulfide bonds. The question is how they pair up.

Although adjacent Cys residues are common in protein sequences, they are, with rare exceptions (11,77), not disulfide crosslinked to each other (87). Such a crosslink requires that the peptide bond between the adjacent Cys amino acids assumes a *cis* conformation. The other possibility is that Cys-192 and Cys-193 are crosslinked to Cys-128 and Cys-142 in either of the two possible arrangements. We did not, however, detect any incorporation of [^{3}H]MBTA into Cys-128 or Cys-142 after reduction of the binding-site disulfide. Furthermore, when we tagged remaining free SH with [^{14}C]-*N*-ethylmaleimide,following mild reduction with 0.2 m*M* dithiothreitol and affinity labeling of the binding site with [^{3}H]MBTA, all ^{14}C, as well as all ^{3}H, was incorporated into the CNBr fragment containing Cys-192 and Cys-193 (Fig. 6). It appears, therefore, that only Cys-192 and Cys-193 are reduced by mild reduction. Furthermore, we have demonstrated that the two CNBr fragments, one containing Cys-128 and Cys-142 and the other Cys-192 and Cys-193, are not disulfide crosslinked to each other and that each contains an internal disulfide bond (48a). Therefore, Cys-192 and Cys-193 form one disulfide bond and Cys-128 and Cys-142 form another.

The results of site-directed mutagenesis of the α subunit (64) and the binding of CTX to proteolytic fragments of α (71,92) are consistent with the disulfide crosslinking pattern determined above.

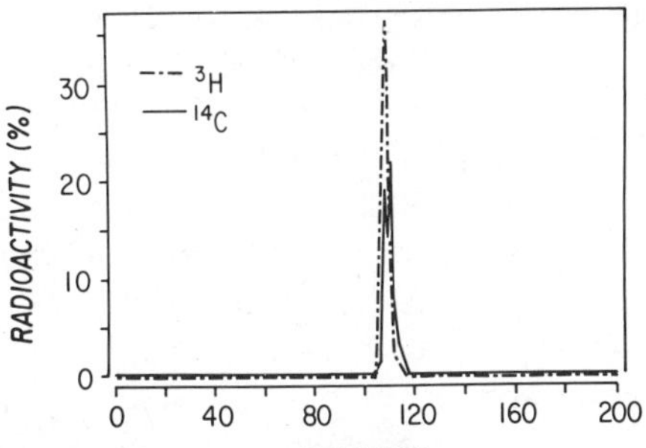

Figure 6. High-pressure liquid chromatography of CNBr peptides of the α subunit containing half-cystinyl residues that are alkylated by [^{3}H]MBTA and [^{14}C]*N*-ethylmaleimide. Purified receptor (5.6-mg and 40-nmol sites) in Triton X-100 solution was alkylated with 1 m*M* *N*-ethylmaleimide, dialyzed, and reduced with 0.2 m*M* dithiothreitol at pH 8.0 and 25° C, for 20 min. The dithiothreitol was removed by dialysis under argon. The mildly reduced receptor was alkylated at 25° C,pH 7.0, for 10 min, with a mixture of 55 nmol [^{3}H]MBTA and 1100 nmol [^{14}C]-*N*-ethylmaleimide in a final volume of 4 ml. Under these conditions, 50% of the binding sites were blocked. The α chain, isolated electrophoretically, contained 92% of recovered ^{3}H and 72% of recovered ^{14}C. It was cleaved with 200 m*M* CNBr in 70% formic acid for 24 hr, and the peptides were separated on a 25-cm Vydac C-18 column. ^{3}H, solid line; ^{14}C, dashed line.

We have located two atoms, the sulfurs of Cys-192 and Cys-193, which are about 10 Å distant from the negative subsite of the ACh binding site *in the folded structure* of the α chain. The identification of the ACh-binding-site disulfide provides an important constraint on models of the folding of the α sequence to form this site. We now need to identify other residues in the ACh binding site.

3. NONCOMPETITIVE INHIBITOR SITES

There exists a class of inhibitors of AChR function that do not compete with the binding of ACh. These are called *noncompetitive inhibitors* (NCIs). Like the better understood class of effectors, the agonists, NCIs affect the transitions between functional states of the receptor and bind with different affinities to the different states.

Four functional states of the receptor have been distinguished: resting, active (open-channel), rapid-onset desensitized, and slow-onset desensitized (19,20,37,41,54,70,83,90). The binding of agonist to the resting state is diffusion controlled; the rate of transition to the active, open-channel state is of the order of 10^4 sec^{-1}; the rate of transition to the rapid-onset desensitized state is of the order of 1 - 10^2 sec^{-1}; and the rate of transition to the slow-onset desensitized state is of the order of 0.01–0.1 sec^{-1}. The agonist-occupied AChR flickers between the resting and open-channel states until the agonist dissociates or the AChR desensitizes. The affinity of the ACh binding sites for agonists is relatively low in the resting state, intermediate in the rapid-onset desensitized state, and high in the slow-onset desensitized state (reviewed in ref. 18).

Noncompetitive inhibitors have been observed to (1) produce a new closed state of the channel (1,69,82); (2) decrease the rate of channel opening (2,85); and (3) increase the rate of rapid (19,20,31,53,61,62,68,76) and slow (14,38,62,76,84) desensitization. The new closed state is hypothesized to arise when the NCI binds within and occludes the open channel (1,69,82). The available evidence, however, does not exclude an allosteric mechanism for the generation of this state. Different NCIs have different spectra of effects. For example, at low concentrations, QX222 (68) and bupivacaine (7,45) appear to block open channels without causing desensitization, quinacrine appears to both block open channels and promote desensitization (3, 4,32,88); the phenothiazines chlorpromazine (17,61) and trifluoperazine (20) mainly promote desensitization; and histrionicotoxin (HTX) slows channel opening, hastens channel closing, and does not promote desensitization (14,38,85).

The multiple effects of NCIs on the AChR may be due at least in part to their binding to different sites. To account for the findings that HTX blocks the induction of a high-affinity state for agonists (correlated with desensitization) by some NCIs but not others and that the binding of NCIs to receptor-rich membrane is complex, Heidmann et al. (38) have postulated two classes of NCI sites (see also ref. 56). (Note, however, that despite the fact that HTX competes with desensitizing NCIs, it is an ineffective desensitizer.) Heidmann et al. (38) present evidence that there is one "high-affinity" site (HTX site) per receptor monomer (*T. marmorata*) and 10–30 "low affinity" sites. Others have found 0.5–1.8 "high-affinity" NCI sites per receptor (*Torpedo* species) (see ref. 47 for additional references). Sine and Taylor (84) inferred that there are at least two high-affinity HTX sites per receptor in BC3H1 cells.

The binding of NCIs not only affects the state of the AChR, but also is dependent on the functional state. Several NCIs bind with higher affinity in the presence of agonists and competitive antagonists (see ref. 47 for additional references) under conditions such that the AChR is desensitized and the channel is closed. Also, a transient state of the AChR appears to bind NCIs more rapidly than does the resting state or the slow-onset desensitized state: the rates of binding of HTX (8) and phencyclidine (76) are enhanced immediately after the addition to the AChR of agonists, but not competitive antagonists.

Noncompetitive inhibitor binding sites on AChR in *Torpedo* membrane have been affinity labeled with either electrophilic or photolyzable derivatives of NCIs or with the NCIs themselves photoactivated by ultraviolet (UV) light. In most cases, the receptor was labeled while in stable, closed-channel states; that is, either resting state or slow-onset desensitized state or the stable state of the receptor favored in the presence of competitive antagonists. With all labels, the extent of incorporation was increased and the distribution of label among the receptor subunits altered when the ACh binding sites were occupied with either agonist or competitive antagonist (36,47,57,66,74,75). The particular chains labeled, however, were different with different labels and different species of *Torpedo*. For example, in the presence of agonist, trimethisoquin azide was found to label predominantly the δ chain (74); triphenylmethylphosphonium the β and δ chains (67); phencyclidine the δ chain (75); and phencyclidine azide, variably, the α, β, and δ chains (36). Chlorpromazine was photocrosslinked by UV irradiation in the presence of agonist to all four chains (75). In one case, the agonist-induced enhancement of the labeling of α and β by quinacrine mustard was correlated with increased inhibition of channel function (47). It should be emphasized that in none of these cases was the channel actually open during the labeling: both agonists (e.g., ACh) and competitive antagonists (e.g., *d*-tubocurarine) produced the same effects on the labeling by NCIs, and, when agonists were used, the time scales of the reactions were such that the receptor was desensitized and the channel was closed.

One motivation for labeling the NCI sites is based on the hypothesis that NCIs bind to and occlude the open channel. Thus reactive NCIs might tag a site in the cation-conducting channel of the receptor and permit the localization of an important functional domain. Given the rates of receptor activation and rapid desensitization, however, fast reaction techniques are required if we are to label the receptor when it is predominantly in the active, open-channel state. The demonstration that the receptor in the active state was particularly susceptible to labeling by NCIs would not prove that the site labeled was in the open channel, but the absence of particular susceptibility of the active state would indicate that an open channel was not required for NCI binding and reaction.

In membrane from Torpedo electric tissue the receptor enters the first phase of desensitization within 100 msec (90) or less (38) of agonist binding. To label the receptor in the potentially open-channel state, it is thus necessary to mix membrane, effectors, and photolabel and to irradiate the mixture within milliseconds of mixing. We built a continuous-flow, rapid-mixing, photolysis instrument, in which membrane in one syringe and photolabel and effectors in another syringe are continuously mixed and the mixture passes through a delay-tube (minimum delay 20 msec), through a quartz tube on which filtered light (>300 nm) from a 1000-W xenon–mercury arc lamp is focused (typically the transit time through the light is 2 msec), and finally into a collecting tube containing quenching agents (23,51).

In addition to rapid mixing, appropriate photolabels are needed. The photolabel should be a derivative of an NCI that has the characteristics of an open-channel blocker. It also would be advantageous for the NCI not to enhance desensitization. We have used quinacrine azide (QA) (65). Quinacrine is a potent NCI; however, it promotes desensitization in addition to acting like an open-channel blocker (3,4,32,88). Both quinacrine and QA have been shown to block voltage-gated channels (78). Of importance for time-resolved photolabeling, QA has a large molar extinction coefficient in the visible range (ϵ = 13,100 at 359 nm) (65). In the instrument described above, QA is completely photolyzed during a 2-msec exposure to the light.

We synthesized [^{3}H]QA and characterized its photoreaction with receptor in membrane isolated from electric tissue of *Torpedo californica* (23,24,51). We rapidly mixed membrane, [^{3}H]QA, and effectors and after delay times of 20 msec or greater, irradiated the mixture for 2 msec. The labeled membrane was dissolved in SDS solution, and the proteins were separated by polyacrylamide gel electrophoresis. The incorporation of [^{3}H]QA into the individual subunits of the receptor was determined by counting tritium in gel slices.

On the basis of the effects of agonists, competitive antagonists, and NCIs, added together with [^{3}H]QA, on the incorporation of [^{3}H]QA into the subunits of the receptor, we distinguish three classes of sites subject to photolabeling by QA: NCI sites, ACh sites, and ''nonspecific'' sites. The NCI sites are protected against QA by NCIs (proadifen, HTX, bupivacaine); the ACh sites are protectd against QA by agonists (ACh and carbamylcholine) and competitive antagonists (*d*-tubocurarine and hexamethonium); and the nonspecific sites are not protected by any of the effectors.

After 20–200 msec delays, in the absence of effectors, [^{3}H]QA was incorporated into α and, to a much smaller extent, into β, γ, and δ (23,51). Competitive inhibitors *d*-tubocurarine and hexamethonium considerably decreased the labeling of α. Furthermore, some ACh binding sites, measured by CTX binding, were blocked by QA in the absence of effectors, and this functional block of ACh sites was completely prevented in the presence of the competitive inhibitors. Therefore, in the absence of effectors, QA labels ACh binding sites on α subunits. [This is consistent with the block of ACH binding by quinacrine (88)]. There was very little change from the labeling obtained in the absence of effectors when NCIs were added. Therefore, few NCI sites are labeled in the absence of effectors. In the absence of effectors, therefore, QA labels ACh binding sites and nonspecific sites on α. The incorporation of [^{3}H]QA in the absence of effectors into the other subunits was also not decreased by NCIs.

In the presence of ACh, the incorporation of [^{3}H]QA into α was much greater than into the other subunits and slightly greater than that obtained in the absence of effectors. In the presence of ACh, however, there was no block of ACh binding sites assayed by CTX binding. Furthermore, the additional presence of the NCI proadifen with the ACh resulted in incorporation of [^{3}H]QA into α much less than in the presence of ACh alone and equal to that obtained with hexamethonium alone; that is, the difference between the labeling of α in the presence of ACh and that in the presence of hexamethonium was eliminated by proadifen. Thus, in the presence of ACh, [^{3}H]QA labels NCI sites and nonspecific sites.

Of the other subunits, only the labeling of β was increased by ACh over that obtained in the absence of effectors or in the presence of hexamethonium, and this increase was completely blocked by proadifen added together with the ACh. Therefore, β also contributes to an NCI site. The labeling of γ and δ was changed little or decreased by the addition of ACh. In summary, α and β contain NCI sites that are susceptible to photolabeling by QA only in the presence of agonist.

On the basis of our finding that 400 μ*M* hexamethonium optimally suppressed the labeling by QA of α in the absence of all other effectors (24), we henceforth took as the *specific* incorporation of [^{3}H]QA the difference between the labeling in the presence of ACh (plus 400 μ*M* hexamethonium in some cases) and the incorporation in the presence of 400 μ*M* hexamethonium alone; in other words

(NCI sites + nonspecific sites) − (nonspecific sites).

In most of these experiments, the delay between mixing and photolysis was in the range of 20–200 msec. Although competitive inhibitors like hexamethonium (74) and *d*-tubocurarine (24,39,47) induce slow changes in the AChR, including an increase in affinity for NCIs, on the timescale of 20–200 msec the predominant effect of hexamethonium is to protect the ACh binding sites. [Paradoxically, hexamethonium, in the absence of all other effectors, inhibited the incorporation of [^{3}H]QA not only into α but also into the other subunits (24). Nevertheless, hexamethonium did not appear to block NCI sites since the incorporation of [^{3}H]QA into α and β promoted by saturating concentrations of ACh (1 m*M*) was the same in both the presence and the absence of 400 μ*M* hexamethonium.)

All the specific labeling, as well as all the NCI-inhibitable labeling, was in the α and β chains. The *maximum* inhibition of the specific labeling by 2 μ*M* [^{3}H]QA of both α and β was 100% by proadifen and 50% by HTX or bupivacaine, all at saturating concentrations. The maximum extents of inhibition by proadifen and HTX were the same after 20- and 200-msec delay. Proadifen (14), histrionicotoxin (25,84,85), and bupivacaine (7,45) are noncompetitive inhibitors that appear to have overlapping sites of action (14,38,45). Proadifen, however, is a more potent blocker of receptor activation and promoter of desensitization in Torpedo membrane than is

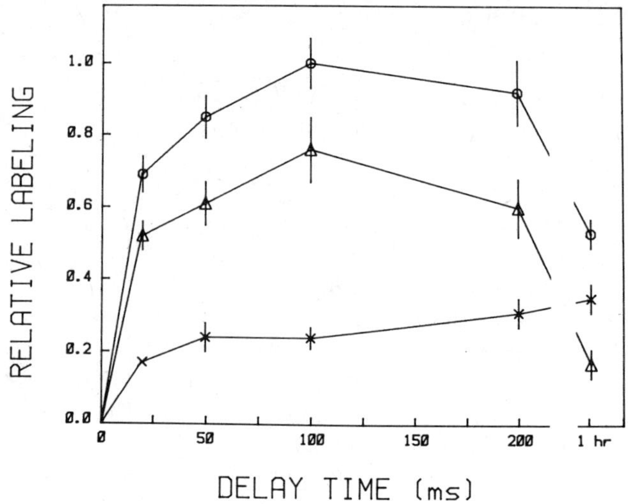

(a)

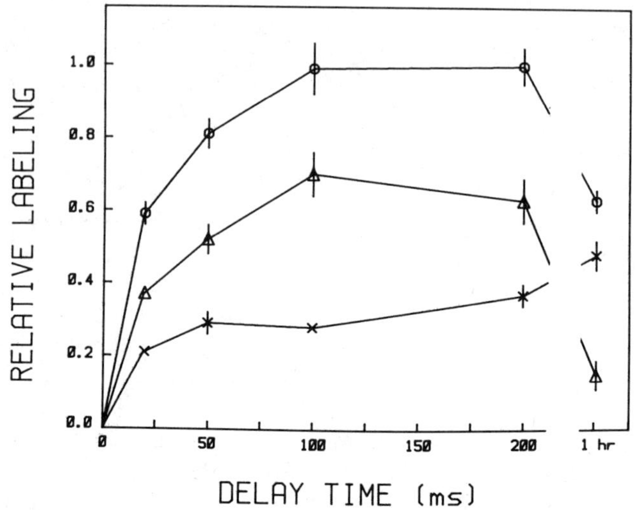

(b)

Figure 7. Incorporation of [^{3}H]QA into the α and β subunits as a function of the delay time between mixing and photolysis (from ref. 24). For each point, 150 μl of receptor-rich membranes containing 60 pmol of toxin-binding sites were mixed with 150 μl of 4 μ*M* [^{3}H]QA plus 800 μ*M* hexamethonium plus or minus 2 m*M* ACh. The protocol for the long delay time points was as follows. Receptor was first desensitized by incubation with ACh (and 200 μ*M* diisopropylfluorophosphate) for 30 min, then irradiated 20–200 msec after mixing with photolabel plus hexamethonium plus ACh. Alternatively, receptor was incubated with hexamethonium for 30 min, mixed with photolabel plus hexamethonium, and irradiated 20–200 msec later. In all cases, the photolabel was present only during the delay time of 20–200 msec. The delay time between mixing and photolysis was varied by changing the length of the delay line between the T mixer and the photolysis tube. The time in the light beam was always 2 msec. The averages and average errors of duplicate determinations are plotted. Symbols: crosses (X), hexamethonium alone; circles, hexamethonium plus ACh; triangles, difference between hexamethonium plus ACh and hexamethonium alone. Part *a*, α subunit; part *b*, β subunit.

HTX (14), and bupivacaine acts as a channel blocker and does not promote desensitization (7,45).

It is apparent that QA labels NCI sites on α and β rendered susceptible specifically by the binding of ACh at its sites. The question is what state of the receptor induced by the binding of ACh is most susceptible to QA labeling. We will examine two kinds of evidence: the time course of the development of the susceptible state and the dependence of its development on ACh concentration.

We investigated the incorporation of [^{3}H]QA at two different concentrations, as a function of the delay time between mixing and photolysis. (Note that it was delay time between mixing and photolysis that was varied, not the duration of exposure of the mixed reactants to light, which was always 2 msec.) The two concentrations of photolabel, 2 μ*M* (Fig. 7) and 3.5 μ*M*, gave similar time courses of specific incorporation. In the presence of hexamethonium alone, the incorporation of [^{3}H]QA into both α [Fig. 7a; crosses (X) and β (Fig. 7b; crosses) increased slowly and slightly. In contrast, the specific incorporation of [^{3}H]QA reached a maximum at 100 msec and declined thereafter (Figs. 7*a*, 7*b*; triangles). The half-time for the rising phase was less than 20 msec. After long incubation times, when the receptor had entered the slow-onset desensitized state, the specific incorporation was about 15% of maximum. The kinetics of the ACh-induced incorporation were the same regardless of whether hexamethonium was also present.

It is clear from these results that a transient state of the receptor is particularly susceptible to labeling by QA. It seems unlikely, however, that it could be the active state that is exclusively susceptible. After rapid mixing of the receptor with 1 m*M* ACh, as in these experiments, a steady state of open channels should have been attained in less than 1 msec; yet the specific labeling did not reach its maximum until 100 msec after the addition of ACh. The kinetics did not change on increasing the QA concentration from 2 to 3.5 μ*M*; hence the association of QA with its site of reaction does not appear to be rate limiting [consistent with the association rate of quinacrine at the frog muscle endplate of about 10^8 liters mole^{-1} sec^{-1} (4)]. The time course of the ACh-induced state susceptible to QA labeling resembles the time course of the ACh-induced change in the fluorescence of QA bound to Torpedo membrane, which, Grunhagen et al. (33) suggested, reflected either slow activation or accelerated desensitization.

We investigated the incorporation of [^{3}H]QA into α and β as a function of ACh concentration. Acetylcholine was always added simultaneously with 400 μ*M* hexamethonium, so that at all concentrations of ACh, the ACh binding sites were fully occupied and protected from reaction with QA. The delay was 20 msec. For 2 μ*M* [^{3}H]QA, the specific incorporation of photolabel saturated at an ACh concentration of about 1 m*M* (Fig. 8). The concentration of ACh producing half-maximal specific incorporation was 116 μ*M* for the α subunit (Fig. 8*a*) and 209 μ*M* for the β subunit (Fig. 8*b*). The K_I for hexamethonium is 50 μ*M*, which is eight fold lower than the concentration used in these experiments; hence the apparent dissociation constant K_{app} for ACh was about 13 μ*M* (= 116 μ*M*/9) for specific incorporation into the α chain and 23 μ*M* (= 209 μ*M*/9) for specific incorporation into β.* A Hill plot of the data (i.e., log $[Z/(Z_{max} - Z)]$ vs. log [ACh], where Z is the *specific incorporation* of [^{3}H]QA and Z_{max} is the maximum specific incorporation) yielded a straight line with a slope of 1 for both the α and β subunits. At a higher concentration of [^{3}H]QA (5 μ*M*), the dose response curve was shifted to the left. The K_{app} (corrected for hexamethonium competition) for specific incorporation into α was 3 μ*M* and into β, about 6 μ*M*. As with 2 μ*M* [^{3}H]QA, the Hill plot of the data for 5 μ*M* [^{3}H]QA was a straight line over the range of concentrations tested, with a slope of 1 for both the α and β subunits.

By comparison, the K_{app} for the ACh-induced permeability increase in Torpedo membrane was found to be about 150 μ*M* (10,70). Depending on the model assumed for receptor activation, different equilibrium dissociation constants can be derived from K_{app}, but the dissociation constant is not likely to be less than $K_{app}/2.4 = 60$ μ*M*.† By comparison, the dissociation constant of the intermediate affinity state (presumably the rapid-onset desensitized state) was found to be about 1 μ*M* (13,37). Thus, the ACh-induced incorporation of [^{3}H]QA is characterized by an apparent affinity for ACh that falls between that of the resting state and that of the intermediate state.

A second consideration is that the specific incorporation of [^{3}H]QA shows a lack of cooperativity in its dependence on ACh concentration; the Hill coefficients were equal to 1 both for α and for β. However, the Hill plot of the receptor-mediated permeability response in *Torpedo* membrane (37,70,90), as well as in several other preparations (reviewed in ref. 50), has a slope close to 2 (but cf. ref. 10). In contrast to channel opening, both fast and slow desensitization are characterized by Hill coefficients of 1 (refs. 40, 90; but cf. ref. 37). The interpretation of the Hill coefficients in the present case is complicated by the possibility discussed below that there might be two different sites of specific labeling by [^{3}H]QA, so that the dose–response curves (Fig. 8) might be compound. A second possible complication is the effect of the simultaneous addition to the membrane of hexamethonium and ACh.§

*Based on the equation for agonist (A) binding in the presence of a competitive inhibitor (I), namely

$$Y = \frac{1}{1 + \dfrac{K_A}{[A]}\left(1 + \dfrac{[I]}{K_I}\right)}$$

†In the two-independent-sites model, in which both ACh sites must be occupied for the channel to open, the relative response is given by

$$\frac{Z}{Z_{max}} = \frac{1}{(1 + K/[A])^2} \quad \text{and} \quad \frac{Z}{Z_{max}} = \frac{1}{2} \quad \text{at } [A] = 2.4K$$

§We recently redetermined the dependence on ACh concentration of [^{3}H]QA incorporation into α and β (DiPaola and Karlin, in preparation). In this case, however, we took as the specific incorporation the difference between the incorporation 20 ms after the addition of ACh and the incorporation 20 ms after the addition of both ACh and 100 μ*M* HTX. This specific incorporation was characterized by a K_{app} of 10 μ*M* and a Hill coefficient of 1.7 ± 0.1. This result indicates that the induction by ACh of the susceptible state, like channel opening, is positively cooperative.

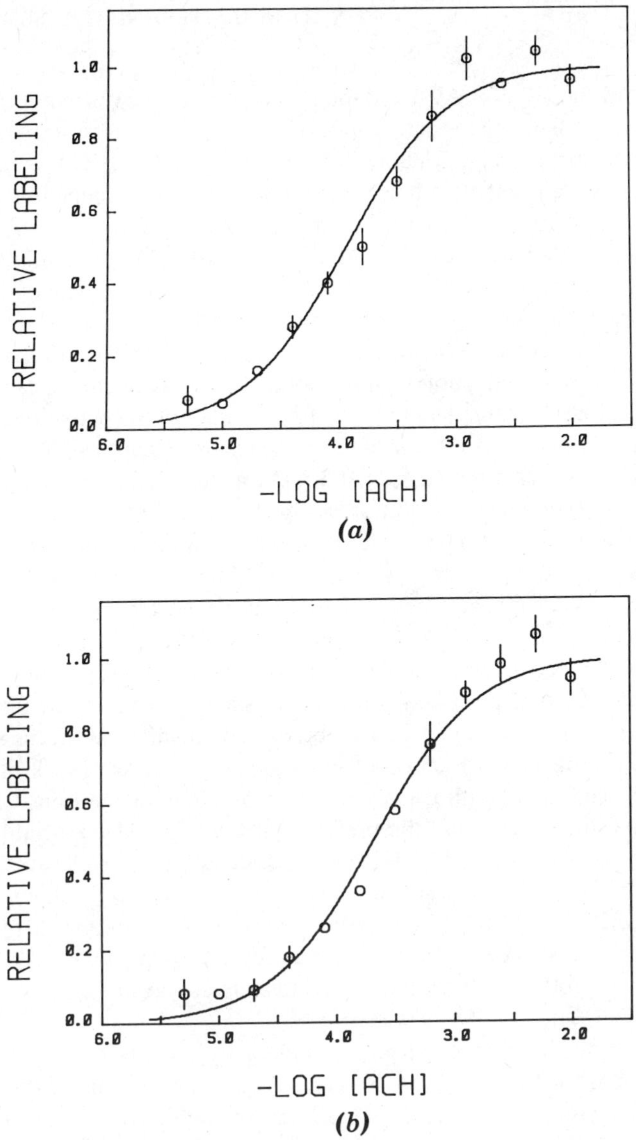

Figure 8. Incorporation of [^{3}H]QA into the α and β subunits as a function of ACh concentration (from ref. 24). 150 μl of membranes (60 pmol of toxin-binding sites) were mixed with 150 μl of 4 μM [^{3}H]QA plus 800 μM hexamethonium (hence final concentrations of 2 and 400 μM) plus or minus ACh, at twice the concentrations listed on the abscissa. Each point represents the difference, normalized to the saturating value, between the incorporation observed in the presence of ACh plus hexamethonium and that observed in the presence of hexamethonium alone. The delay time was 20 msec, and the photolysis time was 2 msec. The data are the average of three experiments, each involving duplicate determinations of each point. The data are fitted by the equation 1/(1 + K/[ACH]) with K = 116 μM for α and 209 μM for β. Part *a*, α subunit; part *b*, β subunit.

On the basis of the K_{app} for ACh and the Hill coefficients, we cannot rule out the possibility that QA is labeling the active state of the receptor, but it is unlikely that it is labeling this state exclusively. Rather, it seems likely that [^{3}H]QA is specifically incorporated into a transient state that is initiated by agonist occupation of a single ACh binding site that arises less than 20 msec after binding and slowly subsides after 100 msec. Of the characterized functional states of the receptor, this state most closely resembles the rapid-onset desensitized state. The development of this state in *Torpedo* membrane has been characterized by rates that range from 5 sec^{-1} (90) to 70 sec^{-1} (37), which correspond to half-times for onsets of 10–140 msec.

In another time-resolved photolabeling study, Heidmann and Changeux (39) concluded that chlorpromazine labeled the open channel of the active receptor. The photoincorporation of chlorpromazine in the presence of ACh into all four subunits of the receptor was characterized by a K_{app} of 30 μM and by a Hill coefficient of 1.3 ± 0.2. The shortest delay between addition of ligands and photolysis was 100 msec; however, these same authors have found that the half-time for the transition to the intermediate state (rapid-onset desensitized state) is as little as 10 msec (37). Finally, chlorpromazine does not appear to be a simple open-channel blocker; rather, it mainly promotes desensitization (17,61), as does the similar trifluoperazine (20). It seems likely, therefore, that the receptor labeled by chlorpromazine was predominantly in a desensitized state rather than in an open-channel state. Our inference that QA and chlorpromazine preferentially label a transient, closed-channel state of the receptor is consistent with mechanisms of action of NCIs other than direct block of the open channel and, in particular, with the acceleration by NCIs of the transitions to the rapid-onset desensitized state and the preferential binding of NCIs to this state.

The specific incorporation of [^{3}H]QA is undoubtedly into NCI sites, but how many classes of NCI sites? As the specific labelings of α and β were characterized by different K_{app} values for ACh, there may be at least two structurally different sites. Given the different spectra of effects of proadifen and bupivacaine, the fact that proadifen can block 100% of the specific labeling and bupivacaine only 50% suggests that QA may label functionally distinct sites.

We are exploring the sites of specific labeling by [^{3}H]QA at the level of the primary structure of the AChR. We have reacted AChR in membrane with [^{3}H]QA 20 msec after mixing with 1 mM ACh, or 400 μM hexamethonium (HEX), or 1 mM ACh plus 100 μM proadifen (PRO), or 1 mM ACh plus 200 μM bupivacaine (BUP). [Subsequently the AChR was solubilized in sodium dodecyl sulfate (SDS) and labeled with [^{14}C]-*N*-ethylmaleimide, and the ^{14}C was used to normalize the ^{3}H for recovery.] In each case, the α subunit was isolated by electrophoresis and cleaved with CNBr, and the CNBr peptides were separated by reverse-phase HPLC (Fig. 9; DiPaola, Kao, and Karlin, in preparation). There are five major peaks of ^{3}H, marked *A* through *E* in Fig. 9. These radioactive peaks are complex, and the labeled peptides have been only tentatively identified by comparison of their elution times with those of unlabeled CNBr peptides. All peaks are lower in HEX than in ACh, with *A* and *E* showing the greatest relative change. The difference between ACh and ACh plus PRO is also greatest in peaks *A* and *E*. Comparison between ACh and ACh plus BUP

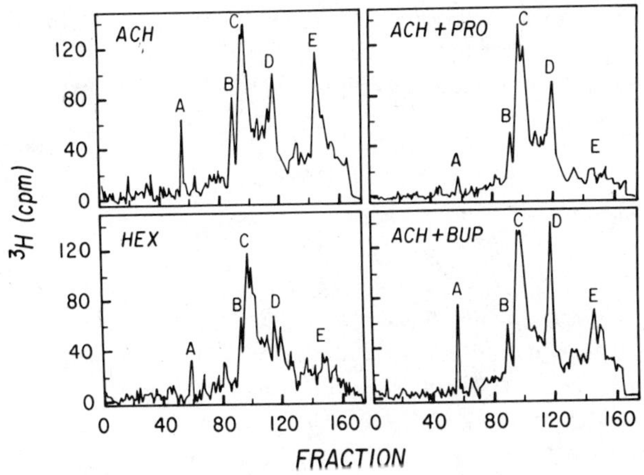

Figure 9. High-pressure liquid chromatography of CNBr peptides of the α subunit that are photolabeled by [^{3}H]QA. In addition to 1.5 μ*M* [^{3}H]QA, the effectors added to the membrane were: ACH (- as in Figure), 1 m*M* ACh; HEX, 0.4 m*M* hexamethonium; ACH + PRO, 1 m*M* ACh and 100 μ*M* proadifen; ACH + BUP, 1 m*M* ACh and 200 μ*M* bupivacaine. The delay between mixing and photolysis was 20 msec, and time in the light beam was 2 msec. The photolabeled membrane was dissolved in 2% SDS and reacted with [^{14}C]-*N*-ethylmaleimide. The ^{3}H radioactivity in the HPLC fractions was normalized by the total ^{14}C recovered in the fractions.

reveals that bupivacaine mainly protects *E*. By comparison with other maps, we know that *E* eluted close to the position of a CNBr fragment containing one of the hydrophobic subsequences postulated to span the membrane.*

Proadifen and bupivacaine differ in their mechanisms of inhibition of AChR function (albeit determined at much lower concentrations than used here): proadifen is a potent desensitizer (14, 62), and bupivacaine acts mainly as an open-channel blocker (7, 45). They both protect a peptide fragment in peak *E*; thus they have at least partially overlapping sites of action. We can ask now, in what functional state peak *E* is preferentially labeled, and, in particular, what the kinetic characteristics of the bupivacaine-protectible labeling of the fragment(s) are in peak *E*. In addition, we are sequencing the fragments in each peak and should be able to identify the amino acid residues with which [^{3}H]QA reacts specifically. We should then be able to compare the positions of the ACh and NCI binding sites in the primary and presently approximate tertiary structure of the α subunit.

ACKNOWLEDGMENTS

We thank Dr. Richard Hudson for a gift of 3-azido-7-methoxy-9-chloroacridine and Dr. J. W. Daly for a gift of histrionicotoxin, and Ms. Niyati Yodh for expert technical

*The major component of peak E is the CNBr fragment of α, α-208 to α-243, containing the first putative membrane-spanning fragment, M1.

assistance. This research was supported in part by research grants from NINCDS and the Muscular Dystrophy Association. Mario DiPaola is a Research Fellow of the New York Heart Association. Peter N. Kao was supported by Medical Scientist Training Grant GM07367

REFERENCES

1. Adams, P. R. (1976). *J. Physiol.* (London), **260**, 531–552.
2. Adams, P. R. (1977). *J. Physiol.* (London), **268**, 291–318.
3. Adams, P. R. and Feltz, A. (1980). *J. Physiol.* (London), **306**, 261–281.
4. Adams, P. R. and Feltz, A. (1980). *J. Physiol.* (London), **306**, 283–306.
5. Anderson, D. J. and Blobel, G. (1981). *Proc. Natl. Acad. Sci.* (USA), **78**, 5598–5602.
6. Anholt, R., Lindstrom, J., and Montal, M. (1985). In Martonosi, A. N., Ed., *The Enzymes of Biological Membranes*, Plenum, Vol. 3, pp. 335–401.
7. Aracava, Y., Ikeda, S. R., Daly, J. W., Brookes, N., and Albuquerque, E. X. (1984). *Molec. Pharmacol.*, **26**, 304–313
8. Aronstam, R. S., Eldefrawi, A. T., Pessah, I. N., Daly, J. W., Albuquerque, E. X., and Eldefrawi, M. E. (1981). *J. Biol. Chem.*, **256**, 2843–2850.
9. Barnard, E. A., Miledi, R., and Sumikawa, K. (1982). *Proc. Roy. Soc. Lond. B*, **215**, 241-246.
10. Blanchard, S. G., Dunn, S. M. J., and Raftery, M. A. (1982). *Biochemistry*, **21**, 6258-6264.
11. Bodanszky, M. and Stahl, G. L. (1974). *Proc. Natl. Acad. Sci.* (USA), **71**, 2791-2794.
12. Bon, F., Lebrun, E., Gomel, J., Van Rapenbusch, R., Cartaud, J., Popot, J.-L., and Changeux, J.-P. (1984). *J. Molec. Biol.*, **176**, 205-237.
13. Boyd, N. D., and Cohen, J. B. (1980) *Biochemistry*, **19**, 5344-5353.
14. Boyd, N. D., and Cohen, J. B. (1984). *Biochemistry*, **23**, 4023-4033.
15. Brisson, A., and Unwin, P. N. T. (1984). *J. Cell Biol.*, **99**, 1202-1211.
16. Brisson, A. and Unwin, P. N. T. (1985). *Nature*, **315**, 474-477.
17. Carp, J. S., Aronstam, R. S., Witkop, B., and Albuquerque, E. X. (1983). *Proc. Natl. Acad. Sci.* (USA), **80**, 310-314.
18. Changeux, J.-P., Devillers-Thiery, A., and Chemoulli, P. (1984). *Science*, **225**, 1335-1345.
19. Clapham, D. E., and Neher, E. (1984). *J. Physiol.*, **347**, 255-277.
20. Clapham, D. E., and Neher, E. (1984). *J. Physiol.*, **353**, 541–564.
21. Claudio, T., Ballivet, M., Patrick, J., and Heinemann, S. (1983). *Proc. Natl. Acad. Sci.* (USA), **80**, 1111–1115.
22. Conti-Tronconi, B. M., Gotti, C. M., Hunkapiller, M. W., Lindstrom, J. M., and Raftery, M. A. (1982). *Proc. Natl. Acad. Sci.* (USA), **79**, 6489–6493.
23. Cox, R. N., Kaldany, R.-R., Brandt, P. W., Ferren, B., Hudson, R. A., and Karlin, A. (1984). *Anal. Biochem.*, **136**, 476–486.
24. Cox, R. N., Kaldany, R.-R.J., DiPaola, M., and Karlin, A. (1985). *J. Biol. Chem.*, **260**, 7186–7193.
25. Daly, J. W., Karle, I., Myers, C. W., Tokuyama, T., Waters, J. A., and Witkop, B. (1971). *Proc. Natl. Acad. Sci.* (USA), **68**, 1870–1875.
26. Devillers-Thiery, A., Giraudat, J., Bentaboulet, M., and Changeux, J.-P. (1983). *Proc. Natl. Acad. Sci.* (USA), **80**, 2067–2071.
27. Dwyer, T. M., Adams, D. J., and Hille, B. (1980). *J. Gen. Physiol.*, **75**, 469–492.

28. Fairclough, R. H., Finer-Moore, J., Love, R. A., Kristofferson, K., Desmeules, P. J., and Stroud, R. M. (1983). *Symp. Quant. Biol.*, **48**, 9–20.

29. Finer-Moore, J. and Stroud, R. M., (1984). *Proc. Natl. Acad. Sci.* (USA), **81**, 155–159.

30. Forster, T. (1949). *Z. Naturforsch.*, **4a**, 321–327.

31. Gage, P. W. and Wachtel, R. E. (1984). *J. Physiol.* (London), **346**, 331–339.

32. Grunhagen, H. H. and Changeux, J.-P. (1976). *J. Molec. Biol.*, **106**, 497–516.

33. Grunhagen, H. H., Iwatsubo, M., and Changeux, J.-P. (1977). *Eur. J. Biochem.*, **80**, 225–242.

34. Guy, H. R. (1984). *Biophys. J.*, **45**, 249–261.

35. Hamilton, S. L., McLaughlin, M., and Karlin, A. (1979). *Biochemistry*, **18**, 155–163.

36. Haring, R., Kloog, Y., Kalir, A., and Sokolovsky, M. (1983). *Biochem. Biophys. Res. Commun.*, **113**, 723–729.

37. Heidmann, T., Bernhardt, J., Neumann, E., and Changeux, J.-P. (1983). *Biochemistry*, **22**, 5452–5459.

38. Heidmann, T., Oswald, R., and Changeux, J.-P. (1983). *Biochemistry*, **22**, 3112–3127.

39. Heidmann, T., and Changeux, J.-P. (1984). *Proc. Natl. Acad. Sci.* (USA), **81**, 1897–1901.

40. Hess, G. P., Cash, D. J., and Aoshima, H. (1983). *Annu. Rev. Biophys. Bioeng.*, **12**, 443–473.

41. Hess, G. P., Pasquale, E. B., Walker, J. W., and McNamee, M. G. (1982). *Proc. Natl. Acad. Sci.* (USA), **79**, 963–967.

42. Holtzman, E., Wise, D., Wall, J., and Karlin, A. (1982). *Proc. Natl. Acad. Sci.* (USA), **79**, 310–314.

43. Huang, L.-Y. M., Catterall, W. A., and Ehrenstein, G. (1978). *J. Gen. Physiol.*, **71**, 397–410.

44. Huganir, R. L., Schell, M. A., and Racker, E. (1979). *FEBS Lett.*, **108**, 155–160.

45. Ikeda, S. R., Aronstam, R. S., Daly, J. W., Aracava, Y., and Albuquerque, E. X. (1984). *Molec. Pharmacol.*, **26**, 293–303.

46. Johnson, D. A., Voet, J. G., and Taylor, P. (1984). *J. Biol. Chem.*, **259**, 5717–5725.

47. Kaldany, R.-R., and Karlin, A. (1983). *J. Biol. Chem.*, **258**, 6232–6242.

48. Kao, P. N., Dwork, A. J., Kaldany, R.-R. J., Silver, M. L., Wideman, J., Stein, S., and Karlin, A. (1984). *J. Biol. Chem.*, **259**, 11662–11665.

48a. Kao, P. N., and Karlin, A. (1986). *J. Biol. Chem.* **261**, 8085–8088.

49. Karlin, A. (1969). *J. Gen. Physiol.*, **54**, 245s–264s.

50. Karlin, A. (1980). In Cotman, C. W., Poste, G., and Nicolson, G. L., Eds., *The Cell Surface and Neuronal Function*, Elsevier, Amsterdam, pp. 191–260.

51. Karlin, A., Cox, R., Kaldany, R.-R., Lobel, P., and Holtzman, E. (1983). *Cold Spring Harbor Symp. Quant. Biol.*, **48**, 1–8.

52. Karlin, A., Holtzman, E., Yodh, N., Lobel, P., Wall, J., and Hainfeld, J. (1983). *J. Biol. Chem.*, **258**, 6678–6681.

53. Karpen, J. W., Aoshima, H., Abood, L. G., and Hess, G. P. (1982). *Proc. Natl. Acad. Sci.* (USA), **79**, 2509–2513.

54. Katz, B. and Thesleff, S. (1957). *J. Physiol.* (London) **138**, 63–80.

55. Kistler, J., Stroud, R. M., Klymkowsky, M. W., Lalancette, R. A., and Fairclough, R. H. (1982). *Biophys. J.*, **37**, 371–383.

56. Koblin, D. D., and Lester, H. (1979). *Molec. Pharmacol.*, **15**, 559–580.

57. Kotzyba-Hibert, F., Langenbuch-Cachat, J., Jaganathen, J., Goeldner, M., and Hirth, C. (1985). *FEBS Lett.*, **182**, 297–301.

58. Lindstrom, J., Anholt, R., Einarson, B., Engel, A., Osame, M., and Montal, M. (1980). *J. Biol. Chem.*, **255**, 8340–8350.

59. Lindstrom, J., Merlie, J., and Yogeeswaran, G. (1979). *Biochemistry*, **18**, 4465–4470.

60. Lobel, P., Kao, P. N., Birken, S., and Karlin, A. (1985). *J. Biol. Chem.*, **260**,10605–10612.

61. Magleby, K. L. and Pallotta, B. S. (1981). *J. Physiol.* (London) **316**, 225–250.

62. Maleque, M. A., Souccar, C., Cohen, J. B., and Albuquerque, E. X. (1982). *Molec. Pharmacol.*, **22**, 636–647.

63. Mishina, M., Kurosaki, T., Tobimatsu, T., Morimoto, Y., Noda, M., Yamamoto, T., Terao, M., Lindstrom, J., Takahashi, T., Kuno, M., and Numa, S. (1984). *Nature*, **307**, 604–608.

64. Mishina, M., Tobimatsu, T., Imoto, K., Tanaka, K., Fujita, Y., Fukuda, K., Kurasaki, M., Takahashi, H., Morimoto, Y., Hirose, T., Inayama, S., Takahashi, T., Kuno, M., and Numa, S. (1985). *Nature*, **313**, 364–369.

65. Mueller, D. M., Hudson, R. A., and Lee, C. P. (1982). *Biochemistry*, **21**, 1445–1453.

66. Muhn, P. and Hucho, F. (1983). *Biochemistry*, **22**, 421–425.

67. Muhn, P., Fahr, A., and Hucho, F. (1984). *Biochemistry*, **21**, 2725–2730.

68. Neher, E. (1983). *J. Physiol.* (London), **339**, 663–678.

69. Neher, E. and Steinbach, J. H. (1978). *J. Physiol.* (London) **277**, 153–176.

70. Neubig, R. R., Boyd, N. D., and Cohen, J. B. (1982). *Biochemistry*, **21**, 3460–3467.

71. Neumann, D., Barchan, D., Safran, A., Gershoni, J. M., and Fuchs, S. (1986). *Proc. Natl. Acad. Sci.* (USA) **83**, 3008–3011.

72. Noda, M., Takahashi, H., Tanabe, T., Toyosato, M., Furutani, Y., Hirose, T., Asai, M., Inayama, S., Miyata, T., and Numa, S. (1982). *Nature*, **299**, 793–797.

73. Noda, M., Takahashi, H., Tanabe, T., Toyosato, M., Kikyotani, S., Furutani, Y., Hirose, T., Takashima, H., Inayama, S., Miyata, T. and Numa, S. (1983). *Nature*, **302**, 528–532.

74. Oswald, R. and Changeux, J.-P. (1981). *Biochemistry* **20**, 7166–7174.

75. Oswald, R., and Changeux, J.-P. (1981). *Proc. Natl. Acad. Sci.* (USA) **78**, 3925–3929.

76. Oswald, R. E., Heidmann, T., and Changeux, J.-P. (1983). *Biochemistry*, **22**, 3128–3136.

77. Ovchinnikov, Y. A., Lipkin, V. M., Shuaeva, T. M., Bogachuk, A. P., and Shemyakin, V. V. (1985). *FEBS Lett.*, **179**, 107–110.

78. Oxford, G. S., and Hudson, R. A. (1982). *Biochem. Biophys. Res. Commun.*, **104**, 1579–1584.

79. Popot, J.-L. and Changeux, J.-P. (1984). *Physiol. Revs.*, **64**, 1162–1239.

80. Raftery, M. A., Hunkapiller, M. W., Strader, C. D., and Hood, L. E. (1980). *Science*, **208**, 1454–1457.

81. Reynolds, J. A. and Karlin, A. (1978). *Biochemistry*, **17**, 2035–2038.

82. Ruff, R. L. (1977). *J. Physiol.* (London), **264**, 89–124.

83. Sakmann, B., Patlak, J., and Neher, E. (1980). *Nature*, **286**, 71–73.

84. Sine, S. M. and Taylor, P. (1982). *J. Biol. Chem.*, **257**, 8106–8114.

85. Spivak, C. E., Maleque, M. A., Oliveira, A. C., Masukawa, L. M., Tokuyama, T., Daly, J. W., and Albuquerque, E. X. (1982). *Molec. Pharmacol.*, **21**, 351–361.

86. Sumikawa, K., Houghton, M., Smith, J. C., Bell, L., Richards, B. M., and Barnard, E. A. (1982). *Nucl. Acid Res.*, **10**, 5809–5822.

87. Thornton, J. M. (1981). *J. Molec. Biol.*, **151**, 261–287.

88. Tsai, M.-C., Oliveira, A. C., Albuquerque, E. X., Eldefrawi, M. E., and Eldefrawi, A. T. (1979). *Molec. Pharmacol.*, **16**, 382–392.

89. Vandlen, R. L., Wu., W. S. C., Eisenbach, J. C., and Raftery, M. A. (1979). *Biochemistry*, **18**, 1845–1854.

90. Walker, J. W., Takeyasu, K., and McNamee, M. G. (1982). *Biochemistry*, **21**, 5384–5389.

91. Weill, C. L., McNamee, M. G., and Karlin, A. (1974). *Biochem. Biophys. Res. Commun.*, **61**, 997–1003.

92. Wilson, P. T., Lentz, T. L., and Hawrot, E. (1985). *Soc. Neurosci. Abstr.*, **11**, 169.

93. Wise, D. S., Karlin, A., and Schoenborn, B. P. (1979). *Biophys. J.*, **28**, 473–496.

94. Wu, W. C. S., and Raftery, M. A. (1979). *Biochem. Biophys. Res. Commun.*, **89**, 26–35.

95. Young, E. F., Ralston, E., Blake, J., Ramachandran, J., Hall, Z. W., and Stroud, R. M. (1985). *Proc. Natl. Acad. Sci.* (USA), **82**, 626–630.

96. Zingsheim, H. P., Barrantes, F. J., Frank, J., Hanicke, W., and Neugebauer, D.-C. (1982a). *Nature*, **299**, 81–84.

97. Zingsheim, H. P., Neugebauer, D.-C., Frank, J., Hanicke, W., and Barrantes, F. J. (1982b). *EMBO J.*, **1**, 541–547.

CHAPTER 4

TOPOLOGICAL MAPPING AND THE IONIC CHANNEL IN AN ACETYLCHOLINE RECEPTOR

ROBERT MICHAEL STROUD
Department of Biochemistry and Biophysics
University of California
San Francisco, California

Acetylcholine receptors (AChRs) are found at the junctions between nerve and muscle cells. Located in the plasma membrane of the target cell, they translate the binding of ACh into conductivity through an ion-conducting channel (11,14,19,16,20). We have determined the three-dimensional structure of this molecule from ACh-rich membrane fractions prepared from the electric fish *Torpedo californica* (13). The structure, at 25-Å resolution, was obtained by a combination of electron-microscopic studies on tubular crystalline arrays of the native AChR (12) and by x-ray diffraction of AChR-rich membranes (5). Recent studies of an extensively proteolyzed version of the AChR from *T. marmarota* show that the remaining core appears more fivefold symmetrical (2,3) and packs into a different lattice than does the undegraded receptor (12). The structure consists of five transmembranous subunits, each shown to transpose the membrane by the binding of antibodies or by susceptibility to proteolysis from either side of the membrane.

The center of the molecule, which is densely stained by negative stains such as uranyl acetate, contains these electron dense uranyl ions throughout 80–90% of the entire length of the AChR in a cylinder that is 7 Å in diameter in the narrowest transmembranous part of the molecule (13) (Fig. 1). It was shown by x-ray diffraction that neurotoxins that compete with ACh for binding sites located primarily on the α chains are located on the top crest of the AChR (5). Thus ligands binding on the top crest must signal their effect through at least 60 Å to the entrance to the ion channel.

X-ray diffraction shows that the AChR contains 12–30 α helices oriented predominantly perpendicular to the membrane plane. Alpha helices provide a means of protecting hydrogen bonding carbonyl and amide NH groups during cotranslational insertion into the membrane and provide properties to the helix characteristic of the side chains that are all external to this feature. There is precedent

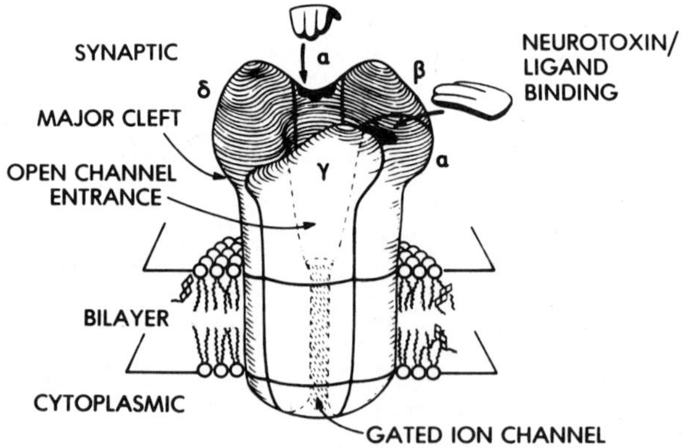

Figure 1. Three-dimensional structure of AChR derived from electron microscopy and x-ray diffraction. The cytoplasmic domain must be larger than seen in this electron-microscope structure. The subunit locations are tentative. The ion channel probably contains ions in the resting state based on TB^{3+} binding experiments (6). Residues Cys-208 and Cys-209 lie at the top crest where toxin and cholinergic ligands bind (10).

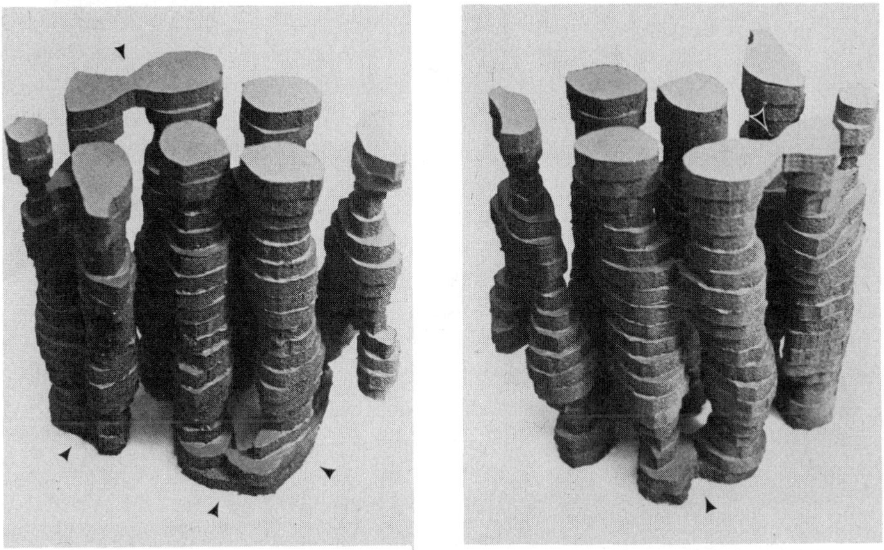

Figure 2. The three-dimensional structure of bacteriorhodopsin refined by Agard and Stroud (1) shows predominantly helical rods perpendicular to the membrane planes. These are probably α helices.

for α helices that interface between protein and lipid as seen, for example, in the photoreaction center complex (4) or inferred for bacteriorhodopsin and many other proteins (1) (Fig. 2).

Amino acid sequences, deduced from the sequences of cDNA clones, show a homology between all four different gene (α β γ δ) products from *N* terminal to *C* terminal (18). Each sequence contains four very hydrophobic sequences that transpose the membrane. The topology of each subunit is most probably identical in light of this strong sequence homology. Using antibodies, generated against peptides contained within the sequence of different chains, we were able to establish that residues around 392 and 500 both lie on the cytoplasmic side of the membrane while the sequence 1–228 is known to lie on the synaptic surface (see Fig. 3) (21). The latter contains the amino terminal, generated by cleavage of the signal sequence in the lumen of the endoplasmic reticulum, and regions involved in the binding of ligands (10) as well as the sites of N-glycosylation (18). The topology indicated in Fig. 3 and proposed for all subunits implies an even number of crossings between residues 392 and 500. This sequence contains one highly hydrophobic crossing per subunit and a most unusual sequence that presents an alternation of charge and hydrophobic residues, which display a periodicity expected for sequences in α-helical structure. Such periodicities can be analyzed by Fourier transformation of the properties of amino acids (7). Such Fourier transforms quantitatively demonstrate the degree of amphipathic quality expected from this secondary structure and provide a powerful tool in the analysis of sequence (7,15,20). Applied to chains of the AChR, this method shows that the *N*-terminal domain includes a long sequence of 100 amino

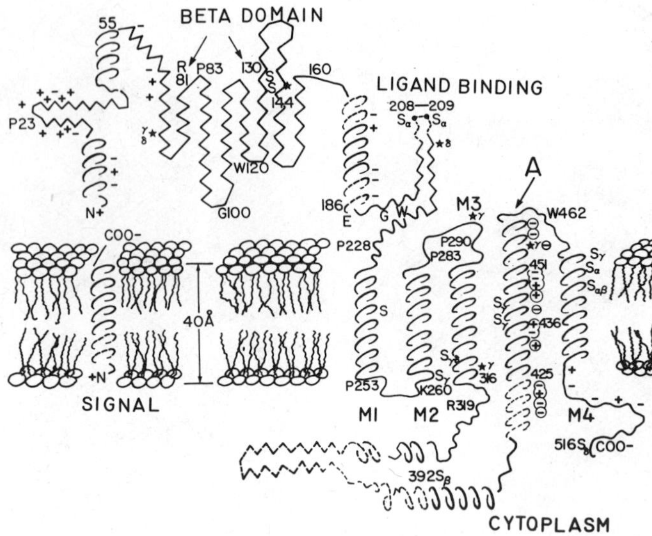

Figure 3. Topology of the sequence in AChR subunits deduced from antibody labeling and electron microscopy. Stars indicate sites of N-glycosylation. Numbers and letters refer to the consensus aligned sequence numbers due to Stroud and Finer-Moore (20).

acids that alternate with the frequency expected for β sheets, and that from the locations of turn-forming residues we expect to form one eight-stranded, antiparallel β-pleated sheet structure per subunit. This analysis also focuses on the strongly amphipathic helix A immediately preceding the final hydrophobic sequence, close to the carboxyl terminal of each chain. Topological studies indicate that the amino acid sequence must cross the membrane in this region, and the issue is now whether this sequence, one copy per subunit, could provide the lining for the water-filled ion-conducting channel, which we have shown to be located in the center of the molecule (13). One may thus ask whether an ionic channel, constructed in this way, between a fivefold symmetrical arrangement of these sequences within the five subunits, could create the characteristics of specificity and flux observed by electrophysiological studies for AChR.

To evaluate these possibilities, a model was constructed in which these sequences were arranged and close packed around the ion channel. The sequences presented toward the channel are shown in Fig. 4. This model is built with only two parameters, rotation around the axis of one helix and the tilt angle with respect to the vertical of the helical stretch. The former was determined to bring the hydrophilic chains in toward the center of the molecule, and the latter was determined for optimum packing of side chains located between adjacent helices. Energy calculations in which a sodium ion is placed at each position throughout the length of the ion-conducting channel permit an estimation of the energy barriers to conductivity of that ion (20). These studies show that the energy barriers to conduction of a hydrated monovalent cation such as Na^+ are on the order of 3–4 kcal/mole, close to that estimated by electrophysiologists from

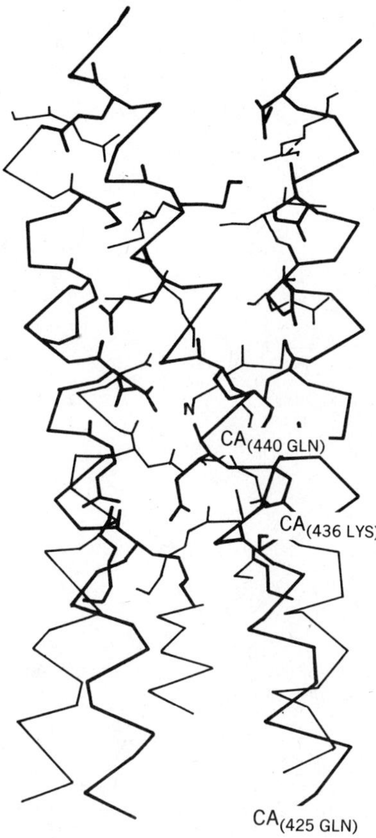

Figure 4. Five-helix model of the ion channel in AChR. One amphipathic helix (*A* in Fig. 3) derives from each subunit.

ion-conductivity measurements (Fig. 5) (9). The size of the model channel is sufficient to provide passage for larger ions such as isopropyl ammonium, which are known to pass through the channel. Although the entry to the ion channel is negatively charged, it provides a negative free energy, hence a stable attractive space charge for cations, and a repulsion for anions such as chloride. It thus can explain the observed specificity for cations rather than ions in AChR, as revealed in the energy plots (see Fig. 6).

This model thus stands the test of reasonableness alongside the known parameters of conductivity of ions. However, it raises the question as to why the central well in this model should carry charge at all. It is clear from similar energy calculations in which the charged side chains are replaced by alanine, or by calculations in which the side chains are simply polar rather than charged, that the barriers to ion transduction by such models would also be similarly small. The (unproved) proposal for alamethecin action (8) depends on an uncharged arrangement of protein helices that

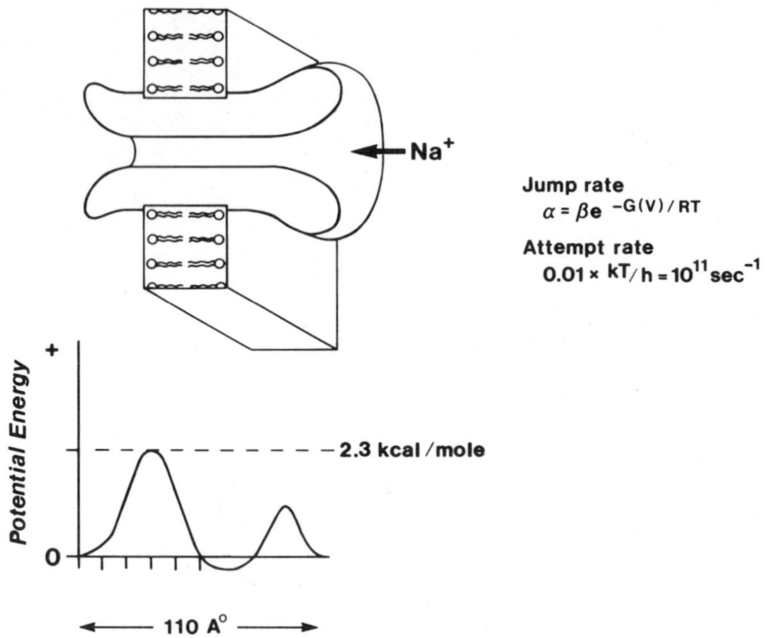

Figure 5. Sketch of the energy profile for passage of a sodium ion through the AChR channel, deduced from electrophysiological experiments (9).

contain only a single (negative) charge around a central channel also. Why, then, would the amphipathic helices in AChR carry charge? The answer seems to be that the high charge density within the channel would require hydration in the center of the molecular complex. Thus it would stabilize the specific ion channel against collapse caused by repacking of the more hydrophobic helices. In summary, this model suggests the following tentative conclusions:

1. The specificity for cations rather than anions arises from the predominantly negatively charged entry to the transmembranous region of the ion channel.
2. The channel carries an almost equally balanced set of adjacent positive and negative charged side chains that preserve a water-filled volume but do not provide deep energy wells for a charged ion, since this would impede conductivity rather than enhance it.
3. The size presented by the five-helix model, linking the number of homologous subunits in any similar channel, is adequate to explain the size limitations for conductivity of organic and inorganic hydrated ions.

The question as to where the ion gating system might lie within the ion channel is addressed most directly in experiments by Fairclough et al. (6) in which terbium ions have been localized within the resting state of AChR. Terbium ions compete with

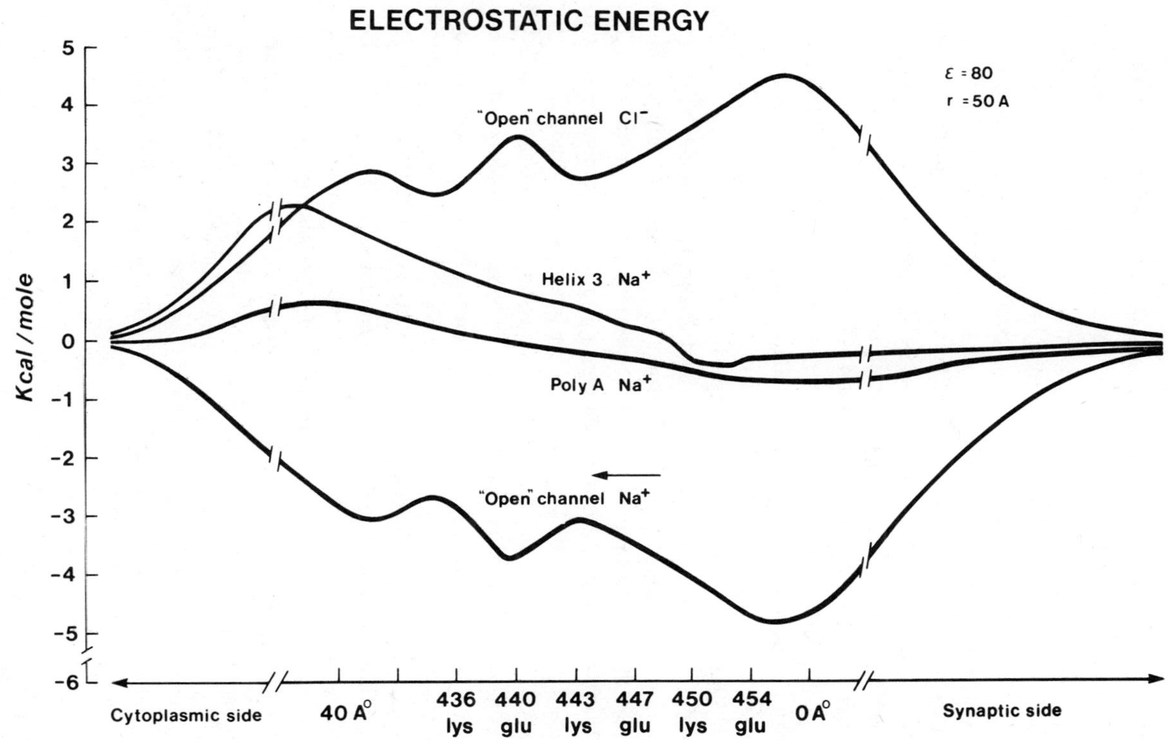

Figure 6. Energy profiles for a Na^+ and a Cl^- ion as a function of distance from the extracellular side of the membrane are shown; they display cation selectivity. Calculations are for the model shown in Fig. 5 and for similar models constructed with hydrophobic helices M3 or for polyalanine helices. In this case conductivity would also be high, although the channel could easily collapse without need for solvation.

calcium ions for sites on the AChR, and it has been shown that approximately 50% of these ion sites are removed by the binding of ligands such as carbamyl choline. The Tb^{3+} binding sites were located perpendicular to the membrane by anomalous x-ray diffraction analysis. This technique is especially sensitive to lanthanide ions such as terbium with unfilled *d* orbitals where the wavelength-dependent dispersion is particularly large. One-dimensional diffraction data perpendicular to the membrane plane were recorded at five different wavelengths. Thus the experiment is extremely well controlled and allows determination of the terbium ion positions, or rather, the probability of Tb^{3+} occupation, relative to the membrane plane. In this case three distinct terbium ion locations were found in the transmembranous region of the protein. If it is assumed that these ion positions must be located in the most ionophilic region of the transmembrane complex, it is implied that terbium ions are bound inside the ion-conducting channel in the *resting* state of AChR, at positions that are approximately 11 Å apart as revealed from the difference Fourier map and refinement analysis by Fairclough et al. (6). This distribution is consistent with the notion that there are stable positions for ions in the channel at each second turn of α helix, surrounded by rings of negatively charged side chains, and suggests that the resting state of AChR may contain bound ions in these locations *in vivo*.

Since the Tb^{3+} or Ca^{2+} sites displaced by ACh binding have not yet been assigned, no direct conclusion about mechanisms of channel gating can be inferred from this distribution; however, the minimum probability density of finding terbium ions lies 15 Å from the cytoplasmic surface of the protein, thus suggesting that this region is a possible focus for attention in the search for a gating mechanism. Kao et al. (10) have localized the region of Cys-208–209 as being at the binding site for activating ligands. The neurotoxin binding site, which is most probably the ACh binding site, is located on the top crest of AChR (5). Thus the effect of ligand binding must be relayed to the channels. Although it once seemed that a large-scale twisting of all subunits could be involved, this kind of scheme does not answer the question at atomic level, or the level at which residues generate the "gate." Various possible gating mechanisms are still being tested in part by the elegant experiments of Mishina et al., (17), who have demonstrated that expression of protein followed by functional reconstitution in frog oocytes can permit the trial of different deletions or substituants in given regions of the protein structure using site-directed deletion and mutagenesis. Their studies pinpoint the functional domains in AChR and consequences for structure and/or function of microscopic changes of side chains in the sequence of this complex protein. These experiments, and those of Kao et al. (10), have illuminated the contribution of specific regions of the sequence in the structure to the mechanisms of the AChR at the microscopic level.

ACKNOWLEDGMENT

Work reported in these studies was carried out under the support of the National Institute of Health, grants GM24485 and GM32079 and the NSF grant PCM83-16401.

REFERENCES

1. Agard, D. A. and Stroud, R. M. (1982). *Biophys. J.*, **37**, 589–602.
2. Brisson, A. and Unwin, P. N. T. (1984). *J. Cell Biol.*, **99**, 1202–1211.
3. Brisson, A. and Unwin, P. N. T. (1985). *Nature*, **315**, 474–477.
4. Desenhofer, J., Epp, O., Miki, K., Huber, R., and Michel, H. (1985). *Nature*, **318**, 618–624.
5. Fairclough, R. H., Finer-Moore, J., Love, R. A., Kristofferson, D., Desmeules, P. J., and Stroud, R. M. (1983). *Cold Spring Harbor Symp. Quant. Biol.*, **48**, 9–20.
6. Fairclough, R. H., Miake-Lye, R. C., Stroud, R. M., Hodgson, K. O., and Doniach, S. (1986). *J. Molec. Biol.* **189**, 673–680.
7. Finer-Moore, J. and Stroud, R. M. (1984). *Proc. Natl. Acad. Sci.* (USA), **81**, 155–159.
8. Fox, R. O. and Richards, F. M. (1982). *Nature*, **300**, 325–330.
9. Horn, R. and Stevens, C. F. (1980). *Comments Mol. Cell Biophys.*, **1**, 57–68.
10. Kao, P. N., Dwork, A. J., Kaldany, R.-R. J., Silver, M. L., Wideman, J., Stein, S., and Karlin, A. (1984). *J. Biol. Chem.*, **259**, 11662–11665.
11. Karlin, A. (1983). *Neurosci. Commun.*, **1**, 111–123.
12. Kistler, J. and Stroud, R. M. (1981). *Proc. Natl. Acad. Sci.* (USA), **78**, 3678–3682.
13. Kistler, J., Stroud, R. M., Klymkowsky, M. W., Lalancette, R. A., and Fairclough, R. H. (1982). *Biophys. J.*, **37**, 371–383.
14. Lindstrom, J. (1983). *Neurosci. Commun.*, **1**, 139–156.
15. Liscum, L., Finer-Moore, J., Stroud, R. M., Luskey, K. L., Brown, M. S., and Goldstein, J. L. (1985). *J. Biol. Chem.*, **260**, 522–530.
16. McCarthy, M. P., Earnest, J. P., Young, E. F., Choe, S., and Stroud, R. M. (1986). *Ann. Rev. Neurosci.*, **9**, 383–413.
17. Mishina, M., Tobimatsu, T., Imoto, K., Tanaka, K., Fujita, Y., et al. (1985). *Nature*, **313**, 364–369.
18. Noda, M., Takahashi, H., Tanabe, T., Toyosato, M., Kikyotani, S., et al. (1983). *Nature*, **302**, 528–532.
19. Popot, J.-L. and Changeux, J.-P. (1984). *Physiol. Rev.*, **64**, 1162–1239.
20. Stroud, R. M. and Finer-Moore, J. (1985). *Ann. Rev. Cell Biol.*, **1**, 369–401.
21. Young, E. F., Ralston, E., Blake, J., Ramachandran, J., Hall, Z. W., and Stroud, R. M. (1985). *Proc. Natl. Acad. Sci.* (USA), **82**, 626–630.

CHAPTER 5

COORDINATE REGULATION OF EXPRESSION OF SYNAPTIC PROTEINS IN SKELETAL MUSCLE: STUDIES WITH ACETYLCHOLINE RECEPTOR

J. P. MERLIE[1]
A. BUONANNO[1,4]
B. CARLIN[1]
J. COVAULT[3]
C. M. CROWDER[1,4]
T. DIETZ[1,4]
D. FRAIL[1]
K. ISENBERG[1,2]
J. LAWRENCE[1]
J. MUDD[1]
J. SANES[3]
V. SHAH[1]
M. M. SMITH[1]
L. L. THIO[1,4]

Departments of Pharmacology[1], Psychiatry[2], and Anatomy and Neurobiology[3], and the Division of Biomedical Sciences[4]
Washington University Medical School
St. Louis, Missouri

The neuromuscular junction is highly specialized in its molecular composition, containing high concentrations of several cytoskeletal, membrane, and extracellular matrix components that are virtually absent from extrasynaptic portions of the muscle fiber surface. The best studied of these synapse-specific proteins is the acetylcholine receptor (ACh). Acetylcholine receptors are present throughout the plasma membrane of embryonic myotubes but become concentrated at synapses and are lost from extrasynaptic areas after formation of synapses. However, denervation or paralysis of adult muscle induces synthesis and insertion of new AChRs in extrasynaptic areas. These alterations in receptor levels and distribution parallel at least in part those of several other synapse specific molecules (Table 1) and correlate with the ability of the muscle fiber to form new synapses. Therefore, it seems likely that the mechanisms that regulate the synthesis of synapse-specific molecules are, in fact, important components of the program that regulates the formation and maintenance of synapses.

To understand the intricacies of this program, we have begun to study the mechanisms of regulation of AChR levels in developing muscle cells, normal adult muscle, and denervated muscle.

1. cDNA CLONES FOR MOUSE MUSCLE ACh RECEPTOR

To have the reagents necessary for analysis of regulation of AChR synthesis at the level of subunit gene transcription and mRNA processing, we began to prepare and characterize cloned cDNAs for each mouse muscle subunit mRNA. Our starting material for this cloning project was the mouse cell line BC3H-1, which we had shown by cell-free translation experiments to be a rich source of α subunit mRNA (48). Our first α subunit cDNA clone, pA59, was identified in a library prepared in

TABLE 1. SYNAPSE-SPECIFIC MOLECULES THAT MAY BE COORDINATELY REGULATED IN MUSCLE

	AChR (18)	N-CAM (13)	JS-1 (46)	AChE (26)	Cyto-skeletal Elements (4)
Concentrated in or near postsynaptic membrane	+	+	+	+	+
In extrasynaptic membrane in embryo	+	+		+	
Increased after denervation	+	+		+/−	+
Suppressed by activity; increased by paralysis	+	+	+	−	
Increased by brain extract	+		+	+	

pBR322 by hybrid selected translation (32) and subsequently confirmed by partial nucleic acid sequence analysis (34). Recently, clones that cover the mRNA sequence from nine bases 5′ of the initiator methionine codon to the poly(A) sequence at the 3′ end have been isolated from cDNA libraries prepared in the bacteriophage λgt11. The composite sequence of the α-subunit cDNA and predicted amino acid sequence of the pre-α-subunit polypeptide are shown in Fig. 1. Comparison of our sequence with the BC3H-1 α-subunit sequence (5) has indicated no differences within the amino acid coding region except that our sequence is 36 base pairs longer at the 5′ end and our sequence predicts a cysteine residue rather than a serine at amino acid 8.

The δ-subunit clone that we have used was isolated by LaPolla et al. (24) from a λgt10 library by screening with a homologous *Torpedo* cDNA. It is approximately 1900 base pairs in length, extending from the initiator codon through approximately 400 base pairs of 3′ noncoding sequence. Since no poly(A) addition signal was found and the most abundant RNA species that hybridized with this probe was 3300 bases in length, there may be significant stretches of 5′ and 3′ noncoding sequences not represented in the clone. Nevertheless, this δ-subunit cDNA has proved useful for many different types of experiments, including analysis of DNA and RNA by blot hybridization, transcription run on assays (see the following paragraphs), and expression of functional receptor in frog oocytes (57).

A strategy for the isolation of BC3H-1 β and γ cDNA clones has been devised on the basis of use of a synthetic oligonucleotide that was predicted to be homologous to all four subunits. The 25-base sequence between nucleotides 402 and 426 (see ref. 40 for sequence conventions) was found by sequence alignment to be highly conserved among all subunits of all species examined (see Fig. 2). The β, γ, and δ subunits of the chick, bovine, human, and mouse are most conserved, whereas all α subunits and Torpedo β, γ, and δ show the most divergence from the consensus sequence. We have no explanation for the conservation of nucleotide sequence in this region. Most of the codons represented have one or more synonyms, so that conservation of functional domains of the polypeptides cannot fully explain the conservation of nucleotide sequence. It is interesting to speculate that this region of the subunit sequences may participate in a process for which conservation of nucleotide sequence may be necessary, such as transcription or RNA processing.

Screening of 30,000 recombinants with the 25-mer 402-426 oligonucleotide resulted in approximately 68 positive signals. Of these, 60 melted at 72°C in 6× SSC and were thus predicted to include the less homologous α- and δ-subunit clones. These were set aside for future characterization. Eight isolates were found to hybridize at 72°C in 6× SSC, and these were chosen for sequence analysis. Two of the eight have been identified as full-length β-subunit clones by sequence alignment with the five bovine cDNAs. We are in the process of sequencing the remaining six clones in the expectation that γ cDNAs will have the same sequence in the 402-426 region.

It is significant that the sequence for oligonucleotide 402-426 is conserved in the novel ϵ subunit recently described by Takai et al. (54). Thus it seems likely that screening with this synthetic oligonucleotide will prove useful for identifying AChR-related sequences from many species and tissues.

```
       -60       -50       -40       -30       -20       -10         1        11        21
CACAGCCCATGGAGCTCTCGACTGTTCTCCTGCTGCTAGGCCTCTGCTCCGCTGGCCTTGTTCTGGGCTCCGAACATGAGACGCGTCTG
          MetGluLeuSerThrValLeuLeuLeuLeuGlyLeuCysSerAlaGlyLeuValLeuGlySerGluHisGluThrArgLeu

        31        41        51        61        71        81        91       101       111
GTGGCAAAGCTCTTTGAAGACTACAGCAGTGTAGTCCGGCCAGTGGAGGACCACCGTGAGATTGTACAAGTCACCGTGGGTCTACAGCTG
ValAlaLysLeuPheGluAspTyrSerSerValValArgProValGluAspHisArgGluIleValGlnValThrValGlyLeuGlnLeu

       121       131       141       151       161       171       181       191       201
ATCCAGCTTATCAATGTGGATGAAGTAAATCAGATTGTGACAACCAATGTACGTCTGAAACAGCAATGGGTCGATTACAACTTGAAATGG
IleGlnLeuIleAsnValAspGluValAsnGlnIleValThrThrAsnValArgLeuLysGlnGlnTrpValAspTyrAsnLeuLysTrp

       211       221       231       241       251       261       271       281       291
AATCCAGATGACTATGGAGGAGTGAAAAAAATTCACATCCCCTCGGAAAAGATCTGGCGGCCGGACGTCGTTCTCTATAACAACGCAGAC
AsnProAspAspTyrGlyGlyValLysLysIleHisIleProSerGluLysIleTrpArgProAspValValLeuTyrAsnAsnAlaAsp
                *
       301       311       321       331       341       351       361       371       381
GGCGACTTTGCCATTGTCAAATTCACCAAGGTGCTCCTGGACTACACCGGCCACATCACCTGGACACCGCCAGCCATCTTTAAAAGCTAC
GlyAspPheAlaIleValLysPheThrLysValLeuLeuAspTyrThrGlyHisIleThrTrpThrProProAlaIlePheLysSerTyr

       391       401       411       421       431       441       451       461       471
TGTGAGATCATTGTCACTCACTTTCCCTTCGATGAGCAGAACTGCAGCATGAAGCTGGGCACCTGGACCTATGACGGCTCTGTGGTGGCC
CysGluIleIleValThrHisPheProPheAspGluGlnAsnCysSerMetLysLeuGlyThrTrpThrTyrAspGlySerValValAla

       481       491       501       511       521       531       541       551       561
ATTAACCCGGAAAGTGACCAGCCCGACCTGAGTAACTTCATGGAGAGCGGGGAGTGGGTGATCAAGGAAGCTCGGGGCTGGAAGCACTGG
IleAsnProGluSerAspGlnProAspLeuSerAsnPheMetGluSerGlyGluTrpValIleLysGluAlaArgGlyTrpLysHisTrp

       571       581       591       601       611       621       631       641       651
GTGTTCTACTCCTGCTGCCCCACCACTCCCTACCTGGACATCACCTACCACTTCGTCATGCAGCGCCTGCCCCTCTACTTCATTGTCAAC
ValPheTyrSerCysCysProThrThrProTyrLeuAspIleThrTyrHisPheValMetGlnArgLeuProLeuTyrPheIleValAsn

       661       671       681       691       701       711       721       731       741
GTCATCATTCCCTGCCTGCTCTTCTCCTTCTTAACCAGCCTGGTGTTCTACCTGCCCACAGACTCAGGGGAGAAGATGACGCTGAGCATC
ValIleIleProCysLeuLeuPheSerPheLeuThrSerLeuValPheTyrLeuProThrAspSerGlyGluLysMetThrLeuSerIle

       751       761       771       781       791       801       811       821       831
TCTGTCTTACTGTCCCTGACCGTGTTCCTTCTGGTCATTGTGGAGCTAATCCCTTCCACCTCCAGCGCTGTGCCCCTGATCGGGAAGTAT
SerValLeuLeuSerLeuThrValPheLeuLeuValIleValGluLeuIleProSerThrSerSerAlaValProLeuIleGlyLysTyr

       841       851       861       871       881       891       901       911       921
ATGTTGTTCACCATGGTCTTTGTCATTGCGTCCATCATCATCACCGTCATCGTCATCAACACACACCACCGTTCGCCCAGCACCCACATC
MetLeuPheThrMetValPheValIleAlaSerIleIleIleThrValIleValIleAsnThrHisHisArgSerProSerThrHisIle

       931       941       951       961       971       981       991      1001      1011
ATGCCCGAGTGGGTGCGGAAGGTTTTTATCGACACTATCCCAAACATCATGTTTTTCTCCACAATGAAAAGACCATCCAGAGATAAACAA
MetProGluTrpValArgLysValPheIleAspThrIleProAsnIleMetPhePheSerThrMetLysArgProSerArgAspLysGln

      1021      1031      1041      1051      1061      1071      1081      1091      1101
GAGAAAAGGATTTTTACAGAAGACATAGATATATCTGACATCTCTGGAAGCCGGGTCCTCCACCTATGGGCTTTCACTCTCCGCTGATC
GluLysArgIlePheThrGluAspIleAspIleSerAspIleSerGlyLysProGlyProProProMetGlyPheHisSerProLeuIle

      1111      1121      1131      1141      1151      1161      1171      1181      1191
AAGCACCCTGAGGTGAAAAGCGCCATCGAGGGCGTGAAGTACATTGCAGAGACCATGAAGTCAGACCAGGAGTCCAATAACGCCGCTGAG
LysHisProGluValLysSerAlaIleGluGlyValLysTyrIleAlaGluThrMetLysSerAspGlnGluSerAsnAsnAlaAlaGlu

      1201      1211      1221      1231      1241      1251      1261      1271      1281
GAATGGAAGTATGTTGCCATGGTGATGGATCACATCCTCCTCGGAGTCTTTATGCTGGTGTGTCTCATCGGGACGCTGGCTGTGTTTGCA
GluTrpLysTyrValAlaMetValMetAspHisIleLeuLeuGlyValPheMetLeuValCysLeuIleGlyThrLeuAlaValPheAla

      1291      1301      1311      1321      1331      1341      1351      1361      1371
GGTCGGCTCATTGAGTTACATCAACAAGGATGAGCAGAGGCTGAGCTAAGCCTACCTCTGTCCCAGCCATAGCCATCGCTAGGAAAGATG
GlyArgLeuIleGluLeuHisGlnGlnGly*op

      1381      1391      1401      1411      1421      1431      1441      1451      1461
GAAGAGAGGAAGGTCTGTCTCCTTGAATCCTTTCACACTTACCAAACATGCAGTGTTCTACATGTCCTACATGTTAATGAGAGTGATCTC

      1471      1481      1491      1501      1511      1521      1531      1541      1551
TGCTCACACGGCTGTATTCTTGAAGTGTCTCCCCTTTGCTTCTGCTTTTAACACTATGGGCCTCCTTAAAGGGCGAACCCTTTGAGTAAA

      1561      1571      1581      1591      1601      1611      1621      1631      1641
TAAAAGTGAGCCCTCAAAAGAAGTGTTTGCTTCTAAATGGCCCCTGGGAGAGTTTTGCTTGGATACTCAAGGTTTTCTGTTTGTATTGCC

      1651      1661      1671      1681      1691      1701      1711      1721      1731
ATGGCTAGTTGTTTTTGTTTTCTTTCCTTTAATAAATATAATTGTACTTATAAAAAAAA
```

Figure 1. Nucleotide and predicted amino acid sequence of BC3H-1 α-subunit cDNA. The complete nucleotide sequence of both coding and noncoding DNA strands was determined as follows. Plasmid clone pA59 (32) extends from approximtely N = 850 to 1550; it was sequenced by the method of Maxam and Gilbert (27). Clone 14 extending from N = -69 to 1600 and clone 11 also extending from N = 800–1703, both from a γgt11 library, were subcloned in both orientations into M13 mp10 and sequenced by the dideoxy chain termination method (51), using synthetic oligonucleotide primers. The first nucleotide of the serine codon is predicted by comparison with the *Torpedo* α-subunit sequence to be the first amino acid of the mature polypeptide and has been designated as N = 1 (41).

Oligonucleotide		C	T T C	C C C	T T T C	G A C	T G G	C A G	A A C	T G C
TORPEDO	ALPHA	T	T T C	C C [A]	T T T	G A [T]	[C][A][A]	C A [A]	A A [T]	T G C
	BETA	C	T T C	C C C	T T C	G A [T]	T G G	C A [A]	A A C	T G C
	GAMMA	C	T T C	C C [T]	T T C	G A C	T G G	C A G	A A C	T G [T]
	DELTA	T	T T [T]	C C [A]	T T T	G A C	T G G	C A G	A A C	T G C
BOVINE	ALPHA	C	T T [T]	C C C	T T T	[C] A C	[G][A][A]	C A G	A A C	T G C
	BETA	C	T T C	C C C	T T T	G A C	T G G	C A G	A A C	T G C
	GAMMA	C	T T C	C C C	T T C	G A C	T G G	C A G	A A C	T G C
	DELTA	C	T T C	C C C	T T C	G A C	T G G	C A G	A A C	T G C
	EPSILON	C	T T C	C C [T]	T T C	G A C	T G G	C A G	A A C	T G C
CHICKEN	GAMMA	C	T T C	C C C	T T T	G A C	T G G	C A G	A A C	T G C
	DELTA	C	T T C	C C C	T T T	G A C	T G G	C A G	A A C	T G C
MOUSE	ALPHA	C	T T [T]	C C C	T T C	G A [T]	[G][A] G	C A G	A A C	T G C
	DELTA	C	T T C	C C [T]	T T C	G A C	T G G	C A G	A A C	T G C
HUMAN	ALPHA	C	T T [T]	C C C	T T T	G A [T]	[G][A][A]	C A G	A A C	T G C
	GAMMA	C	T T C	C C C	T T C	G A C	T G G	C A G	A A C	T G C

Figure 2. Nucleotide sequence comparison in the region of N = 402–426 of all published AChR subunit sequences. The oligonucleotide sequence listed at the top was chosen as the region most homologous among the bovine, chicken, mouse, and human β, γ, δ subunits. Since the nucleotides at the tenth position seemed to vary randomly between T and C, a mixture containing both T and C at this position was made. The squares indicate positions that vary from the synthetic oligonucleotide.

2. DEVELOPMENTAL CHANGES IN STEADY-STATE mRNA LEVELS

Northern blot analysis has been used to investigate the changes in steady-state levels of mRNA for receptor α and δ subunits during terminal differentiation of myoblasts in tissue culture. As shown in Fig. 3, the amount of hybridizable mRNAs specific for α or δ subunits increases significantly after 4–6 days in culture in both mouse muscle cell lines BC3H-1 (47) and C2 (58). The increase in mRNA was approximately two-fold for α subunit in BC3H-1, ten-fold for δ in BC3H-1 and α in C2, and thirty-fold for δ in C2. These changes are qualitatively similar to what others have found for a number of muscle-specific mRNAs (44). However, a significant difference may exist between receptor subunit mRNAs and skeletal muscle α actin. As shown in Fig. 3, for both cell lines, growing myoblasts have detectable levels of both α- and δ-subunit mRNAs, whereas α-actin mRNA is undetectable. We have previously established that growing populations of BC3H-1 contain significant quantities of translatable α-subunit mRNA, although they have undetectable levels of α bungarotoxin binding sites (42). Further studies of mRNA changes and the precise mechanisms that are responsible should lead to a resolution of the apparent differences observed between actin and receptor subunit mRNA accumulation.

3. DEVELOPMENTAL REGULATION OF RECEPTOR SUBUNIT GENE TRANSCRIPTION

The pronounced increase in α- and δ-subunit mRNAs observed in C2 cells suggests that regulation of these genes during myogenesis occurs at the level of transcription as has been shown for myosin heavy-chain gene (28). To test this possibility, we have measured transcription rates by measuring the number of nascent RNA transcripts associated with α- and δ-subunit genes in isolated nuclei of undifferentiated and differentiated C2 cells. The nuclei were incubated with [α^{32}P]UTP, after which nuclear RNA was purified and hybridized, to an excess of cDNAs fixed to nitrocellulose filters. Transcription rates were determined simultaneously for actin genes and receptor α- and δ-subunit genes. Figure 4 shows that the rates of transcription of both the α- and δ-subunit genes are increased from levels that we could not detect because of the degree of nonspecific hybridization (0.15 ppm as measured with pBR322 or M13 RF DNAs) to levels that are approximately tenfold above background. Thus the accumulation of α- and δ-subunit mRNA correlates well with a minimum tenfold increase in the rates of transcription of these genes in differentiating C2 cells. It should be possible to accurately determine both relative rates of transcription and steady-state mRNA levels for all four subunits as the remaining cDNAs become available.

For a positive control for the quality of the RNA prepared from undifferentiated cells in Fig. 4, we used a β-actin cDNA clone (52) (provided by B. M. Spiegleman)

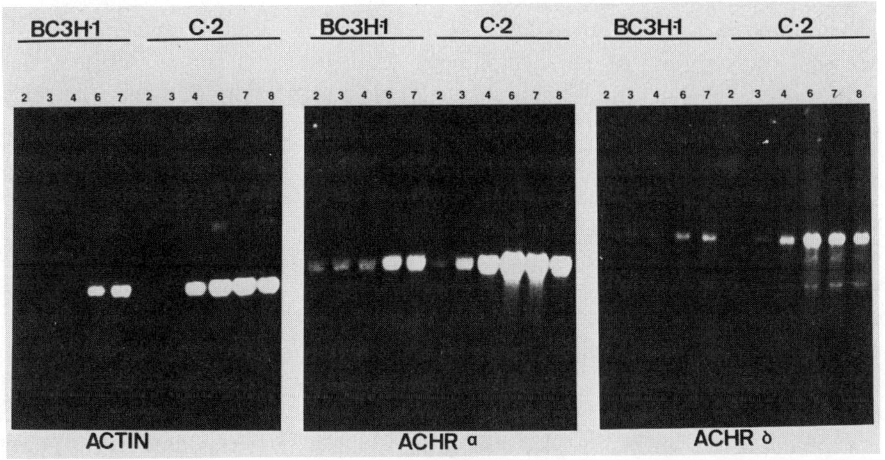

Figure 3. Steady-state levels of Actin and AChR mRNAs in developing muscle cells. The BC3H-1 and C2 cell lines were grown as described previously (22,31). The age (in days) of the cultures after trypsinization and replating is indicated above each gel lane. Ribonucleic acid was extracted, purified, fractionated on formaldehyde-containing agarose gels (10 μg of total RNA per lane), and hybridized with nick translated cDNAs as described previously (34). The cDNAs used were p91-1 for α actin (7) (provided by M. Buckingham), pA59 (32), and p6H (24) (provided by N. Davidson).

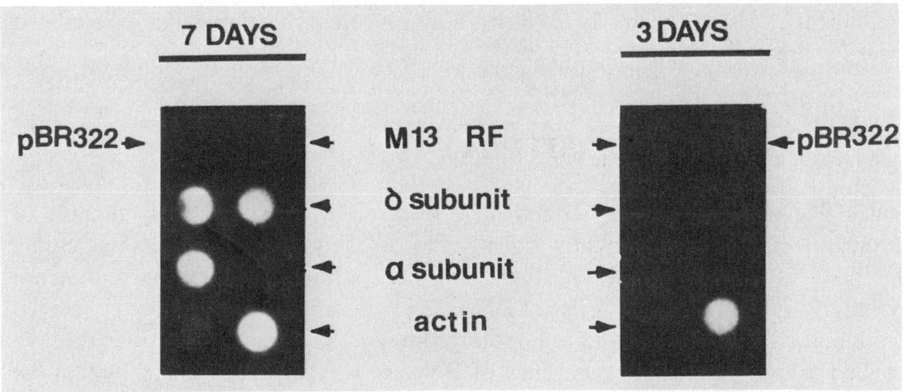

Figure 4. Transcription run-on of receptor α- and δ-subunit and actin genes. Nuclei from undifferentiatea or differentiated C2 mouse muscle cell line were isolated essentially as described previously (55). Nuclear run-on transcription was carried out at 25°C with approximately 2×10^7 nuclei in a 200-μl reaction (25), and the ^{32}P-UMP labeled RNA was purified on CsCl gradients. Hybridization of the labeled RNA to excess cDNA spotted on filters was performed as described by Darnell (15), and filters were exposed to prefogged Kodak XAR-5 film with intensifying screens. In the autoradiography shown above, each dot contains 5 μg of immobilized DNA. From left to right: row 1, pBR322 and replicative form of bacteriophage M13; row 2, noncoding and coding strand of the δ-subunit clone (24) in M13; row 3, noncoding and coding strand of an α-subunit partial clone (32) in M13; row 4, a mouse β-actin clone (52) in pBR322.

that included sufficient coding sequence for detection of all actin-related species (ref. 7). Therefore, the signal observed with ^{32}P-RNA from the 3-day-old nuclei probably represents transcription of the cytoplasmic actin genes, β and γ, whereas that observed with ^{32}P-RNA from 7-day-old nuclei includes transcription of α skeletal as well as α cardiac genes that have been shown to be expressed only after differentiation (37). Similar controls have been performed with other genes known to be expressed in growing myoblasts; thus experiments using mouse histone (50) (provided by D. Sittman) and mouse glyceraldehyde-3-phosphate dehydrogenase (52) (provided by B. Spiegelman) have indicated that these, like β actin, are transcribed in dividing myoblasts, and that the absence of detectable transcription of receptor subunit genes in undifferentiated C2 myoblasts is specific.

Surprisingly, when we measured transcription from the opposite (noncoding) strand of the δ-subunit gene, we found the signal to be of an intensity equal to that from the coding strand, and these transcripts were also developmentally regulated. In the experiment shown in Fig. 4, full-length δ cDNA (24) was used, so it is not possible to determine whether the opposite strand transcripts actually cover all or part of the δ gene. In contrast, the α cDNA (32) used in Fig. 4 corresponded to the 3′ half of the coding sequence and no transcripts were detected from the opposite strand. In further experiments using full-length α cDNA, however, we have observed opposite strand transcription of the α gene (not shown). It seems likely, therefore, that mapping of α and δ gene opposite strand transcripts will reveal that they do not cover the full length of the structural genes. It is interesting to consider that such transcripts may be involved in regulation of subunit expression, but we have no evidence to support this possibility, nor is it known how prevalent developmentally regulated opposite strand transcription may be.

4. DNAase I SENSITIVITY

Genes that are actively transcribed have been shown to be within domains of chromatin that are preferentially sensitive to digestion with deoxyribonuclease I (DNAase I) (56). In addition, there are sites within these domains that are hypersensitive to DNAase I. Sites of DNAase I hypersensitivity have been observed to correlate in many cases with regions of promoter and enhancer function (17). We decided to investigate the occurrence of DNAase I-hypersensitive sites within the region of AChR subunit genes as a possible approach to identifying sequences involved in regulation of expression. Figure 5 shows the results of an experiment in which hypersensitive sites were observed within BC3H-1 BamHI restriction fragments that hybridize with the δ cDNA. More recently similar results have been obtained with C2 cells. Of the three hypersensitive sites observed in Fig. 5, two occur in myoblasts and thus precede the period during which we observe increased expression of receptor genes (and in C2 cells, at a time that precedes detectable transcription). The third site was not detected in dividing myoblasts, but appeared in differentiated cells, correlating well with increased receptor expression. Several other hypersensitive sites have been detected in Hind III digests of the same chromatin. All sites are tissue specific in that they were not detected in chromatin from 10T1/2 mouse fibroblasts (11) (provided by Andrew Lassar, the Fred Hutchinson Cancer Research Center, Seattle, WA).

Preliminary information concerning the location of some of the DNAase I-hypersensitive sites has been obtained by mapping with RNA probes specific for the 3′ and 5′ ends of the gene, prepared by *in vitro* transcription of the δ-subunit cDNA using T7 and SP6 polymerases. In addition, these sites have been localized to a δ subunit containing genomic clones and mapped with respect to restriction fragments that hybridize to synthetic oligonucleotides specific for exons 1 and 12 (see ref. 39 for exon numbering of the chicken δ gene). Thus we have tentatively mapped one hypersensitive site within 1–2 kilobases of the 5′ end of the δ-subunit gene and three others toward the 3′ end; a location that, if homologous to the chicken gene arrangement (39), will be near the 5′ end of the linked γ-subunit gene.

Precise mapping and nucleotide sequencing are currently under way to characterize further the hypersensitive sites nearest the 5′ end of the δ gene.

5. DENERVATION RESULTS IN INCREASED LEVELS OF RECEPTOR mRNA

In contrast to what we had found in tissue culture of muscle cell lines, we detected very low levels of AChR α- and δ-subunit mRNA in adult innervated muscle tissue (13, 34, 35). Denervation of the mouse lower leg resulted in a rapid, approximately hundredfold increase in the steady-state levels of these mRNAs, while actin mRNA changed little (Fig. 6). The relative abundances of receptor mRNAs observed after denervation are similar to that observed in the cultured cell lines. Indeed, the large increases in mRNA levels observed after denervation are reminiscent of the large increases associated with differentiation of the cultured myoblasts. However, there may be an important difference; the increased expression observed in differentiating myoblasts is paralleled with increases in expression of α actin (and other muscle genes), whereas the increase in receptor mRNA observed after denervation is completely uncoupled from α-actin mRNA (Fig. 6). We believe that this indicates two separate mechanisms of regulation of receptor genes: one by which receptor is coregulated with other muscle genes during development and a second that effects a down regulation of receptor gene expression in response to nerve-induced activity and dramatic up regulation in response to denervation. Implicit in this view is the assumption that the latter, activity-linked, mechanism also operates at the level of transcription. We are in the process of testing this assumption by nuclear run on experiments.

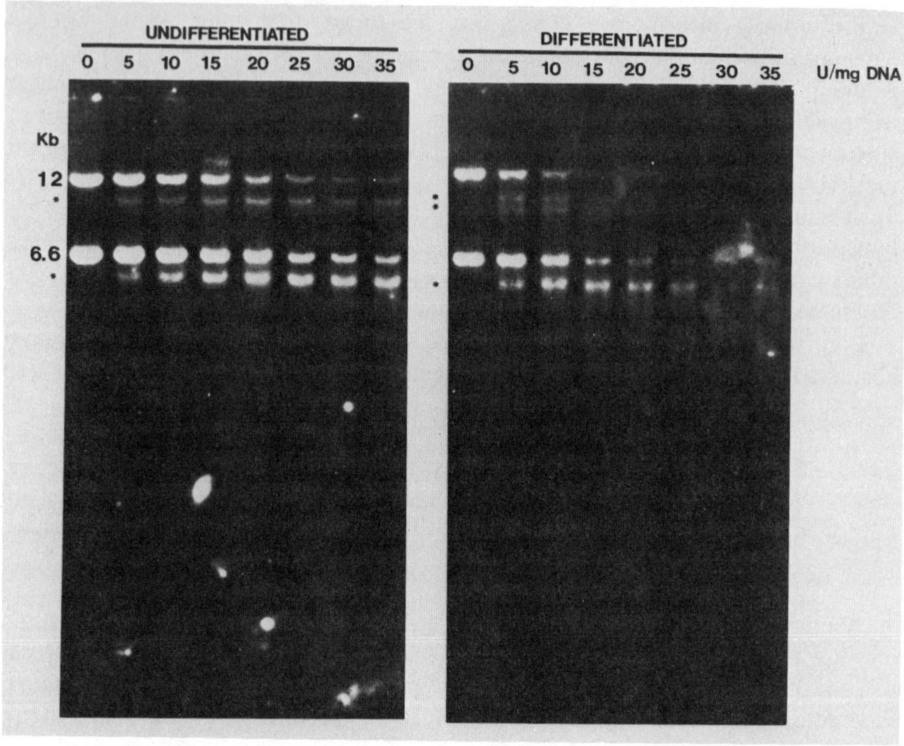

Figure 5. Southern blot autoradiograph of BamH1 cut, DNAase 1 digested BC3H-1 DNA probed with a radioactively labeled mouse δ cDNA. Nuclei were isolated from undifferentiated (2-day-old) and differentiated (7-day-old) BC3H-1 cells (56) and resuspended to a DNA concentration of 1 mg/ml. The DNAse I digestions were performed at 37°C for 10 min with the DNAse I concentrations given above each lane. Deoxyribonucleic acid was purified from each sample and restricted with BamH1. Aliquots of 5 μg were fractionated on a 1% agarose gel and transferred to Gene-Screen Plus. The membranes were hybridized with p6H, δ-cDNA probe (24). The blots were washed at high stringency (0.2 × SSC at 65°C) and autoradiographed. Fragment lengths are relative to Hind 3 cut λ standards. The stars indicate the position of fragments generated by digestion at DNAase I-hypersensitive sites.

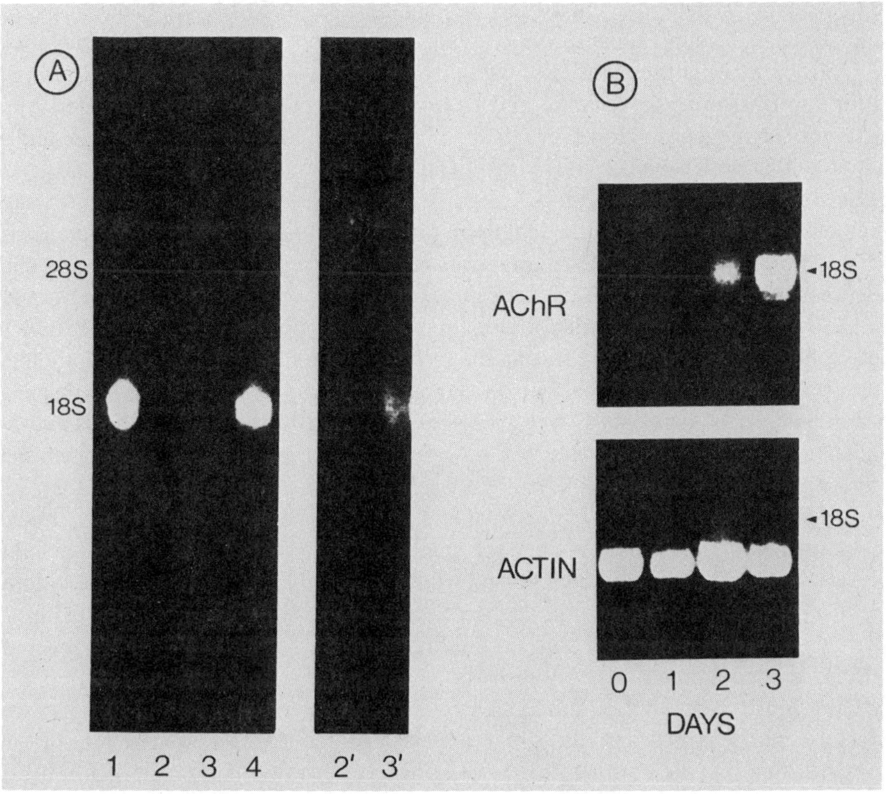

Figure 6. Steady-state levels of actin and AChR α-Subunit mRNA in innervated and denervated mouse leg muscle. (*a*) Ribonucleic acid was fractionated by gel electrophoresis, transferred to nitrocellulose, hybridized with ^{32}P-A59-DNA specific for receptor α subunit, and exposed to x-ray film. Lane 1: 10 μg of total RNA from a mouse muscle cell line, BC3H-1. Lane 2: 10 μg of total RNA from liver. Lane 3: 3 μg of poly A^+ RNA from innervated skeletal muscle. Lane 4: 3 μg of poly A^+ RNA from denervated skeletal muscle. Lanes 2′ and 3′ show autoradiographs exposed tenfold longer than lanes 2 and 3 (0 vs. 4 hr). (*b*) Total RNA from innervated muscles, or from muscles denervated for 1, 2, or 3 days, was fractionated and transferred as described above, then incubated with ^{32}ρ-DNA specific for receptor α subunit (top panel) or for skeletal muscle actin (bottom panel). The ~ 2-kilobase region of autoradiographs exposed for 4 days is shown. In this experiment, 10 μg of RNA was applied to each lane. (From ref. 34, with permission.)

6. CONCENTRATION OF AChR mRNA NEAR SYNAPTIC REGIONS OF ADULT MUSCLE

An important consequence of the hypothesis that activity suppresses transcription of AChR genes is that one must then explain how high AChR concentrations are maintained at the adult synapse. Since synaptic receptors turn over (3, 45), they must be replaced by ongoing synthesis. One possibility is that synaptic receptors, which differ subtly from extrasynaptic and embryonic receptors (21), are encoded by a different set of genes. However, there is growing evidence, at least for α and δ subunits, that both synaptic and extrasynaptic AChRs are encoded by the same genes in the mouse (23, 24, 32, 39, 40, 49). A second possibility is that receptors are synthesized uniformly along the length of the multinucleate fiber as a consequence of a basal level of subunit mRNA transcription. In this case, one might imagine that posttranslational steps, such as assembly of subunits or insertion into membrane, would be more efficient in synaptic areas. In fact, some posttranslational regulation is detectable in tissue culture (43), and the greater metabolic stability of synaptic than extrasynaptic AChR (3) probably contributes to their concentration at synaptic sites. A third and, to us, particularly intriguing possibility is that receptor synthesis may be higher near synapses. The fact that muscle fibers are multinucleate suggests that transcription of receptor mRNA might itself be regulated differentially along the length of the muscle—adult muscle fibers typically have several hundred nuclei distributed along their lengths, but a few nuclei (three to five or about 1% of the total) are invariably situated directly beneath the postsynaptic membrane. One might, therefore, imagine that synaptic and extrasynaptic nuclei could transcribe different sets of genes. In support of this possibility we have found recently that receptor mRNA is concentrated near synapses in adult muscle.

Our experiments have exploited the convenient anatomy of the mouse diaphragm. Most of the synapses and the great majority of the AChRs in this muscle are concentrated in its central third. Diaphragms were dissected into central, synapse-rich and distal, synapse-free portions. Poly A^+ mRNA was prepared from each, fractionated on agarose gels, and then transferred to nitrocellulose and probed with ^{32}P-cDNAs specific for α and δ subunits of mouse receptor and with a cDNA specific for mouse skeletal muscle actin. The relative abundance of mRNA in the synapse-rich and synapse-free samples was determined from these Northern blots as described previously (34).

Figure 7 shows that whereas actin mRNA was equally distributed between synapse-rich and synapse-free samples (ratio 1.1 ± 0.2 mean ± SEM in six experiments), α- and δ-subunit mRNAs were severalfold more abundant (2.9- ± 0.5-fold, $n = 6$, for α subunit; 9- and 14-fold in two experiments on δ subunit) in synapse-rich than in synapse-free regions of the diaphragm. Although the precise distribution of receptor mRNAs and the processes by which they become preferentially concentrated near synapses remain to be elucidated, we have adopted the following as a working hypothesis: in the adult muscle fiber, AChR genes are

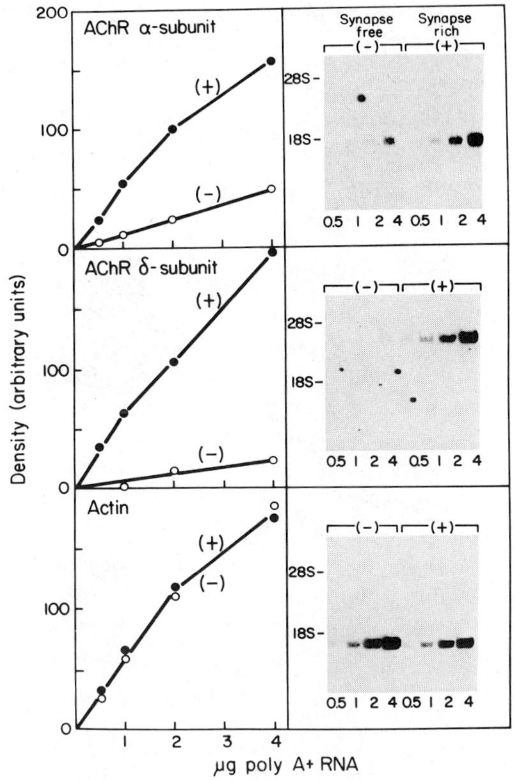

Figure 7. Abundance of AChReceptor α subunit, AChR δ subunit, and actin mRNAs in synapse-rich and synapse-free segments of diaphragm muscle. Graphs on left display results obtained by densitometry of autoradiographs of Northern blots, shown on right. The gels contained samples of 0.5–4 μg of total poly A^+ RNA extracted from synapse-rich and synapse-free portions of adult mouse diaphragms. Autoradiographs were exposed for 7 days with intensifying screens for receptor subunit mRNAs and 12 hr without intensifying screen for actin mRNA. (From ref. 35, with permission.)

preferentially transcribed in nuclei located closest to the synapse. Synapse-associated nuclei and the cytoplasm that surrounds them are morphologically distinguishable from the nonsynaptic nuclei and cytoplasm. Since only about 1% of all muscle fiber nuclei are synaptic, these nuclei might transcribe receptor mRNAs at the maximal rates characteristic of nuclei in denervated and embryonic muscle. How could transcription of receptor genes be maintained at a high level in synaptic nuclei while transcription from extrasynaptic nuclei is efficiently suppressed? Transcription at synaptic nuclei might be stimulated by the localized action of a trophic factor released by the motor neuron (see, e.g., ref. 6). Alternatively, synaptic nuclei may escape the intracellular mediator of activity-linked down regulation, because of their location in a unique region of the muscle cytoplasm.

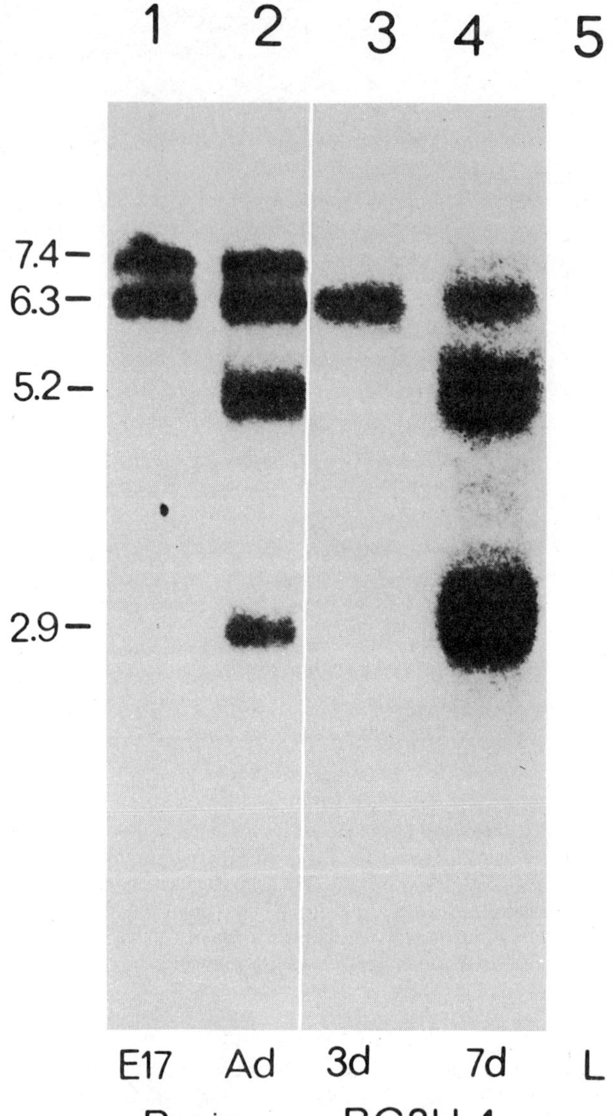

Figure 8. N-CAM RNA in developing brain and muscle, detected with an N-CAM cDNA. Ribonucleic acid isolated from E17 or adult rat brain, from undifferentiated (3-day-old) or differentiated (7-day-old) cultures of BC3H-1 cells, and from rat liver was fractionated by gel electrophoresis, transferred to filters, incubated with ^{32}P-pM1.3 DNA specific for N-CAM, and exposed to x-ray film. All lanes contain 5 μg of total RNA.

7. COREGULATION OF ACh RECEPTOR AND N-CAM GENES

If we assume that activity-linked regulation and concentration of AChR mRNAs near synaptic regions are important aspects of the program that regulates the formation and maintenance of synapses, it follows that other genes for synaptic proteins may be coregulated by the same mechanisms. We would like to test this idea by examining the regulation of other genes for synaptic proteins and comparing the mechanisms responsible.

The neural-cell adhesion molecule (N-CAM) is an excellent candidate for a synaptic protein that may exhibit coregulation with AChR. Although N-CAM is not as tissue specific as AChR, as it is found throughout the nervous system as well as elsewhere in embryonic and adult tissues (16), it is concentrated near synapses in adult muscle (12), and extrasynaptic N-CAM levels increase when muscle is denervated or paralyzed (13,14). Thus, similar to AChR, N-CAM exhibits activity-dependent regulation in muscle.

We analyzed N-CAM mRNA levels by Northern blotting techniques as had been done previously for AChR subunit mRNA. The N-CAM-related cDNA clone used for this purpose was pM1.3 described by Goridis et al. (20); pM1.3 has been characterized by monoclonal antibody binding to a fusion protein expressed by λgt11. In this way it was found to encode an epitope located 90 kilodaltons from the *N* terminus of the N-CAM molecule (20). As shown in Fig. 8, there are four N-CAM-related RNA species in total RNA extracted from adult mouse brain. The two largest of these are present in embryonic brain, and all except the largest are found in differentiated BC3H-1 cells (and C2 cells, not shown). Interestingly, the two smallest N-CAM RNA species are clearly developmentally regulated during development in cultured muscle cells. Denervation of rat diaphragm or mouse leg muscle resulted in increases in the levels of the two smallest N-CAM RNAs of the same order of magnitude and with the same time course as ACh α and δ subunits (Fig. 9). The striking similarities in the accumulation of N-CAM and AChR RNAs, and the fact that these proteins are structurally and evolutionarily unrelated, make them excellent candidates for a comparative analysis of the mechanisms involved in their coregulation.

8. POSTTRANSLATIONAL EVENTS IN AChR SYNTHESIS

The posttranslational processes involved in AChR synthesis have been characterized by using subunit-specific monoclonal antibodies and α-bungarotoxin-mediated immunoprecipitation combined with biosynthetic labeling of receptor subunits in

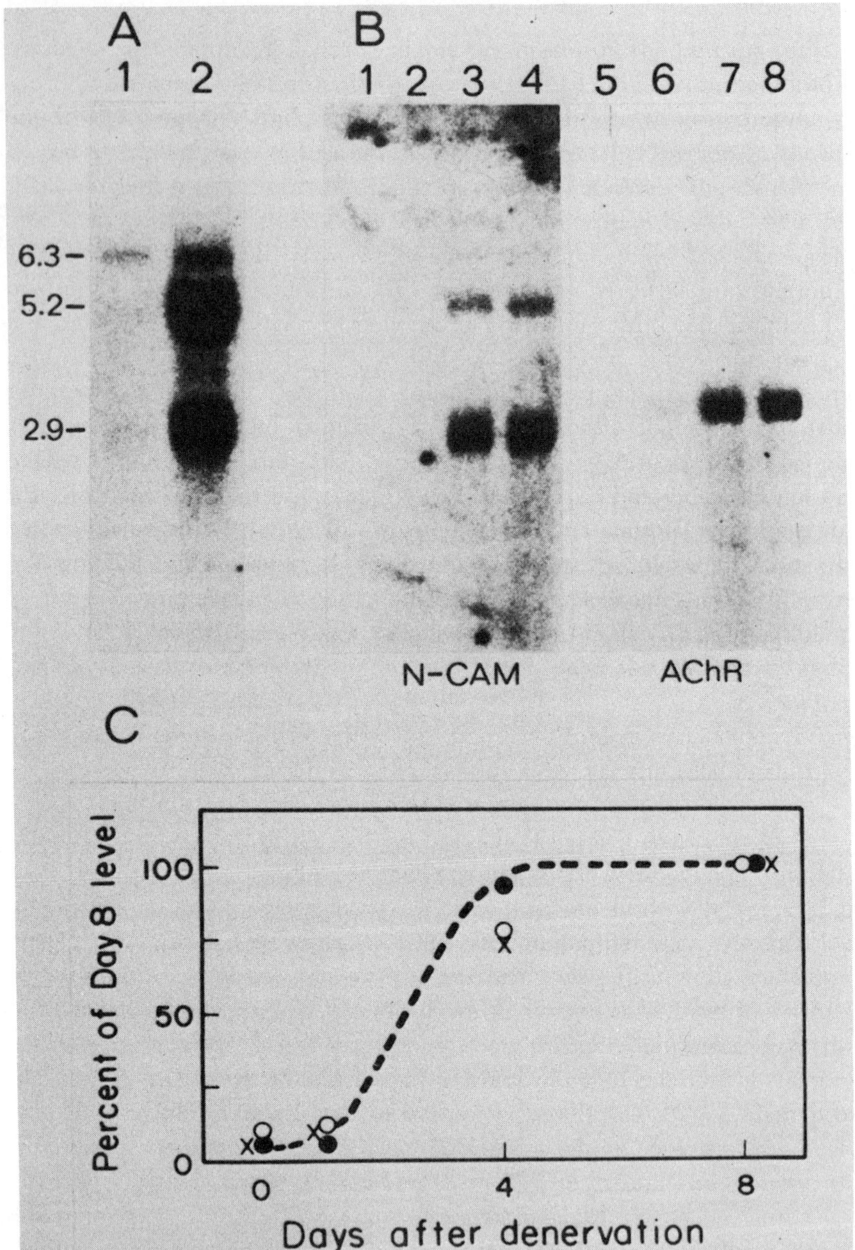

Figure 9. Denervation induces the accumulation of N-CAM RNA. (*a*) RNA from normal adult (lane 1) or 10-day denervated adult (lane 2) rat diaphragm was fractionated by gel electrophoresis, transferred to filters, and hybridized with ^{32}P-pM1.3 DNA specific for N-CAM; (*b*) RNA from normal (lanes 1 and 5), 1-day denervated (lanes 2 and 6), 4-day denervated (lanes 3 and 7), or 8-day denervated (lanes 4 and 8) adult mouse leg muscle, probed with N-CAM cDNA (lanes 1–4) or with a cDNA-specific for the δ subunit of AChR (lanes 5–8); (*c*) time course of the increase in N-CAM and AChR-specific RNAs after denervation of mouse leg muscle. Levels of N-CAM 5.2 kilobases (-•-), 2.9 kilobases (-0-), and AChR δ subunit (-x-) RNAs obtained by densitometric scanning of films such as those shown in part *b* are expressed as a percentage of the value obtained on day 8 for each species. Values for AChR α subunit (---) mRNA are redrawn (34). All lanes contained 5 μg of total RNA.

tissue culture cell lines and cell-free protein synthesizing systems. These studies have been reviewed recently (1,2,8,29,33,36) and are summarized in the form of a working model (Fig. 10). As shown in this model, α subunit is translated from mRNA and undergoes cotranslational insertion into the rough endoplasmic reticulum, where the approximately 20 amino acid signal sequence is cleaved and a single-core oligosaccharide is added at Asn-141. Since these modifications occur cotranslationally, intermediates in these processes cannot be detected by biosynthetic labeling. At early times following pulse labeling, an α-subunit species α_0 can be detected that does not bind α bungarotoxin. This intermediate resembles α subunit synthesized in cell-free translation systems supplemented with dog pancreas microsomes, in that both have undergone signal peptide cleavage and core glycosylation, but do not bind toxin. In transition 1 (Fig. 10), α_0 that has a sedimentation coefficient of 5*S* is converted to α_{TX}, a 5*S* species that binds α bungarotoxin (e.g., Fig. 11). Although the time course of transition 1 in BC3H-1 cells is relatively slow, our recent experiments indicate that it occurs in the endoplasmic reticulum before transport to the golgi (Smith and Merlie, in preparation). This transition appears to require prior oligosaccharide addition and/or disulfide bond formation, as implied by the results of Mishina et al. (38), in which the oligosaccharide linkage site or cysteine residues of α subunit were altered by point mutation. When the altered mRNAs were expressed in *Xenopus* oocytes the mutant α subunit did not bind α bungarotoxin. These results are also consistent with the finding that blockage of core glycosylation with tunicamycin prevented the formation of the toxin binding site (31).

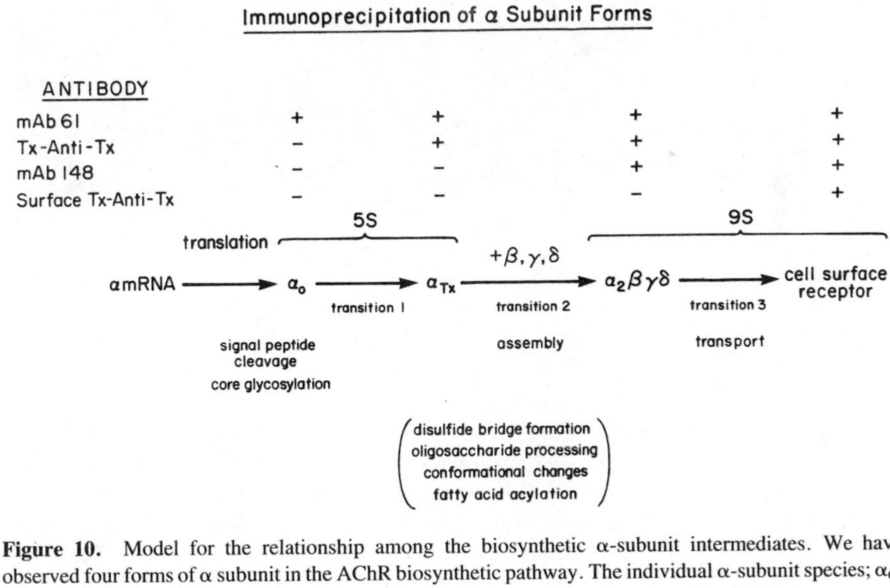

Figure 10. Model for the relationship among the biosynthetic α-subunit intermediates. We have observed four forms of α subunit in the AChR biosynthetic pathway. The individual α-subunit species; α_0, α_{TX}, $\alpha2\beta\gamma\delta$ (assembled intracellular receptor), and cell surface receptor, can be identified by one or more of the antibody precipitations listed at the left, as indicated by the columns of pluses and minuses above each form. See the text for a description of the pathway.

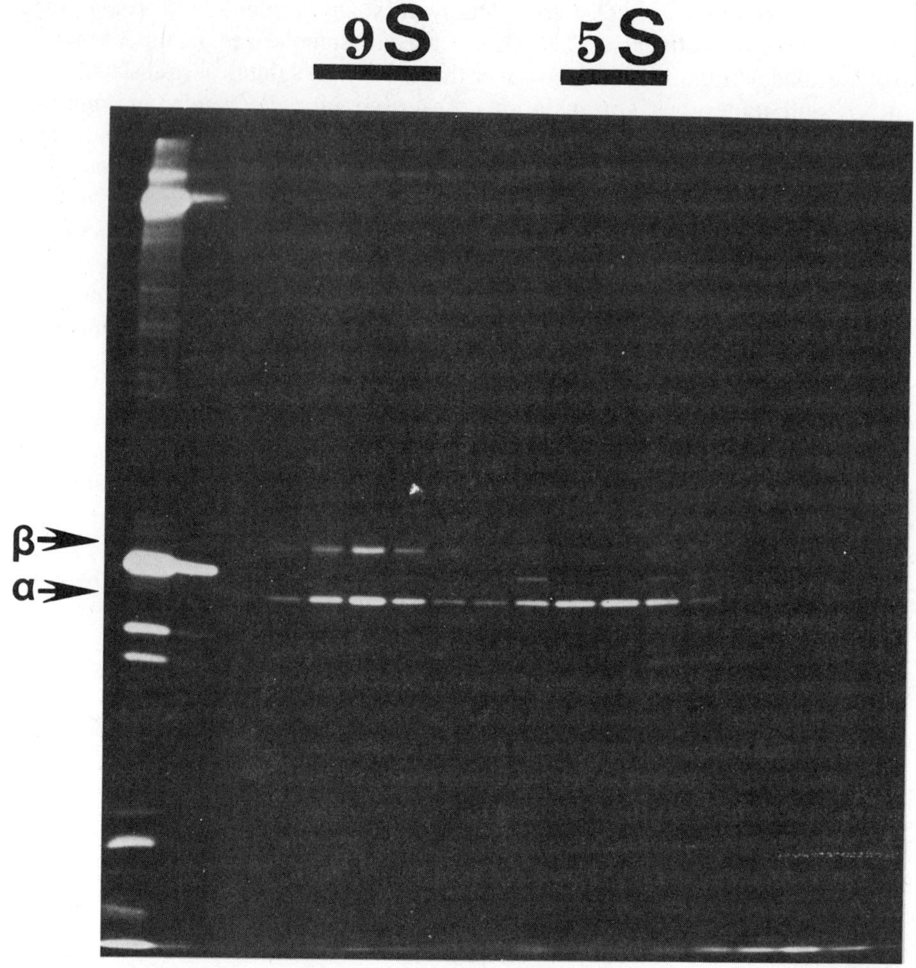

Figure 11. Fractionation by velocity sedimentation of 9*S* AChR α subunit from the 5*S* α_{TX} precursor. Primary rat myotubes were labeled with [^{35}S]-methionine for 0.5 hr and chased for 1 hr, a time predetermined to yield a 1 : 1 mix of labeled α subunit in the 9*S* and 5*S* species. Cells were extracted with Triton X-100, fractionated by velocity sedimentation, and immunoprecipitated with α bungarotoxin and anti-α bungarotoxin as described (9). Immuno-precipitates from each gradient fraction were analyzed by SDS polyacrylamide gel electrophoresis (SDS–PAGE) and exposed to x-ray film.

The model assumes that α subunits remain as monomers until their assembly with β subunits during transition 2. However, at present we cannot exclude the possibility that the 5*S* species consists of dimers of α_2, αγ, or αδ composition and that these associations may form at an earlier time. Assembly with β subunit as detected with subunit-specific monoclonal antibodies was found to occur simultaneously with a shift in *s* value from the 5*S* α subunit precursor to 9*S*, characteristic of the $\alpha_2\beta\gamma\delta$

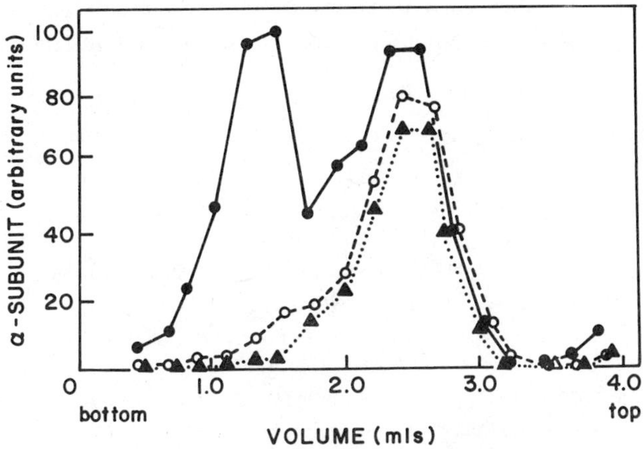

Figure 12. Analysis of curare and decamethonium inhibition of α-bungarotoxin binding to 9*S* ACh receptor and the α_{TX} precursor. [^{35}S] Methionine labeled extracts containing 9*S* and 5*S* species were prepared and fractionated as described in the legend to Fig. 11. Aliquots of each fraction were immunoprecipitated with 10 n*M* α bungarotoxin for 0.5 hr at 4°C in the absence (●) or with 0.5-hr preincubation with 1 m*M* curare (○) or 1 m*M* decamethonium (▲). Immunoprecipitates were analyzed as described in Fig. 12, and the fluorographic image was quantitated by densitometry (from ref. 9).

mature oligomer. However, since γ and δ subunits have been difficult to detect, probably as the result of proteolytic nicking, we cannot be sure of any particular order of subunit assembly. Our efforts to determine whether assembly occurs in the endoplasmic reticulum have not yielded definitive results. At present, our data are consistent with assembly in the endoplasmic reticulum and rapid transport to the golgi, or with assembly in both endoplasmic reticulum and golgi (Smith and Merlie, unpublished observations).

Finally, we assume that all subunits undergo a conformational transition similar to α_0 conversion to α_{TX}. We have observed such a transition for the β subunit, using two monoclonal antibodies: one that precipitates only a form that appears approximately 30 min after synthesis and a second that recognizes both the newly synthesized β_0 as well as the more mature form. Although similar transitions have not been observed for γ and δ subunits, comparison of the amino acid sequences of all four subunits of *Torpedo* electric organ and bovine muscle indicates strong conservation of sites of covalent modification and secondary structure in general (41, 53). It would not be surprising, therefore, if all subunits followed a similar folding pathway.

Recently, we have further characterized the 5*S* α_{TX} intermediate in primary cultures of embryonic rat muscle cells. We have shown that α_{TX} does not bind decamethonium or curare (Fig. 12) (9). Since all the 9*S* receptor does bind these ligands, the small ligand binding site must be acquired coincident with transition 2. Whether the formation of the small ligand binding site is accompanied by further covalent modification or, rather, the result of noncovalent conformational changes is not known (9). In either case it seems as though subunit assembly must be accompanied by further conformational changes in the subunits themselves.

REFERENCES

1. Anderson, D. J. (1983) *Trends NeuroSci.*, **6**, 169–171.
2. Anderson, D. J. and Blobel, G. (1983). *Cold Spring Harbor Symp. Quant. Biol.*, **48**, 125–134.
3. Berg, D. K. and Hall, Z. W. (1975). *J. Physiol.* (London), **252**, 771–789.
4. Bloch, R. J. and Hall, Z. W. (1983). *J. Cell Biol.*, **97**, 217–223.
5. Boulter, J., Luyten, W., Evans, K., Mason, P., Ballivet, M., Goldman, D., Stengelin, S., Martin, G., Heinemann, S., and Patrick, J. *J. Neurosci.*, **5**, 2545–2552.
6. Buc-Caron, M. H., Nystrom, P., and Fischbach, G. D. (1983). *Dev. Biol.*, **95**, 378–386.
7. Caravatti, M., Minty, A., Robert, B., Montarras, D., Weydert, A., Cohen, A., Daubas, P., and Buchingham, M. (1982). *J. Molec. Biol.*, **160**, 59–76.
8. Carlin, B. E. and Merlie, J. P. (in press). In Strauss, D., Kreil, G., and Boime, I., Eds., *Protein Compartmentalization* Springer-Verlag, Berlin.
9. Carlin, B. E., Lawrence, J. C., Jr., Lindstrom, J. M., and Merlie, J. P. (1986). *Proc. Natl. Acad. Sci.* (USA), **83**, 498–502.
10. Chirgwin, J. M., Pryzbla, A. E., MacDonald, R. J., and Rutter, W. J. (1979). *Biochemistry*, **18**, 5294–5298.
11. Constantinides, P. G., Jones, P. A., and Gevers, W. (1977). *Nature*, **267**, 364–366.
12. Covault, J. and Sanes, J. R. (1986). *J. Cell Biol.*, **102**, 716–730.
13. Covault, J. and Sanes, J. R. (1985). *Proc. Natl. Acad. Sci.* (USA), **82**, 4544–4548.
14. Covault, J., Merlie, J. P., Goridis, C., and Sanes, J. R., (1986) *J. Cell Biol.*, **102**, 731–739.
15. Darnell, R. B. (1984). *Molec. Cell. Biol.*, **4**, 829–839.
16. Edelman, G. M. (1985). *Ann. Rev. Biochem.*, **54**, 135–169.
17. Elgin, S. C. R. (1984). *Nature*, **311**, 213–214.
18. Fambrough, D. M. (1979). *Physiol. Rev.*, **59**, 165–226.
19. Glisin, V., Crkvenjakov, R., and Byus, C. (1974). *Biochemistry*, **13**, 2633–2637.
20. Goridis, C., Hirn, M., Santoni, M.-J., Gennarini, G., Deagostini-Bazin, H., Jordan, B. R., Kiefer, M., and Steinmetz, M. (1985). *EMBO J.*, **4**, 631–635.
21. Hall, Z. W., Roizin, M. P., Gu Y., and Gorin, P. D. (1983). *Cold Spring Harbor Symp. Quant. Biol.*, **48**, 263–274.
22. Inestrosa, N. C., Miller, J. B., Silberstein, L., Ziskind-Conhaim, L., and Hall, Z. W. (1983). *Exp. Cell Res.*, **147**, 393–405.
23. Klarsfeld, A., Devillers-Thiery, A., Giraudat, J., and Changeux, J.-P. (1984). *EMBO J.*, **3**, 35–41.
24. LaPolla, R. J., Mixter Mayne, K., and Davidson, N. (1984). *Proc. Natl. Acad. Sci.* (USA), **81**, 7970–7974.
25. Marzluff, W. F. and Huang, R. C. (1984). In Hames, B. D. and Higgins, S. J., Eds., *Transcription and Translation: A Practical Approach*, IRL Press, Oxford, pp. 89–129.
26. Massoulié, J. and Bon, S. (1982). *Ann. Rev. Neurosci.*, **5**, 57–106.
27. Maxam, A. M. and Gilbert, W. (1980). *Methods Enzymol.*, **65**, 499–560.
28. Medford, R. M., Nguyen, H. T., and Nadal-Ginard, B. (1983). *J. Biol. Chem.*, **258**, 11063–11073.
29. Merlie, J. P. (1984). *Cell*, **36**, 573–575.

30. Merlie, J. P. and Sebbane, R. (1981). *J. Biol. Chem.*, **256**, 3605–3608.
31. Merlie, J. P., Sebbane, R., Tzartos, S., and Lindstrom, J. (1982). *J. Biol. Chem.*, **257**, 2694–2701.
32. Merlie, J. P., Sebbane, R., Gardner, S., and Lindstrom, J. (1983). *Proc. Natl. Acad. Sci.* (USA), **80**, 3845–3849.
33. Merlie, J. P., Sebbane, R., Gardner, S., Olson, E., and Linstrom, J. (1983). *Cold Spring Harbor Symp. Quant. Biol.*, **48**, 135–145.
34. Merlie, J. P., Isenberg, K. E., Russell, S. D. and Sanes, J. R. (1984). *J. Cell Biol.*, **99**, 332–335.
35. Merlie, J. P. and Sanes, J. (1986). *Nature*, **317**, 66–68.
36. Merlie, J. P. and Smith, M. M. (1986). *J. Membrane Biol.*, **91**, 1–10.
37. Minty, A. J., Alonso, S., Caravatti, M., and Buckingham, M. E. (1982). *Cell*, **30**, 185–192.
38. Mishina, M., Tobimatsu, T., Imoto, K., Tanaki, K., Fujita, Y., Fukuda, K., Kurasaki, M., Takahashi, H., Morimoto, Y., Hirose, T., Inayama, S., Takahashi, T., Kuno, M., and Numa, S. (1985). *Nature*, **313**, 364–369.
39. Nef, P., Nauron, A., Stalder, R., Alliod, C., and Ballivet, M. (1984). *Proc. Natl. Acad. Sci.* (USA), **81**, 7975–7979.
40. Noda, M., Furutani, Y., Takahashi, H., Toyosato, M., Tanabe, T., Shimizu, S., Kikyotani, S., Kayano, T., Hirose, T., Inayama, S., and Numa S. (1983). *Nature*, **305**, 818–823.
41. Numa, S., Noda, M., Takahashi, H., Tanabe, T., Toyosato, T., Furutani, M., and Kikyotani, S. (1983). *Cold Spring Harbor Symp. Quant. Biol.*, **48**, 57–69.
42. Olson, E., Merlie, J. P., Lindstrom, J., and Glaser, L. (1983). *J. Biol. Chem.*, **259**, 3330–3336.
43. Olson, E. N., Glaser, L., Merlie, J. P., Sebbane, R., and Lindstrom, J. (1983). *J. Biol. Chem.*, **258**, 13946–13953.
44. Pearson, M. L. and Epstein, H. F., Eds. (1982). *Muscle Development, Molecular and Cellular Control*, Cold Spring Harbor.
45. Salpeter, M. M. and Harris, R. J. (1983). *J. Cell Biol.*, **96**, 1781–1785.
46. Sanes, J. R., Feldman, D. H., Cheney, J. M., and Lawrence, J. C., Jr. (1984). *J. Neurosci.*, **4**, 464–473.
47. Schubert, D., Harris, A. J., Devine, C. E., and Heinemann, S. (1974). *J. Cell Biol.*, **61**, 398–413.
48. Sebbane, R., Clokey, G., Merlie, J. P., Tzartos, S., and Lindstrom, J. (1983). *J. Biol. Chem.*, **257**, 2694–2701.
49. Shibahara, S., Kubo, T., Perski, H. S., Takahashi, H., Noda, M., and Numa, S. (1985). *Eur. J. Biochem.*, **146**, 15–22.
50. Sittman, D. B., Travers, R. A., and Marzluff, W. F. (1983) *NAR*, **11**, 6679–6697.
51. Smith, A. J. H. (1980). *Methods Enzymol.*, **67**, 560–580.
52. Spiegelman, B. M., Frank, M., and Green, H. (1983). *J. Biol. Chem.*, **258**, 10083–10089.
53. Takai, T., Noda, M., Furutani, Y., Takahashi, H., Notaki, M., Shimizu, S., Kayano, T., Tanabe, T., Tanaka, K., Hirose, T., Inayama, S., and Numa, S. (1984). *Eur. J. Biochem.*, **143**, 109–115.
54. Takai, T., Noda, M., Mishina, M., Shimizu, S., Furutani, Y., Kayano, T., Ikeda, T., Kubo, T., Takahashi, H., Takahashi, T., Kuno, M., and Numa, S. (1985). *Nature*, **315**, 761–764.
55. Tata, J. R. (1974). *Methods Enzymol.*, **31**, 253–270.
56. Weintraub, H. and Groudine, M. (1976). *Science*, **193**, 848–856.
57. White, M. M., Mayne, K. M., Lester, H. A., and Davidson, N. (1985). *Proc. Natl. Acad. Sci.* (USA), **82**, 4852–4856.
58. Yaffe, D. and Saxel, O. (1977). *Nature*, **270**, 725–728.

CHAPTER 6

SODIUM CHANNEL STRUCTURE–FUNCTION RELATIONS

CHARLES F. STEVENS
Molecular Neurobiology
Yale Medical School
New Haven, Connecticut

A consensus about the molecular mechanism for channel gating has evolved over the three decades since Hodgkin and Huxley (8) published their analysis of sodium channel function. The molecular basis for the gating properties they described is now generally accepted to be conformational changes of the channel.* Starting from a closed, resting state, the sodium channel must pass through a series of distinct closed conformational states to reach the unique open state. Although more complicated possibilities are, of course, entertained, the sequence of conformational states is frequently represented as a linear scheme, the simplest nontrivial version of which is

$$\text{Closed} \rightleftharpoons \text{closed} \rightleftharpoons \text{open}$$

The rate constants that govern these conformational changes, denoted here $a(i,j)$ for the transition from state i to state j, depend on voltage and are coupled to the membrane potential by charges and dipoles that experience the membrane electric field and alter their positions in that field when the channel steps from one conformation to another.

Whereas the general picture just presented is widely accepted, the precise molecular details are not known. How many conformational changes are associated with channel opening, and how much does the protein actually move to open its pore? How many (equivalent) charges move with each of these conformational transitions? What are the molecular structures involved in gating? Since the primary structure of a protein thought to be the complete *Electrophorus* sodium channel (1820 amino acids) has become available (12), several models have been developed (6,7,10,12) that are sufficiently specific to serve as the basis for predictions about the magnitude of movements and the number of charges that move. The goal of this chapter is to place some limits on the size of the protein motions likely to be involved in gating and on the number of charges displaced. These estimates are then compared with the proposed structural theories.

1. MOVEMENT OF ABOUT SIX EQUIVALENT CHARGES ACCOMPANIES GATING

Two methods for estimating charge displacements associated with gating conformational changes are available: gating-current measurements and the determination of voltage dependence for the corresponding conformational transitions. These approaches are discussed in turn.

*Although the passive ion-transport pathways were previously commonly called "pores," some workers objected to this term as too restrictive since permeability properties did not seem to match those expected for macroscopic water-filled pores. As a consequence, the less specific term "channel" came into use. The molecular basis of ion transport has become clearer in the last decade or so, and the term "channel" now generally is used to specify the protein involved in passive ion translocation across the membrane, whereas the hole in the protein through which ions pass is now called (again) the "pore." This terminology is followed here.

Gating currents, when appropriately integrated to separate out contributions not associated with channel function, give a direct measurement of the total number of charges associated with all the channels opening in a membrane (2). If channels are identical and independent, the total charge movement divided by the number of functional channels present gives the number of charges that move as a result of the channel opening. This number is found to be about 6. That is, when a channel progresses from its resting state to being open, about six unitary charges move through the membrane field (2). This number specifies ''equivalent charges'' and places a lower limit on the number of charged side chains of amino acids whose movement might couple conformation to electric field. Two observations are important in this context: (1) the entire charge displacement could, in principle, be composed of dipole reorientation with no charged side chains moving at all (e.g., a membrane spanning α helix with 20 residues would have a dipole moment of 68*D*, and β sheets also can have considerable dipole moments (9); and (2) the number of charged side chains (or their equivalent dipoles) that move may be, and presumably is, much larger than 6. The observed charge displacement can be equally well achieved by six charges moving through the entire membrane field or, for example, 60 charges each moving on the average through one-tenth of the field.

By separating the gating current into distinct kinetic components and determining the charge displacement associated with each, one may estimate the number of charges that move in association with different conformational steps involved in channel opening. The difficulty is that the identification of conformational steps defined in this way with those defined by traditional kinetic analysis is not always clear.

The second way of estimating charge movements is through the connection of charge displacements with gating kinetics. This approach depends on the fact that the free energy of a charge in an electric field is qV (where q is the magnitude of the charge and V the voltage) and the theoretical relation between free energy and rate of a transition:

$$a(V) = a(0) \exp \frac{qfV}{kT}$$

where $a(V)$ is the rate of the conformational change, V is the membrane potential in millivolts, q is the equivalent charge that moves, f is the fraction of the membrane voltage through which the charge q moves as the protein undergoes the conformational change necessary to reach the top of the energy barrier separating two major conformational states, k is Boltzmann's constant, and T is the absolute temperature (see e.g., ref. 13). For a simple two-state process (just open and closed), this sort of relation implies that the fraction of open channels p is given by

$$p(V) = \frac{1}{1 + A \exp(QV/kT)}$$

UNIVERSITY COLLEGE LIBRARY CARDIFF

where A is the ratio of forward and backward rates at 0 voltage, V is the membrane potential, and Q is the total charge moved in the conformational transition. Here, Q is the sum of the charge moved from closed to barrier peak and from barrier peak to open.

When the voltage dependence of the movement from resting to open channel is determined from the fraction of open channels, about six charges are found to be displaced (in agreement with gating-current measurements).

2. ABOUT THREE OR MORE EQUIVALENT CHARGES MOVE WHEN A CHANNEL CLOSES

The preceding discussion has dealt with the total charge displaced as a sodium channel moves through a sequence of conformational steps from resting to open. An extension of this approach can be used to estimate the number of charges that move for specific single-step conformational changes. Using single-channel recording to determine the voltage dependence of sodium channel mean open time, Aldrich and Stevens (1) found the open-to-inactivated transition to be associated with about 0.3 charges moving and the open-to-closed (open to the closed state adjacent to open) transition to displace about three charges. In this instance the individual rate constants for transitions are estimated so the values for q are for the single step: open-to-peak-of-barrier.

According to our observations, inactivation of open channels has little voltage dependence, and thus few charges (about one-third) move during this transition. Armstrong and Bezanilla (3) have also concluded from gating current measurements that inactivation is not voltage dependent and has little or no associated charge movement. Although the conclusions from these two types of experiments may seem the same, they are in fact complementary and not mutually confirmatory. This is an important point because it underlines the differences between at least some versions of the two approaches for estimating charge displacements. When a conformational transition occurs, the protein moves from the first state (state 1) to a transition state (at the top of the energy barrier separating conformations) and then from the transition state to conformation 2. An equivalent charge movement q_1 is associated with the motion state-1-to-transition and q_2 with transition-to-state-2. The voltage dependence of the rate for the 1-to-2 transition is determined by q_1 and the voltage dependence for the reverse transition 2-to-1 by q_2. The single-channel recording experiments have estimated q_1 (where state 1 is open and state 2 is inactivated), and the gating-current measurements determine $Q = q_1 + q_2$.

The steady-state voltage dependence of inactivation (the h_∞ curve), if it could be determined separately for open channels, would, like the gating-current measurements, estimate Q. Generally q_1 and q_2 have the same sign so that $Q > q_1$, but this need not always be so. For the acetylcholine receptor (AChR) channel, for example, q_1 and q_2 have opposite signs so that the voltage dependence of the distribution between open and closed states is smaller than that for the closing rate.[11]

In summary then, about six equivalent charges move when channels undergo several steps from their resting state to the open state, and the individual transitions open-to-closed and open-to-inactivated have about 3 and 0.3 equivalent charges associated with the motion to the transition state. Note that these latter estimates are for a single conformational step, not for the sequence of steps associated with displacing six charges. Gating-current measurements indicate that little charge is moved in the transition-state-to-inactivated step, and kinetic properties of sodium channels reveal that the entire charge movement in the open-to-closed step must be at least three charges.

3. PROTEIN MOTION ASSOCIATED WITH GATING CHARGE DISPLACEMENTS ARE AT LEAST FOUR OR FIVE Å

If the charges that move for a conformational change are associated with a transmembrane helix, a maximum of about 20 could move because this is the number of residues needed to span the hydrophobic core of a membrane. The movement of a single equivalent charge, then, would correspond to each actual charge moving through at least one-twentieth of the membrane field. If the membrane potential gradient from inside to outside occurs mostly over about 30 Å, a single helix bearing the charges would have to be displaced about 9 Å for the movement of six equivalent charges; alternatively, three helices [corresponding to the m^3 term in the Hodgkin–Huxley equations (8)] would each have to move 3 Å. An approximate lower limit for a conformational change associated with channel opening would thus have to be about 3 Å. For the open-to-closed transition studied with the single-channel recording approach, the displacement would have to be at least 4 or 5 Å (since three equivalent charges would move) if the closing step corresponds to a single charged helix moving. Because the primary sequence of what is believed to be the entire sodium channel does not contain a region of 20 consecutive positively (or negatively) charged amino acids, a 3–5 Å motion normal to the membrane plane should be a lower limit for the sort of conformational transition we are considering.

4. GATING CHARGE DISPLACEMENTS ARE ACCOMPLISHED WITH ONLY VERY SMALL CHANGES IN OVERALL CHANNEL STRUCTURE

Additional information about the gating conformational changes comes from experiments on the pressure dependence of gating-current time courses. In these experiments (5) squid axons subjected to high pressures, up to about 600 atm, exhibited gating currents that were simply slowed without changing form. From the relationship between amount of slowing and pressure, an activation volume of 25–30 $Å^3$ was calculated. Although the meaning of this volume can be rather complicated, the simplest view is that it represents the increase in volume of the channel in going from resting to the peak of the energy barrier on the way to channel opening. As the total volume of a sodium channel should be about 0.26×10^6 Å^3 (208,321 daltons $\times$ 1.23 Å^3/dalton), the increase in volume in reaching the barrier peak is about 0.01% of the total. In other words, the motions associated with gating really perturb the channel structure to only negligible extent.

In summary, gating must involve a substantial motion, at least 4 or 5 Å, of some parts of the protein embedded within the membrane (so the membrane field is experienced), but this motion must be so smooth, at least to the peak of the energy barrier separating states, that the remainder of the channel is undisturbed.

5. MOTIONS OF THE REQUIRED MAGNITUDE ARE PLAUSIBLE

Are motions of 5–10 Å reasonable for proteins such as the sodium channel? Although the proteins that have been studied in two conformations crystallographically are not necessarily good models for an integral membrane protein like a channel (the best studied examples are soluble enzymes) they do give some indication of what might be expected (4). Usually the motions of enzymes are represented as the rotation of one domain relative to another around a flexible hinge region. These rotations serve to trap the substrate in the groove that forms the active site and to move functional groups involved in catalysis into the proper position.

Rotations are usually in the range of 10–20° and involve movements in the 5–10-Å range. The rate constants for these conformational changes are often on the order of 1/msec and thus show that motions of the same general size as those expected of channels do occur in other proteins with rates appropriate for channel gating. Because the role of conformational changes in these enzymes is quite different from channel gating, one expects that special types of gating structures and motions may well have evolved for the voltage-gated channels.

6. PREDICTIONS ABOUT SODIUM CHANNEL STRUCTURE FROM PRIMARY SEQUENCE

Four proposals for sodium channel structure based on the primary sequence have appeared (see Table 1 for terminology). The first proposed structure was described in the original paper presenting the amino acid sequence deduced from cDNA believed to code for the complete eel sodium channel (12). These workers identified four highly homologous regions of sequence, domains, each of which in turn contained six segments of sequence. The six segments found within each domain have been denoted S_1, S_2, through S_6. Two of these segments (S_5 and S_6) are highly apolar, three (S_1, S_2, and S_3) are relatively apolar with only a few charged side chains, and the remaining segment (S_4) has the unusual feature that every third residue is positively

TABLE 1. PROPOSED SODIUM CHANNEL STRUCTURES[a]

Model	Designation of Segments								Domain
12	S_1	S_2	S_3	S_4	S_5			S_6	
7	S_1	S_2	S_3	S_4	S_5	S_6	S_7	S_8	
6	S_a	S_b	S_c	S_d	S_e	S_f	S_g	S_h	
12	114–141	150–171	177–197	204–224	244–264			397–402	
7	121–143	147–170	183–203	204–224	243–266	345–357	360–373	378–402	I
6	114–141	150–170	182–203	208–224	244–264	337–355	359–374	379–402	
12	558–585	597–620	626–643	651–671	691–711			767–790	
7	565–587	597–620	629–649	651–671	690–712	733–745	748–761	766–793	II
6	558–585	597–619	628–650	655–671	691–711	728–743	747–759	763–790	
12	992–1019	1033–1057	1062–1079	1092–1112	1132–1152			1236–1264	
7	1000–1021	1033–1056	1065–1085	1092–1112	1131–1153	1197–1209	1216–1229	1240–1267	III
6	992–1019	1033–1056	1064–1091	1096–1112	1132–1152	1199–1212	1215–1227	1232–1264	
12	1314–1341	1353–1376	1381–1398	1417–1437	1454–1474			1544–1567	
7	1321–1343	1353–1376	1384–1404	1417–1437	1453–1475	1510–1517	1521–1527	1543–1570	IV
6	1314–1341	1353–1375	1383–1405	1421–1437	1454–1476	1484–1499	1522–1535	1540–1567	

[a] Numbers at the far left of the table indicate the reference, cited in the bibliography, that identifies the sequences given in the table (numbers in the body of the table follow the amino acid sequence positions presented in ref. 12). The nomenclature adapted for the various segments by each article is displayed at the across the top of the table over the numbers specifying the amino acids included in the segment.

charged (usually an arginine). In the original model, four of these segments were considered to be membrane spanning (S_1, S_2, S_5, and S_6), and the remaining two were placed on the cytoplasmic side of the membrane.

The next model to appear is quite different from the one proposed by the Numa group. Kosower (10) has placed the S_4 segment in the membrane as a 3_{10} helix and suggested that the positive charges of these segments are neutralized within the membrane by negative charges provided by other regions (not any of the homologous segments originally defined by Numa's group). The Kosower model has not made full use of the homologies found in the different domains.

Two recent models (6,7) are extensions of the original six-segment model from the Numa group. These models are very similar to each other, and both differ from the original proposal by making all six segments cross the membrane and also by identifying two more segments between the original S_5 and S_6 that are viewed as at least partially membrane spanning. Guy's proposal is the most specific one and suggests that the coupling of membrane field to gating occurs through a screwlike movement of segment S_4 (recall that S_4 is the segment in which every third residue has a positively charged side chain).

7. STRUCTURE–FUNCTION CORRELATIONS

The following discussion is limited to the three models that exploit the interdomain homologies first noted by the Numa group.

The original model from the Numa group is implausible because an insufficient number of charged residues are placed within the membrane field. The extensions of this model recognize the difficulty and thus propose that the S_4 helix is the gating sensor that spans the membrane and has its positive charges neutralized by surrounding negative charges contributed by other segments. If the motion of each S_4 segment were identified with at least one conformational change involved in gating, at least one to two equivalent charges would have to be displaced by the motion of this helix. As each of the S_4 segments bears—in different domains and with slightly different assignments for the location of this segment in the various models—from four to seven positive charges, a single "step" would move from about one-half to one equivalent charges through the membrane field. A step would be a sliding of the S_4 helix that displaces positive charge into the position of the next positive charge on the helix surface, and would be a 4.5-Å motion in the field (which would fall over about the 30 Å membrane spanning region of the channel) corresponding to the distance along an α helix for a step of three residues. All four S_4 segments would then move a total of two to four equivalent charges, an amount that is less than that suggested by the physiological data. It would seem that two steps might be required to produce the necessary charge displacement, and this might be accomplished by an opposite movement of the structures that bear the negative charges neutralizing the S_4 positive charges.

The single-channel data provide a further challenge for the current models. The closing step is associated with the movement of about three equivalent charges from an open position to the peak of the energy barrier, presumably a distance less than the full displacement necessary for channel closing. It seems implausible that a single S_4 displacement could provide the required charge movement. The S_4/IV bears seven charges, so a step of about 15 Å to reach the "half-closed" position would be necessary to displace three equivalent changes, a movement that seems too large to be plausible. Apparently, then, channel closing involves either more than the movement of an S_4 segment, or the concerted motion of S_4s from different domains.

Because the activation volume for gating is so small, the overall structure must be well preserved during the conformational changes required to open a channel. The helical screw proposal by Guy is attractive in this regard because a relatively large motion of charges in the field can be achieved without the transient separation of helices that would appear for many other sorts of motions. For example, a tilting of transmembrane helices would require that the elements move somewhat farther apart for side chains to pass one another and would presumably be associated with a larger activation volume.

8. SUMMARY

The arguments presented here cannot be made sufficiently precise to place clear limits on models of sodium channel function, but they do raise questions about the validity of simplest interpretations made so far. Thus, simple identification of a motion of each of the four S_4 segments with a separate kinetic step in the sequence that leasts to channel opening seems unlikely to be correct. As more detailed structural and functional information becomes available, attempts to relate structure and function in a quantitative way should indicate which features of the models are least in accord with data, and will help in deciding which among the various models that will arise is most satisfactory.

REFERENCES

1. Aldrich, R. A. and Stevens, C. F. (in press). *J. Neurosci.*
2. Armstrong, C. M. (1981). *Physiol. Rev.* **61**, 644–683.
3. Armstrong, C. M. and Bezanilla, F. (1977). *J. Gen. Physiol.*, **70**, 567–590.
4. Bennett, W. S. and Huber, R. (1984). *Crit. Rev. Biochem.* **15**, 291–384.
5. Conti, F., Inouc, I., Kukita, F., and Stuhmer, W. (1984). *Eur. Biophys. J.*, **11**, 137–147.
6. Greenblatt, R. E., Blatt, Y., and Montal, M. (1985). *FEBS Lett.*, **193**, 125–134.
7. Guy, H. R. and Seetharamulu, P. (1986). *Proc. Natl. Acad. Sci.* (USA), **83**, 508–512.
8. Hodgkin, A. L. and Huxley, A. F. (1952). *J. Physiol.* (London), **177**, 500–554.
9. Hol, W. G. J., Halie, L. M., and Sander, C. (1981). *Nature*, **294**, 532–536.

10. Kosower, E. M. (1985). *FEBS Lett.*, **182**, 234–242.
11. Magleby, K. L. and Stevens, C. F. (1972). *J. Physiol.* (London), **223**, 173–197.
12. Noda, M., Shimizu, S., Tanabe, T., Takai, T., Kayano, T., Ikeda, T., Takahashi, H., Nakayama, H., Kanaoka, Y., Minamino, N., Kangawa, K., Natsuo, H., Raftery, M. A., Hirose, T., Inayam, S., Hayashida, H., Miyata, T., and Numa, S. (1984). *Nature*, **312**, 121–127.
13. Stevens, C. F. (1978). *Biophys. J.*, **22**, 295–306.

CHAPTER 7

STATE-DEPENDENT MODIFICATION OF SODIUM CHANNELS BY LIPID-SOLUBLE AGONISTS

BERTIL HILLE
MARK D. LEIBOWITZ
JEFFREY B. SUTRO
Department of Physiology and Biophysics
University of Washington School of Medicine
Seattle, Washington

JÜRGEN R. SCHWARZ
Institute of Physiology
University of Hamburg
Hamburg, West Germany

GEORGE HOLAN
Commonwealth Scientific and
Industrial Research Organization
Melbourne, Australia

Dr. Sutro's present address is: Department of Physiology and Biophysics, University of California, Irvine, California.

The voltage-gated sodium channel is an essential feature of animal life. Hence it has become the target of many classes of natural toxins evolved to immobilize animals. A large and miscellaneous group of lipid-soluble plant toxins and insecticides exert their lethal effects by modifying the gating of sodium channels. Well-known members of the group include the veratrum alkaloids and aconitine, substances that physiologists have studied for more than a century, as well as batrachotoxin (BTX), grayanotoxins, and the pyrethroid and DDT insecticides. The heterogeneous nature of this group is evident from their chemical structures (Fig. 1). Because these lipid-soluble molecules cause sodium channels to open more easily and to stay open longer than normal, we call them *agonists*. They tend to induce repetitive firing, arrhythmias, depolarizing afterpotentials, or even maintained depolarization of nerve and muscle. They can lead to pain, paralysis, and respiratory or cardiac crisis. These agents have recently become popular again because they offer a useful way to activate isolated sodium channels reconstituted in lipid bilayers and vesicles.

This chapter concerns the mechanism of agonist action. One central issue is whether all agonists act by the same mechanism. We describe some of the early work, which did not suggest much commonality of mechanism, and then consider newer experiments that indicate more unity than was first suspected.

Before discussing how the agonists modify gating, we should remind ourselves how the sodium channel normally gates. In the terminology of the Hodgkin–Huxley (13) model, sodium channels are said to activate and then inactivate in response to membrane depolarizations from rest, giving a transient opening that underlies the generation of action potentials. In a voltage clamp step depolarization, the macroscopic opening of sodium channels follows a sigmoid time course, suggesting that opening is a multistep process as reflected in the following diagram of conformational states of the channel protein:

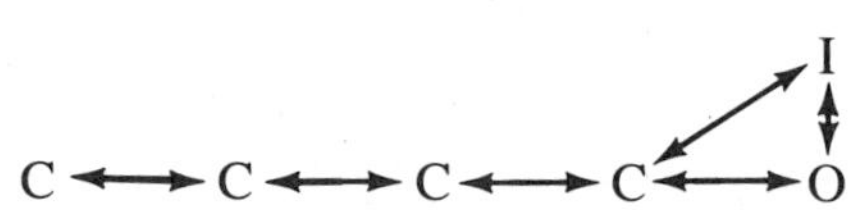

where the Cs are closed states, O is the open state of the pore, and I stands for inactivated states. Depolarization of the membrane causes channels to make conformational transitions from the left-hand states toward the right. In a frog myelinated axon, depolarization to near -25 mV is required to obtain half-maximal opening of channels. During the depolarization, some of the channels enter inactivated states before ever opening and others enter inactivated states after opening. In the myelinated fiber, the inactivation process is relatively complete so that in any strong depolarization the "steady-state" number of sodium channels remaining open after 100 msec must be less than 1% of the peak transient value (8). If the membrane is repolarized in a step to the resting potential (-75 mV), all sodium channels that are open will close within a few hundred microseconds, a rapid reversal of activation that is usually called *deactivation*, and in addition most inactivated channels return slowly (within milliseconds) to the closed resting state.

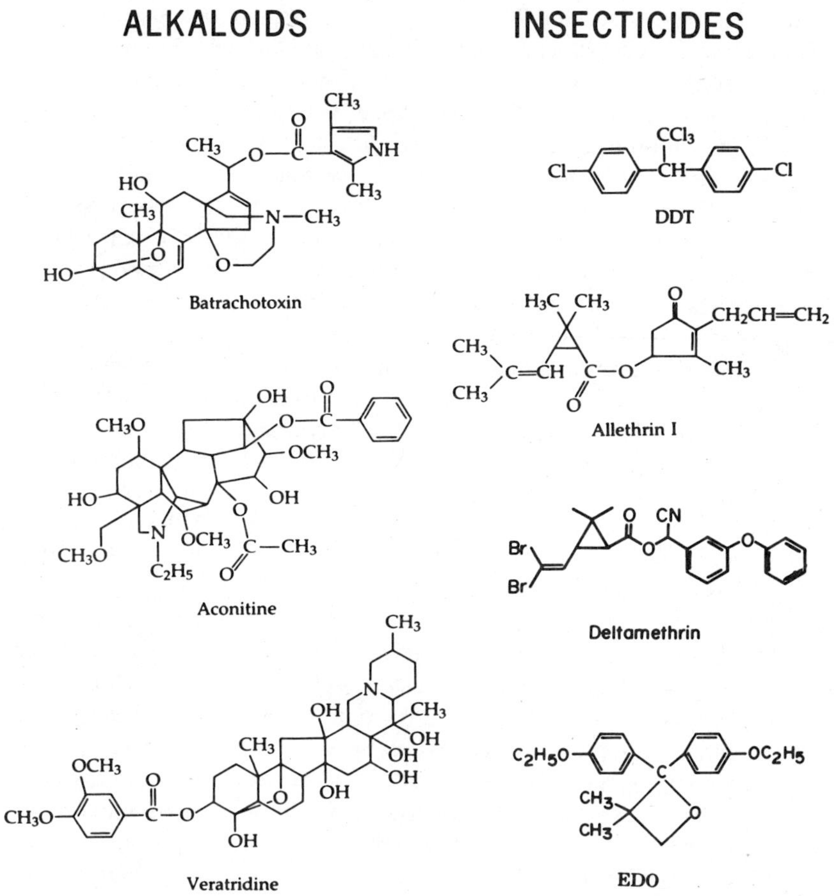

Figure 1. Chemical structures of lipid-soluble agonists of sodium channel gating. Batrachotoxin (BTX), aconitine, and veratridine are natural products. We also used another natural alkaloid, cevadine, purified from veratrum alkaloids, and a synthetic derivative, 3(4-ethoxybenzoyl) veracevine (EBV); EBV differs from veratridine in being the ethoxybenzoyl ester instead of a diethoxybenzoyl ester of veracevine. Cevadine is the 2-methyl-2-butenoyl ester.

1. EARLY BIOPHYSICAL STUDIES WITH VERATRIDINE AND DDT

The first voltage clamp studies with lipid-soluble agonists were done with veratridine applied to nodes of Ranvier of myelinated axons. Ulbricht (32,33) found that veratridine induced a slowly developing, persistent sodium current when the nodal membrane was held depolarized. The new current grew with a time constant of 1.5 sec during a step to 0 mV and turned off again within a few seconds when the membrane was repolarized to -70 mV. The steady-state sodium current–voltage relation of the membrane, traced out by voltage clamping with an extremely slow depolarizing ramp voltage waveform (4 mV/sec), had a shape resembling the peak sodium current–voltage relation of a normal node. Thus at rest there was little sodium current, depolarization to near -25 mV induced a half-maximal persistent sodium conductance, and depolarization to 0 mV or beyond induced nearly maximal persistent sodium conductance. In rapid-flow experiments with short exposures to veratridine, the agonist effects became visible within hundreds of milliseconds of drug application and reversed within seconds after the drug was washed away (34). With long application (minutes) of drug, the effects became more difficult to reverse by rinsing with drug-free solutions.

Ulbricht correctly supposed that veratridine modifies normal sodium channels, changing their gating. Since there is persistent current lasting minutes during a depolarization, the sodium channels no longer inactivate as fully as they do in normal axons. Ulbricht interpreted his observations to mean that veratridine-modified channels open and close (activate and deactivate) over the same voltage range as normal sodium channels but with rate constants that are 100–1000 times slower than normal gating. On the basis of new experiments, we give quite a different interpretation of the gating modification later.

The second lipid-soluble agonist to be studied by voltage clamp was DDT (10,26), which prolongs sodium currents as well but on a time scale of tens of milliseconds rather than seconds and minutes. During a step depolarization, the sodium channels activate in the normal manner but then a fraction of them fail to inactivate during depolarizations lasting up to 30 msec. When the membrane is stepped back to -75 mV, the noninactivating channels close, but with a time constant of approximately 5 msec, much slower than for normal deactivation, which takes only 200 μsec. Hence, like veratridine, DDT alters both deactivation and inactivation gating.

Hille (10) proposed that DDT does not modify sodium channels at rest, but only after they *open*. Thus DDT reacts with normal channels that open during a depolarizing pulse and temporarily converts them to a form that does not inactivate readily during the depolarization and that closes only slowly at the resting potential. At rest DDT would come off the channels again. This precursor–product relation between open normal channels and modified ones was demonstrated kinetically in myelinated nerve by varying the number of channels opening during a test pulse by (1) varying the duration of the pulse, (2) varying the amplitude of the pulse, and (3) adding a depolarizing prepulse before the test pulse to inactivate varying fractions of channels. In each case the number of modified channels produced was linearly proportional to the number of normal channels that opened during the test pulse—a typical example of state-dependent binding of a drug to the channel. Hille (10) also pointed to similarities between the actions of DDT and veratridine and suggested that, like DDT, veratridine might act by holding channels open once they had opened by normal gating.

2. ELECTRICAL EXPERIMENTS WITH "IRREVERSIBLE" AGONISTS

Properties of agonist-modified sodium channel *gating* were first clearly described in the laboratories of H. Schmidt and B. I. Khodorov for aconitine and BTX (15,17,30). The extensive electrophysiological work with BTX has recently been reviewed (16). Unlike veratridine and the insecticides, aconitine and BTX have the advantage that once they modify channels, the modification lasts for hours, at least when the experiments are done in the cold. Hence we call them loosely ''irreversible'' agonists. The progress of the modification can be hastened by stimulating the axon repetitively for several minutes with voltage clamp depolarizations. Khodorov and Revenko (17) found that to accelerate the reaction, these depolarizing pulses had to be sufficient to activate sodium channels and initiated from a holding potential sufficiently negative to prevent channels from remaining inactivated. These experiments can be interpreted to imply that BTX, like DDT, does not react with channels until they are open. As modified channels appear, normal ones disappear. With suitable conditions, nearly the whole complement of sodium channels in the membrane may be modified and then voltage clamp experiments can be done to explore their gating kinetics in detail.

Recall that normal sodium channels of a myelinated axon are closed at -75 mV and activate with a sigmoid time course and then inactivate during a 10-msec depolarizing pulse. Once modified by aconitine or BTX, the channels continue to gate but in a different membrane potential range (15,17,25,30). The *transient* nature of sodium channel opening is eliminated. Half of the modified channels are open at -75 mV. They do not inactivate there but can be closed in 1–3 msec by a hyperpolarization to -120 mV. They reopen quickly on ''depolarization'' back to -75 mV in a process that resembles normal activation but is severalfold slower and no longer appears sigmoid in time course. The results have been summarized by saying that the voltage dependence of activation gating has been shifted from -20 to -40 mV to abnormally negative potentials and that the gating kinetics have only one rate-limiting step as would be expressed by the state diagram $C \longleftrightarrow O$. At the same time as inactivation is removed, sodium channels remain open indefinitely at voltages positive to -75 mV.

In addition to changes in gating, voltage clamp experiments on myelinated nerve revealed that the ionic selectivity of sodium channels is decreased once they are modified by aconitine or BTX (15,25). Methylammonium and Cs^+ become permeant in modified sodium channels, K^+ moves up to being 60% as permeant as Na^+, and NH_4^+ becomes the most permeant ion. These changes suggest that the agonist binding widens the selectivity filter of the channel. At the same time as the peak sodium current decreases, the absolute conductance of each channel is believed to be depressed.

In summary, the experiments with myelinated nerve revealed that (1) the reaction of sodium channels with aconitine and BTX is speeded by depolarizing pulses that open normal sodium channels, (2) the off reaction is very slow, (3) the modified channel hardly inactivates, (4) the modified channel activates at more negative voltages, and (5) the modified channel may have lower conductance and does not discriminate as well against ions larger than Na^+.

These pioneering voltage clamp experiments have subsequently been extended to frog muscle, neuroblastoma cells, squid giant axons, and other excitable membranes (16). The broad strokes are similar throughout, although the amount of shift of activation, the type of change of inactivation, and the extent of the ionic selectivity change vary from membrane to membrane. Batrachotoxin is now also used as the agonist of choice to elicit electrical signals from lipid bilayer membranes containing native or chemically purified sodium channels (see Chapters 8 and 9 in this volume). Such single-channel experiments show that individual BTX-modified channels are open for more than 90% of the time at potentials more positive than -50 mV (18, 24).

3. COMPETITION AMONG AGONISTS

Significant evidence that lipid-soluble agonists share similarities of mechanism, and possibly even a common receptor came from W. Catterall's studies on neuroblastoma cells (refs. 3–5; reviewed in ref. 6). Agonist action was assayed by measuring ^{22}Na uptake into cells on tissue culture dishes. This assay probably reflects equilibrium toxin binding as even the BTX effect could be reversed by washing (with an off time constant of 30 min at 36°C). As the agonist concentration was raised, ^{22}Na flux was stimulated according to a saturable dose–response curve, suggesting binding of toxin to a homogeneous class of saturable sites. The saturating effect (efficacy) differed for different agonists, increasing in the sequence aconitine < veratridine < grayanotoxin < BTX. Batrachotoxin was assumed to give the maximum possible flux, and aconitine at saturating doses gave only 20% of that.

The new evidence for similarity of action of the toxins was the discovery that the apparent affinity of each of the four agonists for their receptor was increased (five- to hundredfold) by a peptide toxin from *Leiurus* scorpions. The peptide toxin was binding to a site on the channel and lowering the agonist concentration needed for half-maximal effect—at another site. *Leiurus* toxin also increased the saturating flux achieved with aconitine, veratridine, and grayanotoxin, bringing it up to the level seen with BTX.

A common receptor for the four lipid-soluble agonists studied was also suggested by an apparent competition among them. Thus when enough aconitine, an agonist of low efficacy, was added to cells already strongly stimulated with BTX, the ^{22}Na flux was depressed to the low saturating level expected for aconitine alone, as if aconitine displaced BTX from the channel. Later experiments with a binding assay for a radioactive BTX analogue confirmed that veratridine and aconitine are competitive inhibitors of BTX binding (7). These results would be expected if BTX, grayanotoxin, veratridine, and aconitine had a common receptor. Conservatively, however, they do not *prove* that there is only one receptor, but they show at least that the receptor for one agonist cannot be occupied when the receptor for another is occupied—mutual exclusion.

Catterall (5) offered a quantitative interpretation of the competition and different efficacies of the lipid-soluble toxins, as well as the potentiation by *Leiurus* toxin. In his model sodium channels have two conformational states, "active" and "inactive," where the active state was implied to correspond to open channels. Lipid-soluble agonists bind to either form, giving the state diagram

where the asterisk (*) represents an agonist molecule. In this model (6, p. 30):

> The principal assumption is that the lipid-soluble toxins bind with a higher affinity to an active state of the sodium channel and thereby shift a preexisting voltage-dependent equilibrium between active states and inactive states. The efficacy . . . depends on the selectivity of binding to the active vs. inactive states. High [binding] selectivity gives high efficacy. The energy of binding is translated into shifts in the voltage-dependence of activation, causing sodium channels to be active at the resting potential.

Finally, since the scorpion peptide α toxins favor the active state, they also promote binding of lipid-soluble agonists.

4. OUR EXPERIMENTS WITH VERATRUM ALKALOIDS

In 1980 we decided to resume voltage clamp experiments on lipid-soluble agonists to see whether a unifying mechanism could be found to explain their seemingly diverse electrophysiological actions. Voltage clamp studies of DDT, veratridine, and BTX–aconitine had led to three apparently different models; however, flux experiments with veratridine, aconitine, grayanotoxin, and BTX had been described in a single model that did not explicitly refer to the classical channel "states" familiar to the electrophysiologist.

Our approach was to look for similarities in the electrophysiological actions of agonists. Using frog skeletal muscle, we asked initially whether veratridine and its relatives, cevadine and 3(4-ethoxybenzoyl) veracevine (EBV), shared properties that had been demonstrated for other agonists: (1) requirement for normal channels to open before they would react with agonist (known for DDT and BTX and suggested for aconitine); (2) drug-modified channels retaining an activation gating process that is, however, shifted to more negative potentials and less sigmoid in time course (known for BTX and aconitine); and (3) drug-modified channels with less strict sodium selectivity than normal (known for BTX and aconitine). As it turned out, each of these properties could be demonstrated for veratridine and its relatives. In addition, we wanted to know in electrophysiological terms how scorpion toxin could potentiate the binding of lipid-soluble agonists.

4.1 Use-Dependent Modification

Consider first the requirement for open channels as a precursor of the modified channel. When sodium currents of a muscle fiber equilibrated with 100 μ*M* veratridine, cevadine, or EBV are recorded with repeated depolarizing test pulses, they seem nearly normal in the first test pulse and become progressively more modified as pulsing continues. As Fig. 2 shows for veratridine, sodium currents in pulse 1 activate and inactivate during the test pulse, nearly returning to zero late in the pulse. With successive depolarizations, the peak transient current shrinks while noninactivating current late in the pulse and persistent current after the pulse increase in size. In the absence of drug, sodium currents do not change during the course of an identical pulse protocol.

The persistent "tail" current at -90 mV is especially diagnostic of the veratridine modified state. It develops pulse by pulse at the expense of channels with normal gating. As had been shown for DDT, the number of channels becoming modified during one test pulse is proportional to the number of normal channels opening during the pulse. For veratridine the proportionality holds when the number of normal channels opening is varied by depolarizing to different test potentials or for different durations or by using depolarizing prepulses to inactivate some of the channels (31). Apparently the rate of reaction with resting or inactivated channels is low and that with open channels is high, leading to a simplified state diagram

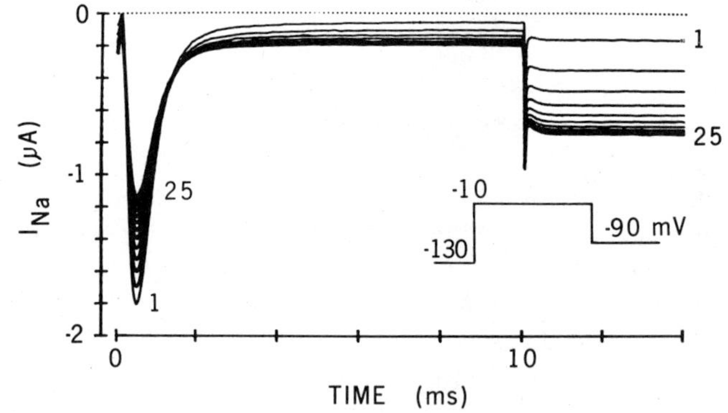

Figure 2. Use-dependent modification of sodium channels by 100 μ*M* veratridine. Sodium currents recorded from a frog skeletal muscle fiber under voltage clamp while the fiber is depolarized repeatedly by a 10-msec pulse to -10 mV. The traces shown are from the first, third, fifth, and so on pulse in a 3.7-Hz train of 25 pulses (T = 9°C). (Modified from ref. 19.)

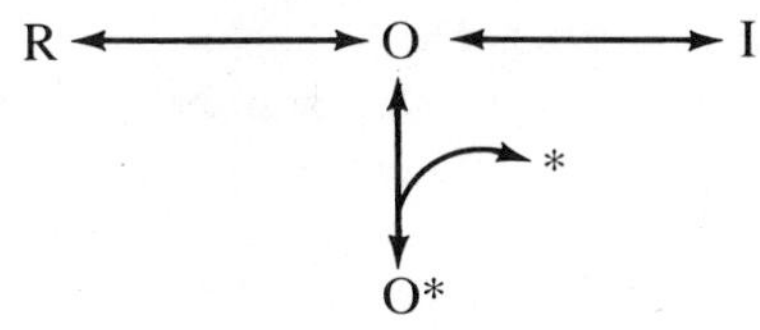

where again the letters represent resting, open, and inactivated states. The initial rate of modification is linearly dependent on the veratridine concentration, as is expected from this scheme, and we estimate that a normal open channel becomes modified with a rate constant of 200 sec^{-1} in 100 μM veratridine (19). Therefore, if a channel remains open for 1 msec, it has a 20% chance of becoming modified.

The modification reaction with veratrum alkaloids reverses when the membrane is returned to negative potentials where *normal* channels would be deactivated. Figure 3 shows that reversal with veratridine occurs with a time constant of approximately 3 sec. The trace shows sodium currents flowing at -90 mV before, during, and after a repetitive train of voltage clamp depolarizations. Before the train there are nearly no modified channels—no standing inward sodium current. As the train is applied, pulse by pulse the modified channels appear. After the train they disappear with an exponential time course, so that within 20 sec there is no remaining evidence of modified channels. Thus modification–unmodification experiments such as Figs. 2 and 3 can be repeated continually. This is an advantage of working with reversible agonists.

We take as a working hypothesis that the physiologically measurable modification–unmodification changes correspond to binding and unbinding of

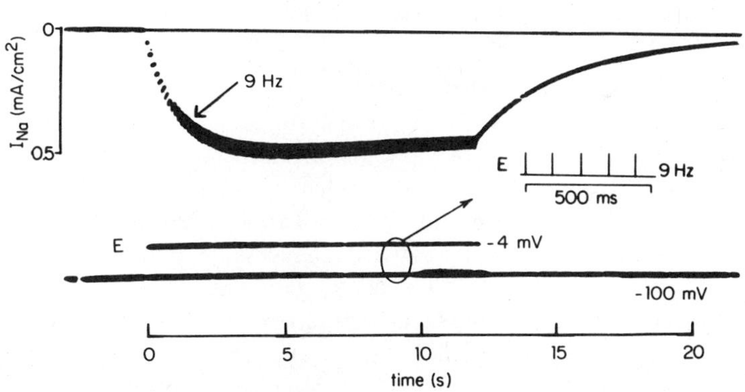

Figure 3. Use-dependent modification of sodium channels by 50 μM veratridine during a pulse train and recovery following the train. The sodium currents at -90 mV are recorded photographically at such a slow sweep speed that the 110-pulse train of depolarizations is seen in one sweep. The individual depolarizations to -4 mV are too brief to be photographed. One can see only the buildup of tail current between pulses and its subsequent decay. In our view, this experiment shows the progressive conversion of normal sodium channels into the agonist-modified form and the subsequent unbinding of agonist from the channel. (Modified from ref. 31.)

agonist at the channel. These drugs are hydrophobic and partition preferentially into the lipid phase of the membrane. When the agonist is not bound, we envision that it diffuses freely within the lipid phase of the membrane at a concentration higher than that in the aqueous phase.

These conclusions permit us to reinterpret Ulbricht's (33) observation that a veratrinized node of Ranvier held depolarized will develop a standing sodium current over 1-sec-long period that decays away in a few seconds when the membrane is repolarized. We would suggest that while most normal sodium channels are inactivated after a few milliseconds of depolarization, a small fraction are always reentering the conducting pool. Suppose that only 0.1% of the channels are open at a time. Then if they react at 200 sec^{-1} with veratridine, 20% of the normal channels would become modified per second. Since the off rate is about 0.3 sec^{-1}, slightly less than 50% of the channels could become modified in the steady state. Thus the slow appearance of inward current with maintained depolarization would not be due to slow gating in modified channels, but rather to a rate of modification of channels that is slow because the precursor, normal open channels, is in short supply.

These conclusions also agree with Catterall's explanation of potentiation by *Leiurus* scorpion toxin. Scorpion toxin greatly slows inactivation gating and thus prolongs the open lifetime of sodium channels. If open channels are the kinetic precursor of agonist-modified channels, scorpion toxin, by holding channels open, should enhance the rate of modification in depolarized cells. This explanation could be further tested by determining whether other agents that slow inactivation gating, such as *N*-bromoacetamide, chloramine-T, or internal proteolytic enzymes, can also hasten the modification by lipid-soluble agonists. Indeed, these three reagents do potentiate agonist action (19,21,22,31), confirming the idea that open channels are the kinetic precursor of agonist modified channels.

4.2 Gating in the Modified State

The next question was whether veratridine-modified channels are simply held open or retain a modified activation gating process, as had been found for BTX and aconitine. The experiment is more complex with a reversible rather than an irreversible agonist for three reasons: (1) to obtain modified channels, a train of ''conditioning'' depolarizations may be required; (2) the test for gating must then be done in the short period before the modified channels have reverted to normal channels; and (3) if hyperpolarization causes a more rapid decay of persistent current than at rest, some criterion must be available to determine whether hyperpolarization actually acted on a voltage-dependent gate or merely hastened the reversal of the modification reaction (unbinding of drug).

With veratridine and its relatives, the result was unambiguous (19,20,31). A hyperpolarization to -170 mV during a period of persistent current decreases the sodium current nearly to zero within 1–2 msec (Fig. 4). ''Depolarization'' from this -170-mV level back to -90 mV brings back the full persistent current in 1–2 msec. Hence there is rapid and reversible gating of modified channels with speeds 1000

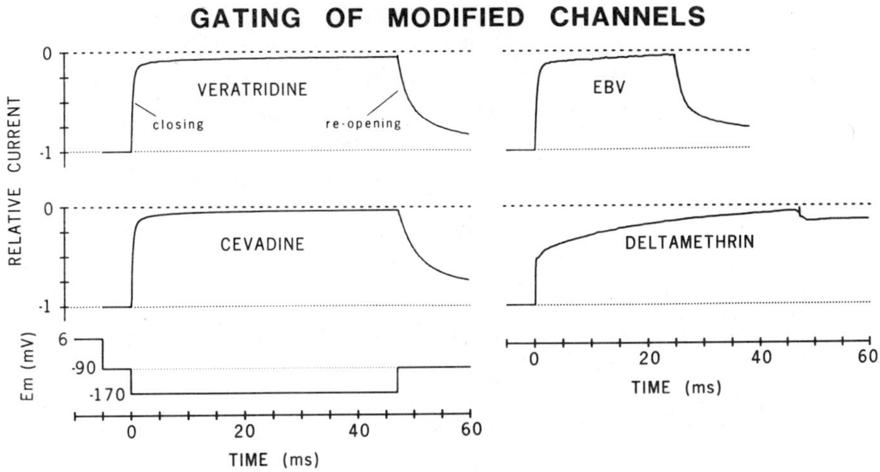

Figure 4. Channel gating observed when modified channels are stepped to −170 mV and back to −90 mV. In each case a population of modified channels is generated before the record begins by applying a train of 10 depolarizing conditioning pulses. This results in a standing inward sodium current through modified channels seen at the beginning of each record. Hyperpolarization to −170 mV closes channels. With the three veratrum alkaloids, channels reopen quickly when the voltage is stepped back up to −90 mV. This is the shifted activation gating typical of modified channels. With deltamethrin, channels do not reopen once they have been closed. (Data from ref. 20.)

times greater than the agonist dissociation reaction. The midpoint for opening of veratridine-modified channels is near -114 mV rather than at -25 mV as in normal channels, so the voltage dependence of activation is shifted by -90 mV when the agonist binds. As with BTX, the time course of activation has lost its sigmoid character as well; however, with veratrine alkaloids, the activation and deactivation time courses have a double-exponential character.

In addition to closing modified channels, we found that hyperpolarization has another interesting effect. Maintained hyperpolarization slows the recovery from the modified state perhaps tenfold (19,20). We suppose, therefore, that while the modified channel remains closed, it traps the agonist molecule at its receptor—closing the channel results in closure of the escape route for the small molecule. The hydrophobic binding site may become submerged within the channel protein.

4.3 Ionic Selectivity in the Modified State

Our last goal was to determine whether veratrine alkaloids alter the ionic selectivity of sodium channels as had been seen with BTX and aconitine. The permeability ratio P_{NH_4}/P_{Na} was determined from reversal potential (zero-current potential) changes when all the Na^+ in the external solution was replaced by NH_4^+. We chose to study NH_4^+ since it was the ion whose permeability was reported to be most profoundly affected by BTX and aconitine (15,25). Our experiment was done with normal sodium channels by measuring the peak of the early transient permeability increase and with alkaloid modified channels by measuring during the persistent permeability increase after a train of pulses. For normal sodium channels the permeability ratio P_{NH4}/P_{Na} was 0.11, and for veratridine and cevadine modified channels it rose to near 0.7 (20). Thus bound veratrine alkaloids do decrease the ionic selectivity of sodium channels. The same result had been obtained by the Leningrad group using veratrine (27).

5. EXPERIMENTS WITH INSECTICIDES

Encouraged by finding several close similarities—gating, selectivity, and use dependence—between the actions of veratrum alkaloids and those of BTX and aconitine, we turned to the other group of reversible agonists, the insecticides. Would they also show the same diagnostic features? Again using frog muscle, we studied DDT, allethrin, deltamethrin, and some related analogues (19,20). The experiments reveal some but not all of the features seen with the other lipid-soluble agonists—there are similarities and differences.

Early experiments (10) demonstrating a requirement for open channels as the precursor for modification by DDT have already been discussed. We confirmed with frog muscle that removal of inactivation by *N*-bromoacetamide enhanced the rate of modification by DDT and allethrin as it had for veratridine and conversely that depolarizing prepulses sufficient to inactivate sodium channels prevent modification by insecticides during a subsequent test pulse (19). Similar experiments had been done in J. van den Bercken's laboratory on nodes of Ranvier and in T. Narahashi's laboratory on squid giant axons and crayfish axons (21,22,35,36).

Do insecticide-modified channels have modified gating? Clearly their gating is not normal since the transient sodium current during a depolarizing pulse is prolonged and residual tail currents appear when the membrane is repolarized. Our experiments with DDT, EDO, and allethrin were complicated by the apparent short lifetime of the agonist–channel complex in frog muscle. The tail current decayed with a time constant of 3–10 msec suggesting that these agonists dissociate from the channel 1000 times faster than veratridine does. In addition, inactivation is not eliminated by these agents, although it is slowed. Repolarization of the membrane at the peak of the transient sodium permeability yielded the expected tail of open, modified channels. Repolarization after much longer (10 msec) time yielded almost no tail. Possibly the *modified* channel inactivates readily—as Campbell found for aconitine-modified muscle channels (2)—or possibly the agonist dissociates during the 10-msec pulse and then the open channel undergoes normal inactivation.

We looked for shifted activation gating by trying to determine whether insecticide-modified channels could be closed reversibly by short hyperpolarizations. As others have found (21,37), the decay of residual tail currents could indeed be greatly accelerated by hyperpolarization (19,20). However, when the hyperpolarization was turned off, sodium current did not redevelop. Therefore, we cannot be sure whether hyperpolarization causes the modified channels to close and that these channels do not reopen or whether hyperpolarization accelerates dissociation of the drug yielding normal channels that do not open at -90 mV. The experiment was particularly ambiguous with DDT, EDO, and allethrin since the drug–receptor complex lasts for only a few milliseconds even at -90 mV, leaving little time to try to close and reopen channels. However, with a cyanopyrethroid, deltamethrin, the residual sodium current lasts for several seconds at -90 mV, and nevertheless we could find no reopening after a hyperpolarization long enough to close most channels (Fig. 4). We favor the hypothesis that hyperpolarization hastens the dissociation of the channel–insecticide complex.

Insecticides also decrease the ionic selectivity of sodium channels (20). Reversal potential measurements show that EDO and deltamethrin raise the permeability ratio P_{NH_4}/P_{Na} from 0.11 to 0.28, a smaller change than was seen with the veratrum alkaloids or with BTX or aconitine but still a significant one.

6. A MODEL FOR THE ACTION OF LIPID-SOLUBLE AGONISTS

This section indulges in several speculations concerning the receptor for agonists. The principal protein subunit of voltage-gated sodium channels has been sequenced, and the most striking feature is that the structure is repetitive (28). There are four complete repeats within the peptide chain—not identical but similar—containing, presumably, four separate voltage-dependent components that would have to be disposed in a symmetrical manner in the membrane, forming a central pore. As S. Numa's laboratory has suggested (28), activation of the channel might require successive conformational changes in each of the repeats, culminating in an open channel when all four have undergone the transition (Fig. 5). This open state also exposes a receptor for lipid-soluble agonists that we suppose faces the hydrophobic interior of the lipid bilayer. We envision it as part of the junction between two major domains, a largely hydrophobic region.

When agonist binds to the exposed receptor, it could form a wedge between domains and stabilize the open state while distorting the macromolecule sufficiently to enlarge the selectivity filter. This stabilization hinders channel closing by either of the two normal mechanisms, deactivation or inactivation. If the modified channel does close, the receptor pocket narrows and presses on the drug. This strain favors reopening of the modified channel even at potentials more negative than usual. It also alters the kinetics of gating so that activation no longer appears like four units making relatively independent transitions, but rather only one of the steps (not the final opening step) remains slow enough to show as rate limiting.

When a channel modified by veratrum alkaloids closes by deactivation, the agonist becomes trapped at its binding site. It leaves more slowly despite the pressure from the receptor closing down around it. On the other hand, the DDT and pyrethroid

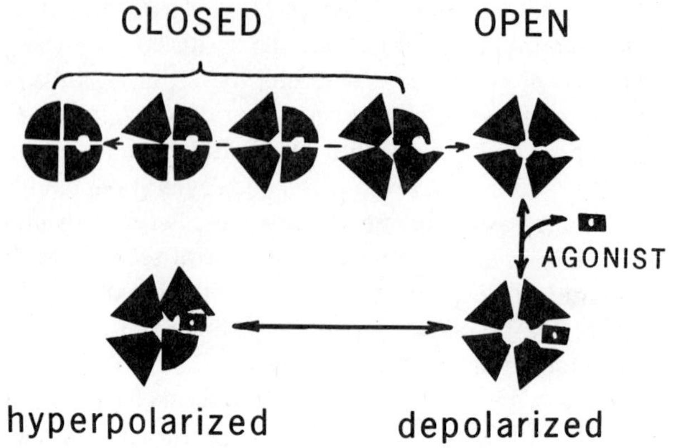

Figure 5. A cartoon of sodium channel gating and modification by lipid-soluble agonists. The channel, viewed frontally, has four equivalent domains whose stepwise conformational changes (top row) lead to opening of a pore and exposure of the agonist binding site to agonist molecules moving freely in the lipid phase. For BTX, aconitine, and veratrum alkaloids, the modified channel can be closed by strong hyperpolarization (bottom row). (Our kinetic experiments show that the closing of modified channels involves more than the single closed state shown here.) For insecticides, which may bind to a different site, the drug may be squeezed off its receptor when hyperpolarization promotes channel closing.

insecticides apparently can leave more easily, perhaps because their binding site is less obscured in the closed channel or perhaps because they are smaller, more flexible, and less polar and lack ionizable groups. The high affinity of BTX and aconitine for their binding sites results in a dissociation so slow that the channel must be able to open and close thousands of times with these compounds bound. They shift activation only half as much as veratrum alkaloids do, so they must be accommodated with less strain into the receptor volume when the channel is closed.

We have supposed that the receptor for lipid-soluble agonists is at the junction between two of the four major domains of the channel, a region whose shape changes considerably during gating. Given the near symmetry of the channel, it is conceivable that there are actually four such sites that could bind agonists but have somewhat different binding specificities and effects. One could imagine that insecticides bind better to one site and BTX, aconitine, and veratridine to another. Our experiments do not distinguish these possibilities. Several published reports of potentiation of BTX binding and of veratridine, grayanotoxin, and BTX action by pyrethroids (1,14) would favor the view that there are several agonist binding sites with different specificity and positive mutual cooperativity.

Inasmuch as sodium channels may have evolved from calcium channels at the time of the evolution of animals, and calcium channels may have diverged from potassium channels at the time of the origin of eucaryotes (11,12), there might be homologous sites for lipid-soluble agonists on calcium and potassium channels. So far the only agonists we know of for these channels are the dihydropyridine analogues that are ''Ca agonists'' (29) (see also Chapters 10, 11, and 13 in this volume). They are not known to change ionic selectivity, shift activation gating, or exhibit use dependence, but they do prolong the open time. They could be agonists that bind and unbind rapidly. It remains to be seen whether any further resemblance to the effects of sodium channels agonists will be noted.

Finally, given that there are agonist binding site(s) on sodium channels, one might wonder whether they play any role in normal physiology. Are there endogenous regulatory agonists? If so, what function would they serve? One function could be to modify sodium channels to suit them better to functioning in spike initiating zones—the axon hillock, primary sensory nerve terminals. The agonist, a modulator of excitability, would render the sodium channels less susceptible to inactivation and more easily opened. The agonist would lower firing threshold and broaden spikes. It would be facilitatory and potentiating. There is evidence for special low threshold and for nonactivating classes of sodium channels (9,23), but there is no evidence at present for an endogenous agonist. However, the possibility should still be considered.

REFERENCES

1. Brown, G. B. and Olsen, R. W. (1984). *Soc. Neurosci. Abstr.*, **10**, part 2, 865.
2. Campbell, D. T. (1982). *J. Gen. Physiol.*, **80**, 713–731.
3. Catterall, W. A. (1975). *Proc. Natl. Acad. Sci.* (USA), **72**, 1782–1786.
4. Catterall, W. A. (1975). *J. Biol. Chem.*, **250**, 4053–4059.
5. Catterall, W. A. (1977). *J. Biol. Chem.*, **252**, 8669–8676.
6. Catterall, W. A. (1980). *Ann. Rev. Pharmacol. Toxicol.*, **20**, 15–43.
7. Catterall, W. A., Morrow, C. S., Daly, J. W., and Brown, G. B. (1981). *J. Biol. Chem.*, **256**, 8922–8927.
8. Conti, F., Hille, B., Neumcke, B., Nonner, W., and Stämpfli, R. (1976). *J. Physiol.* (London), **262**, 699–727.
9. Gilly, W. F. and Armstrong, C. M. (1984). *Nature*, **309**, 448–450.
10. Hille, B. (1968). *J. Gen. Physiol.*, **51**, 199–219.
11. Hille, B. (1984). *Ionic Channels in Excitable Membranes*, Sinauer Associates, Sunderland, MA., 426 pp..
12. Hille, B. (in press). In Edelman, G. M., Gall, W. E., and Cowan, W. M. Eds., *New Insights into Synaptic Function*, Neurosciences Research Foundation; New York.
13. Hodgkin, A. L. and Huxley, A. F. (1952). *J. Physiol.* (London), **117**, 500–544.
14. Jacques, Y., Romey, G., Cavey, M. T., Kartalovski, B., and Lazdunski, M. (1980). *Biochim. Biophys. Acta*, **600**, 882–897.
15. Khodorov, B. I. (1978). In Tosteson, D. C., Ovchinnikov, Y. A. and Latorre, R. Eds., *Membrane Transport Processes*, Vol. 2, Raven, New York, pp. 153–174.
16. Khodorov, B. I. (1985). *Progr. Biophys. Molec. Biol.*, **45**, 57–148.
17. Khodorov, B. I. and Revenko, S. V. (1979). *Neuroscience*, **4**, 1315–1330.
18. Krueger, B. K., Worley, J. F., III, and French, R. J. (1983). *Nature* (London), **303**, 172–175.
19. Leibowitz, M. D., Sutro, J. B., and Hille, B. (1986), *J. Gen. Physiol.*, **87**, 25–46.
20. Leibowitz, M. D., Schwarz, J. R., Holan, G., and Hille, B. Manuscript in preparation.

21. Lund, A. E. and Narahashi, T. (1981). *J. Pharmacol. Exp. Ther.*, **219**, 464–473.
22. Lund, A. E. and Narahashi, T. (1982). *Neurotoxicology*, **3**, 11–24.
23. Matteson, D. R. and Armstrong, C. M. (1982). *J. Gen. Physiol.*, **79**, 739–758.
24. Moczydlowski, E., Garber, S. S., and Miller, C. (1984). *J. Gen. Physiol.*, **84**, 665–686.
25. Mozhayeva, G. N., Naumov, A. P., Negulyaev, Yu. A., and Nosyreva, E. D. (1977). *Biochim. Biophys. Acta*, **466**, 461–473.
26. Narahashi, T. and Haas, H. G. (1968). *Science*, **157**, 438–440.
27. Naumov, A. P., Negulyaev, Yu. A., and Nosyreva, E. D. (1979), *Tsitologia*, **21**, 692–696.
28. Noda, M., Shimizu, S., Tanabe, T., Takai, T., Kayano, T., Ideda, T., Takahashi, H., Nakayama, H., Kanaoka, Y., Minomino, N., Kangawa, K., Matsuo, H., Raftery, M. A., Hirose, T., Inayama, S., Hayashida, H., Miyata, T., and Numa, S. (1984). *Nature*, **312**, 121–127.
29. Nowycky, M. C., Fox, A. P., and Tsien, R. W. (1985). *Proc. Natl. Acad. Sci.* (USA), **82**, 2178–2182.
30. Schmidt, H. and Schmitt, O. (1974). *Pflügers Arch.*, **349**, 133–148.
31. Sutro, J. B. (1986). *J. Gen. Physiol.*, **87**, 1–24.
32. Ulbrucht, W. (1965). *J. Cell. Compar. Physiol.*, **66**, 91–98.
33. Ulbricht, W. (1969). *Ergebn. Physiol.*, **61**, 18–71.
34. Ulbricht, W. (1972). *Pflügers Arch.*, **336**, 187–199.
35. Vijverberg, H. P. M., van der Zalm, J. M., and van den Bercken, J. (1982). *Nature*, **295**, 601–603.
36. Vijverberg, H. P. M., van der Zalm, J. M., Van Kleef, R. G. D. M., and van den Bercken, J. (1982). *Biochim. Biophys. Acta*, **728**, 73–82.
37. Wu, C. H., Oxford, G. S., Narahashi, T., and Holan, G. (1980). *J. Pharmacol. Exp. Ther.*, **212**, 287–293.

CHAPTER 8

VOLTAGE-SENSITIVE SODIUM CHANNELS: AN EVOLVING MOLECULAR VIEW

ROBERT L. BARCHI[1–3]
JAN M. CASADEI[1, 3]
ROBERT D. GORDON[1]
RICHARD H. ROBERTS[1, 3]

Mahoney Institute of Neurological Sciences[1] and Departments of Neurology[2] and of Biochemistry and Biophysics[3]
University of Pennsylvania School of Medicine
Philadelphia, Pennsylvania

Voltage-activated sodium channels play an integral role in the generation of action potentials in nerve and muscle cells. In the excitable membranes of most of these cells, a rapid but transient increase in membrane conductance to sodium ions occurs in response to a triggering membrane depolarization. This conductance increase, mediated by the sodium channel, produces a rapid shift in membrane potential toward the sodium equilibrium potential that results in the rising phase of the action potential. Subsequent closing or inactivation of the sodium channel, often in association with the opening of a separate potassium-specific conductance pathway, terminates the action potential and allows the membrane to repolarize. The unraveling of the ionic currents underlying the action potential and of the time- and voltage-dependent membrane processes that control them represents one of the major single advances in the brief history of molecular neurobiology.

Since the pioneering work of Hodgkin and Huxley on the action potential in the squid giant axon during the late 1940s (25), advancing technology has allowed the development of increasingly sophisticated descriptions of the kinetics of the currents modulated by the sodium channel (see Chapters 6,7, and 9) More recently, attention has focused on the isolation and molecular characterization of the channel protein itself, and progress in this direction has been rapid (4). A molecular view of the channel is now evolving that should, in turn, open the way for the next phase in its characterization, a phase in which direct correlation between the details of protein structure and channel function can be expected.

Various aspects of sodium channel function, especially those dealing with channel kinetics and gating, have been covered in the preceding chapters of this volume and are not discussed further here. In this chapter we consider progress that has been made in the biochemical characterization of the sodium channel protein. For this discussion, we focus on work with the voltage-dependent sodium channel from mammalian skeletal muscle and relate these results to comparable studies of sodium channels from other sources, including eel electric organ, rat brain, and chick heart, to provide a broader perspective. Although the basic similarities among the sodium channels from these various sources is impressive, a number of differences are highlighted whose explanation remains unclear.

1. THE CHANNEL GAME

From the point of view of the biochemist, the isolation and characterization of an integral membrane protein that is responsible for a specific ionic conductance raises two immediate problems: (1) the protein must be removed from its native membrane environment, but this usually results in loss of stability and disappearance of functional identity; and (2) the conflicting demands for rigorous purification and maintenance of functional activity often prove difficult to reconcile. Manipulations that improve purity may sacrifice key binding activity. Compromise solutions to these problems have arisen through trial and error. In this respect much has been

learned from earlier studies with the acetylcholine receptor, (AChR) and work on the biochemical characterization of the sodium channel has moved forward briskly.

In the channel game (Fig. 1), a label is needed to identify the protein once it is removed from its normal membrane environment and a reconstitution system is ultimately required to assess the maintenance of the functional integrity of the protein during the purification. If the isolated protein can simultaneously meet the criteria of purification and preservation of function, questions regarding biochemical characterization can reasonably be asked. In combination with modern molecular biological techniques, this should ultimately lead to experiments correlating biochemical structure with channel function. Work with the voltage-dependent sodium channel has followed roughly along these lines, and it will be useful to keep the game plan in mind while reviewing the most recent results.

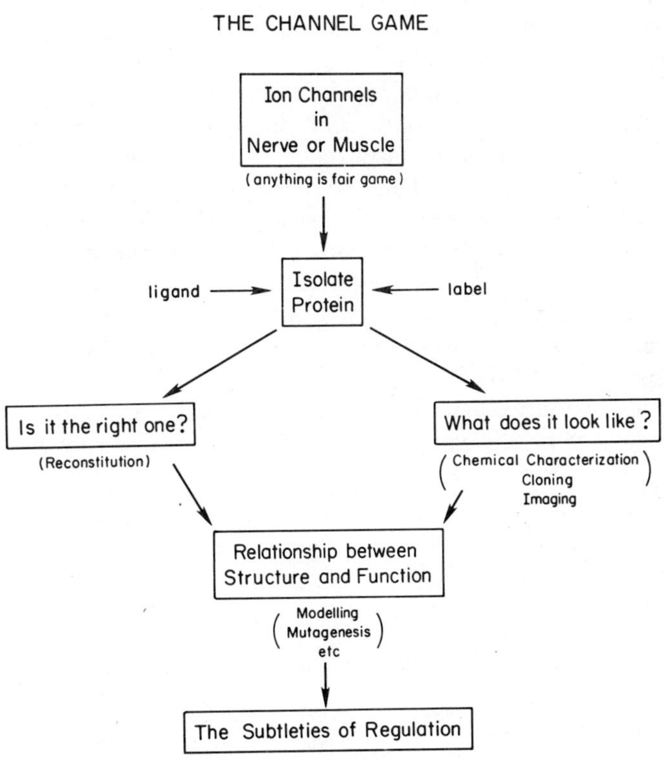

Figure 1. The channel game. Molecular characterization of an ion channel usually begins with the solubilization and purification of a specific ligand receptor or toxin binding protein. If it can be shown through functional reconstitution that this protein is indeed the channel of interest, biochemical characterization can proceed with the ultimate goal of relating channel structure to function.

2. PURIFICATION OF THE SODIUM CHANNEL

For the most part, purification of the voltage-dependent sodium channel has relied on the availability of labeled derivatives (34) of tetrodotoxin (TTX) or saxitoxin (STX) to identify the channel protein once the membrane has been disrupted (1,5,21), although more recently, polypeptide neurotoxins (29) and monoclonal antibodies (11) have also been used for this purpose. While the channel protein can be removed from nerve and muscle membranes by a variety of detergents, its stability appears to be best preserved in nonionic detergents such as NP-40 or Triton X-100; maintenance of low temperature and supplementation of detergent-containing buffers with exogenous phospholipid are also required (2,5).

The solubilized sodium channel is obtained as a mixed micelle in which the protein is associated with detergent and phospholipid. The overall size of the micelle is considerably larger than that of the protein itself, and the properties of the micelle will contribute directly to the behavior of the channel on preparative columns and gradients. The mixed micelles from mammalian skeletal muscle contain about 45% by weight detergent and phospholipid and have a partial specific volume of ~0.83 and an overall molecular weight of approximately 560,000 (6).

A general feeling for the size of the molecule involved can be obtained even in crude preparations from electron inactivation or centrifugation techniques. Early studies using these approaches suggested an overall molecular weight of between 287,000 and 314,000 for the total sodium channel protein from rat, rabbit, and human skeletal muscle (Table 1) (6,27). Similar studies with rat brain suggested a molecular size of 316,000 (23). Electron inactivation studies that specifically examined the rate of loss of the binding site for STX or TTX or for a polypeptide binding site suggested a somewhat smaller limiting size for the components containing these sites ranging in the neighborhood of 260,000 daltons (Table 2) (3,8,28).

Early studies confirmed that the sodium channel, like many intrinsic membrane proteins, was a glycoprotein and was efficiently bound by a number of lectins (13).

TABLE 1. PHYSICAL SIZE ESTIMATES OF THE SODIUM CHANNEL

	Centrifugation	Radiation Inactivation			
		TTX/STX Receptor	Scorpion Toxin Receptor	Other	Reference
Rat brain	316,000	260,000	266,000		3,8, 23
		220,000	263,000	287,000 (BTX)	
Pig brain		229,000			28
Rat sarcolemma	314,000				6
Rabbit T tubule	295,000				27
Human sarcolemma	287,000				35

TABLE 2. MOLECULAR PROPERTIES OF PURIFIED SODIUM CHANNEL

Source	Detergent	Stoke Radius (Å)	Apparent S	Subunits on SDS–PAGE	Reference
Eel electroplax	Lubrol PX	90 85	8	~260,000 ~270,000	1 33
Rat sarcolemma	NP-40	86	9.1	~260,000 38,000	5, 12
Rat brain	Triton X-100	80	11.3	~260,000 38,000 37,000	21, 22
Rabbit T-tubule	NP-40	86	8.6	~260,000 38,000	27
Chick heart	NP-40		8.8	~250,000	29

Of these, wheat germ agglutinin (WGA) has proved the most useful as a tool for channel purification. After the preparation and solubilization of excitable membranes, most methods for the purification of the sodium channel have relied on sequential chromatography on ion-exchange resins and WGA–Sepharose affinity columns followed by one or more additional steps of purification by molecular sieve chromatography or sucrose density-gradient centrifugation. More recently we have used an immunoaffinity column technique to accomplish a similar purification (see the following paragraphs).

From a biochemist's point of view, there is a reassuring degree of similarity in the sodium channel proteins that have now been purified from a variety of sources using these approaches. In each case the purified material contains a large glycoprotein that runs diffusely and anomalously in the high-molecular-weight region of analytical sodium dodecyl sulfate polyacrylamide gel electrophoresis (SDS–PAGE) gels with an apparent molecular weight of 260,000 (Table 1). Current convention refers to this component as the α subunit. This large subunit is particularly sensitive to proteolysis in skeletal muscle, especially from rat, where it is rapidly broken down to a smaller component of ~150,000 daltons through the action of a Ca^{2+}-activated protease (12).

The 260,000 molecular weight α subunit is the only component present in the sodium channel purified from eel electroplax (31,33). In mammalian skeletal muscle and brain most workers find one or more additional smaller subunits in the 37–40 molecular-weight range, designated β subunits (Fig. 2) (11,19,27). In one study, however, the authors contend that no β subunits exist in the mammalian channel (7). The relationship between the α and β subunits remains controversial and are discussed in detail later in this chapter.

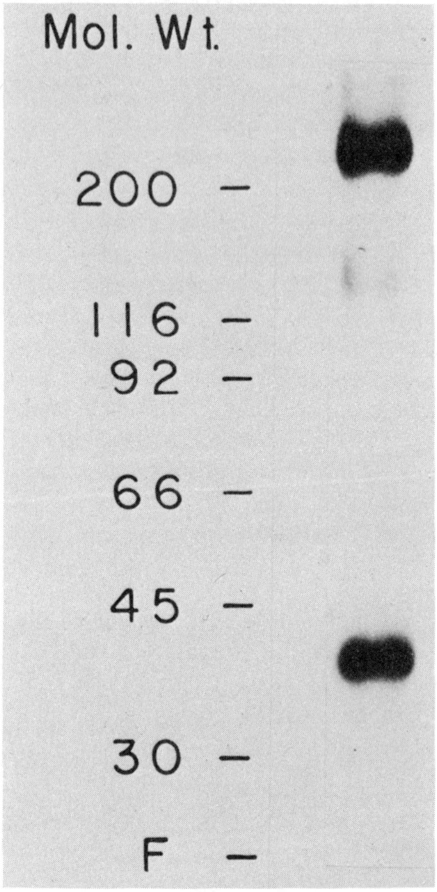

Figure 2. The voltage-sensitive sodium channel purified from rabbit T-tubule membranes contains a 260-kilodalton glycoprotein α subunit and a 38-kilodalton β subunit. The β subunit, also a glycoprotein, is sometimes resolved as two closely spaced bands of 37 and 39 kilodaltons in this preparation.

3. IS IT THE RIGHT PROTEIN?

The voltage-sensitive sodium channel has a number of biophysical and biochemical properties that uniquely define it among membrane ion channels. While STX binding is one such characteristic, the presence of toxin binding in itself does not provide evidence for channel function. In addition to having the appropriate toxin binding sites, the purified protein should retain the ability to select for sodium among monovalent cations, exhibit a characteristic maximal rate of ion flow or single-channel conductance, and be capable of characteristic voltage-dependent channel transitions between conducting and nonconducting states. These criteria for channel function have now been met for the purified protein from rat (41, 42) and rabbit (15,27) skeletal muscle as well as for the sodium channel isolated from rat brain (24,40) and eel electroplax (36,37).

For the mammalian skeletal muscle sodium channel, the purified protein was reconstituted into egg phosphatidyl choline vesicles using Biobeads SM-2 to remove the NP-40 detergent (42). In routine reconstitutions, both the rat and rabbit sodium channels inserted randomly during vesicle formation with about 50% of the channels facing outward.

Reconstitution of the purified channel protein into a lipid bilayer restores the temperature stability usually associated with the sodium channel in nerve or muscle membrane fractions (42). Whereas even brief incubation of the solubilized muscle channel at room temperature will result in the loss of high-affinity toxin binding and functional channel activity, the reconstituted channel can easily withstand incubation at 37°C for several hours without significant loss of activity.

3.1 Pharmacological Activation and Cation Selectivity

Using reconstitution into small phospholipid vesicles, we were able to demonstrate preservation of channel gating in response to pharmacological activation and block for sodium channels purified from rat sarcolemmal membranes (42). Subsequent work with both rat sarcolemmal and rabbit T-tubule membrane sodium channels has confirmed that these purified proteins retain the expected binding sites for STX and

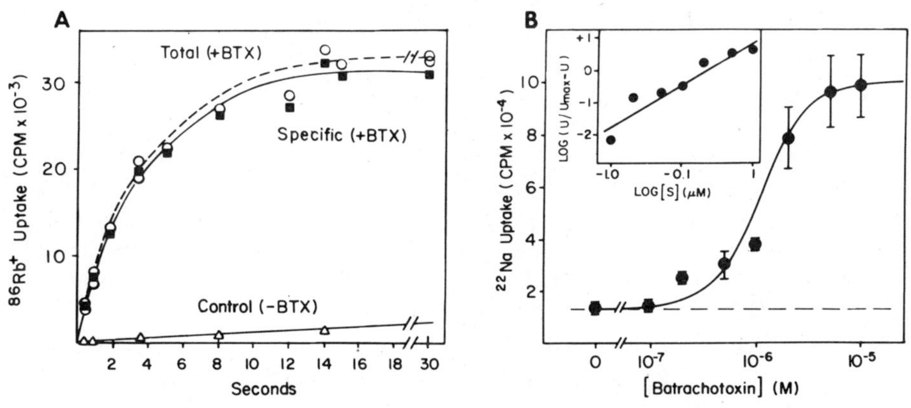

Figure 3. Skeletal muscle sodium channels (here from rabbit T-tubule membranes) gate cation flux in response to pharmacological activation after reconstitution into egg PC vesicles. (*A*) Influx of Rb^+ into vesicles containing the purified sodium channel. Vesicles were incubated at 36°C for 45 min with 5×10^{-6} *M* batrachotoxin (○) in ethanol (0.5% final ETOH concentration in the vesicle suspension) or with ethanol alone (△). Specific BTX-activated influx (■) occurred with a half time of ~3 sec in this example. Fluxes were measured after equilibration of vesicles at 22°C. (*B*) Dose-dependent activation of the purified, reconstituted sodium channel by batrachotoxin. Inset: Hill plot of the same data indicating an apparent K_d of 2 μ*M* and a Hill coefficient of 1.2.

TTX and are activated by batrachotoxin and veratridine at concentrations consistent with the effect of these toxins on the channel *in vivo* (Fig. 3) (27,42). We have not been able to confirm the presence of a functional polypeptide binding site in the purified mammalian skeletal muscle preparations, however.

Other investigators have shown similar results for rat brain and eel electroplax sodium channels after purification and reconstitution into liposomes (36,40). Taken as a whole, these pharmacological results demonstrate a profile for these reconstituted STX or TTX binding proteins that is comparable to the native sodium channel and confirm that the reconstituted proteins modulate membrane sodium permeability in response to alkaloid activation.

Cation selectivity can be demonstrated directly in the reconstituted skeletal muscle channel by use of a modification of a quenched-flow technique to obtain the necessary time resolution. With batrachotoxin to stimulate channel opening, the time course for the influx of various monovalent cations has been measured for both the rat and the rabbit sodium channel after their purification and reconstitution into phosphatidyl choline (PC) vesicles (27,41). This approach allowed clear resolution of K^+, Rb^+, and Cs^+ equilibration rates. The early phase of sodium uptake is still at the limits of the time resolution of the system, but calibration from the initial linear phase of potassium uptake allowed a maximum value to be determined for its equilibration half time (Fig. 4). With this value, selectivity ratios can be calculated for other cations. For the purified rabbit channel, these values were Na^+ (1.0):K^+ (0.13):Rb^+ (0.02):Cs^+ (0.008). Virtually identical results were obtained with the purified channel from rat sarcolemma. Although there are reports that batrachotoxin reduces the ionic selectivity of the sodium channel, the purified reconstituted sodium channel from mammalian skeletal muscle clearly retains much of its native selectivity.

3.2 Voltage-Dependent Toxin Activation

Recently we have extended our observations on the purified reconstituted sodium channel from rabbit T-tubular membrane to the single-channel level (15). Sodium channels were first reconstituted into egg PC vesicles, and these vesicles were then fused with planar bilayers for single-channel measurements. For these experiments, asymmetric solutions were used, with 500 m*M* NaCl on the cis side of the membrane and 200 m*M* NaCl on the trans side.

We found that the sodium channels generally incorporated into the bilayer with their extracellular TTX binding site facing the trans chamber. In the presence of batrachotoxin, voltage-dependent single-channel activity was observed after exposure of the bilayers to vesicles containing the purified sodium channel, but not with control vesicles containing no protein. The single-channel activity was steeply dependent on voltage; channel activity disappeared with large hyperpolarizing potentials but showed rapid voltage-dependent activation over a narrow range of depolarizing potentials (Fig. 5). The half-maximal opening probability was usually found at potentials between -95 and -115 mV. The single-channel conductance averaged 20 +/− 1 pS and was independent of membrane potential (15).

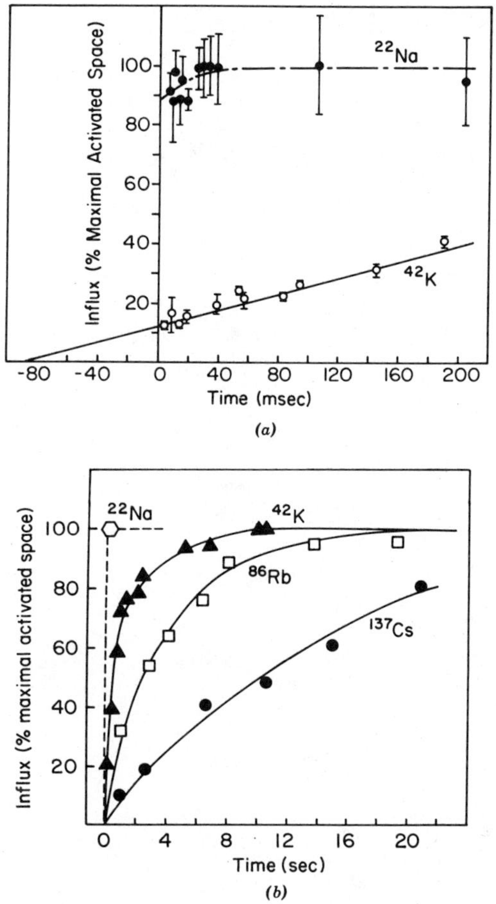

Figure 4. (*a*) Quenched-flow measurements of $^{22}Na^+$ and $^{42}K^+$ influx in batrachotoxin-stimulated vesicles at short incubation periods. Data points represent the specific batrachotoxin-stimulated uptake corrected for control uptake. Each point represents the mean ± SD (standard deviation) of three or more separate measurements after correction for control influx. The effective quenching time for the apparatus was determined by extrapolation of the initial linear uptake rate for $^{42}K^+$ into batrachotoxin-activated vesicles (○), yielding a value 90 msec. The $^{22}Na^+$ influx on this time scale (●) was very rapid, and an initial rate could not be resolved; an upper limit of 50 msec for the half time for $^{22}Na^+$ could be estimated from the measured quenching time of 90 msec for the system. (*b*) The time course of $^{42}K^+$, $^{86}Rb^+$, and $^{137}Cs^+$ uptake into batrachotoxin-activated vesicles as determined by use of the quenched flow apparatus. Specific uptake data from several preparations, corrected for control uptake and plotted as percent maximal activated space, are shown for $^{42}K^+$ (▲), $^{86}Rb^+$ (□), and $^{37}Cs^+$ (●). The maximal space accessible through activated sodium channels was the same for all cations in any given preparation. Maximal $^{22}Na^+$ uptake was already observed at the shortest interval shown on this graph (⬡). All measurements were made at 18°C (41).

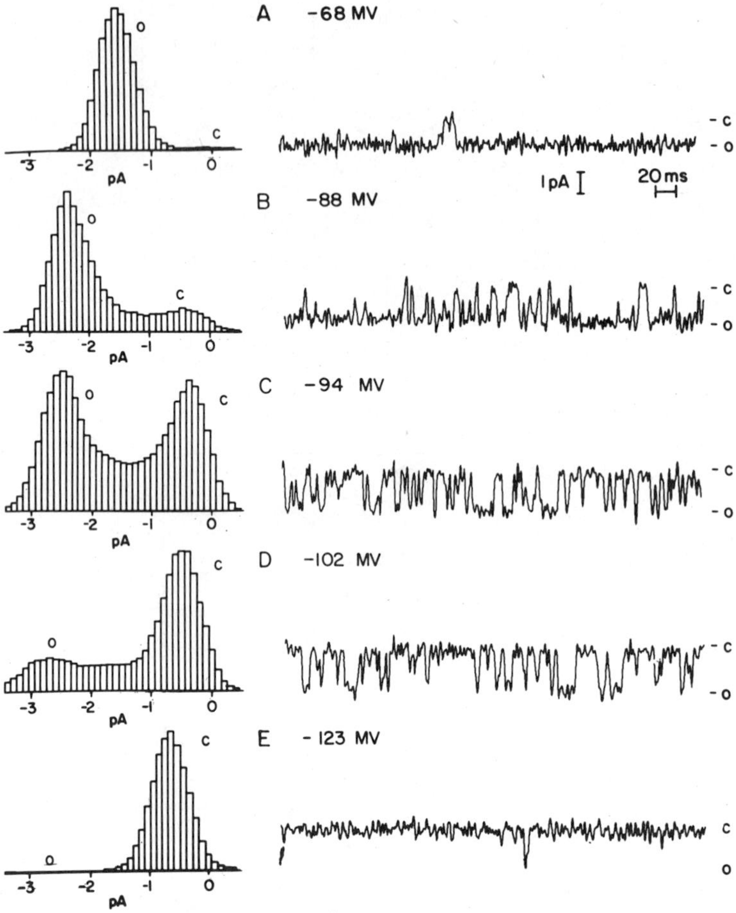

Figure 5. Voltage-dependent gating in a single purified sodium channel. Rabbit T-tubule sodium channels were purified, reconstituted into PC liposomes, and activated with batrachotoxin; activated liposomes were then fused with preformed planar bilayers and voltage clamped at the indicated potentials. The right panel shows representative transmembrane currents from a single trans-oriented sodium channel, and the left panel shows the corresponding current amplitude histograms. The area under each peak is proportional to the time the channel spends in the open (O) or closed (C) states. As the membrane potential is hyperpolarized, the single-channel currents progress from the mostly open state (*A*) through a series of rapid transitions between the open and closed states (*B–D*) to a mostly closed state (*E*). For this channel, the half-maximal opening probability occurred at about -95 mV (*C*). The solutions contained 500 m*M* NaCl on the cis side and 200 m*M* NaCl on the trans side. The current traces were filtered at 300 Hz and digitized at 2 kHz, while the histograms were digitized at 15 kHz. Temperature was 21°C.

Addition of TTX (10^{-6} *M*) to the cis chamber had no effect on either single-channel activity or the distribution between open and closed states when a potential of -77 mV was maintained across the membrane. Application of TTX to the trans chamber resulted in disappearance of single-channel activity within seconds, causing the channel to remain in the closed state more than 95% of the time.

Detailed single-channel measurements have also been reported with the purified rat brain sodium channel reconstituted in a similar planar bilayer system (24), and these results are reviewed in the following Chapter 9. In general, the results are comparable, although the single-channel conductance of the skeletal muscle channel consistently appears slightly smaller than that of the rat brain. Voltage-activated single-channel openings in the absence of batrachotoxin have also been shown for the channel from eel electroplax and by other investigators for the rat brain channel (19,37).

3.3 Ensemble Measurements of Voltage-Dependent Activation

Single-channel analysis of purified sodium channels leaves unresolved the question of whether those few channels observed in the bilayer are truly representative of all the channel protein purified. Since a typical reconstitution of sodium channels from skeletal muscle might reflect 10^{14} molecules and a single flux measurement the activity of 10^{11} or 10^{12} molecules per assay point, results obtained from ensemble measurements of reconstituted channels in vesicles can usually be considered representative of the population of isolated channels as a whole. Correlations drawn between observations on two or three single-ion channels in planar bilayers and the broad properties of the isolated proteins seen on gels, on the other hand, may be more tenuous. In the latter case, trace components that might be undetectable with standard biochemical methods could easily provide the major contribution to the single-channel events observed. We have addressed this issue by measuring the voltage-dependent activation of large populations of purified channels in the reconstituted vesicle system.

Batrachotoxin-activated sodium channels retain their voltage-dependent activation, although the relationship between activation and voltage is shifted about 40 mV in the hyperpolarizing direction (26). Batrachotoxin also slows the channel inactivation process so that channels remain open during prolonged depolarization instead of closing after only a brief open interval. With batrachotoxin, voltage-dependent channel gating should be detectable in vesicles containing the purified reconstituted sodium channel if the vesicle membrane potential can be controlled.

A functionally oriented vesicle population was generated by reconstituting rabbit T-tubule sodium channels into egg PC vesicles in the presence of STX and subsequently removing external STX from the solution by gel filtration (15). With this approach, all inward-facing channels were blocked by the high concentration of intravesicular STX while channels oriented with their binding sites for this polar toxin facing outward were available for normal activation. To establish a transmembrane potential, a potassium concentration gradient was created and vesicles rendered permselective to potassium by the addition of valinomycin. The voltage dependence of these batrachotoxin-activated channels was then determined by measuring sodium uptake at various transmembrane potentials set by altering the external potassium.

With vesicles containing the purified sodium channel, we found that the specific $^{22}Na^+$ influx was maximal at low membrane potentials ($[K]_i = [K]_o$) and completely inhibited at large hyperpolarizing potentials ($[K]_o = O$) (Fig. 6) (15). At intermediate external potassium concentrations the influx increased sharply. The influx stimulated by batrachotoxin in the depolarized vesicles was blocked by the application of external STX. Most importantly, all the influx stimulated by batrachotoxin in depolarized vesicles was blocked by hyperpolarization, indicating that the entire population of reconstituted channels that retained the capacity for STX block and batrachotoxin activation also exhibited the expected voltage dependence for this activation process.

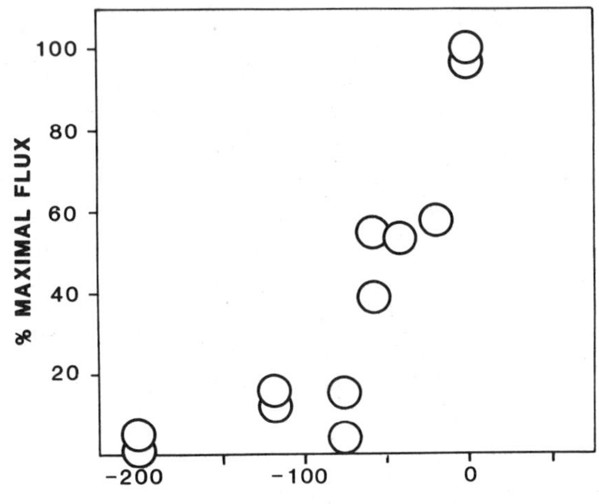

Figure 6. Voltage-activated $^{22}Na^+$ influx into vesicles containing the purified sodium channel. Saxitoxin was trapped within vesicles during the reconstitution to block inward-facing channels, and membrane potential was controlled by imposing a transmembrane K^+ concentration gradient in the presence of valinomycin. Vesicles were diluted 1:2 with buffer containing choline chloride and the desired KCl at constant osmolarity and valinomycin $(10^{-7}\ M)$. Batrachotoxin $(5 \times 10^{-6}\ M$ final concentration) was added from an ethanol stock; comparable amounts of ethanol alone were added to control samples, and all were incubated for 45 min at 37°C. Following incubation, 5-sec ^{22}Na influx measurements were made in quadruplicate on both control and batrachotoxin-stimulated vesicles at each potential. Each point shown here represents the difference of the average of four stimulated and four control measurements. The influx is expressed as a percent of maximal activated space at zero membrane potential in order to combine the results from two reconstitutions.

4. CHARACTERIZATION OF THE SODIUM CHANNEL PROTEIN

At this point most skeptics would agree that the toxin-binding protein purified from mammalian skeletal muscle, rat brain, eel electroplax, and other sources is in fact the voltage-dependent sodium channel. What further can be said concerning the characteristics of the protein itself?

4.1 Relationship Between Large and Small Subunits in Skeletal Muscle

A major unresolved issue is the relationship between the 260-kilodalton α subunit and the one or more small β subunits seen in the purified preparations of mammalian sodium channel, including those from rat skeletal muscle and brain and rabbit T-tubular membrane. In the rat brain channel, Catterall and his co-workers find two smaller subunits, designated β_1 and β_2 (22). These small subunits can be differentiated on the basis of the presence of a disulfide bond covalently linking α and β_2, while β_1 can be dissociated from the complex under denaturing but nonreducing conditions. The β_1 subunit also appears to be removed from the channel complex by

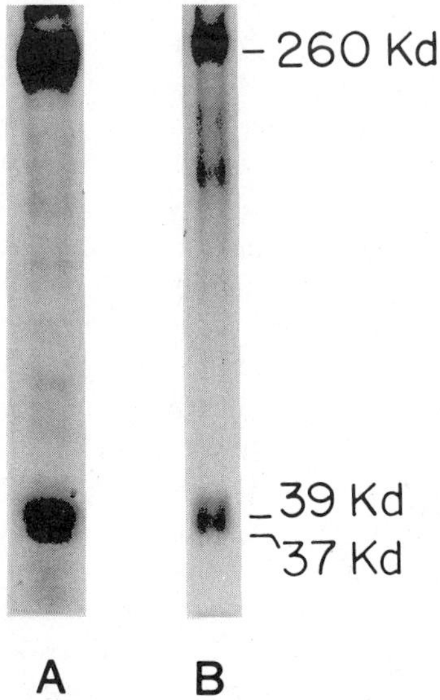

Figure 7. Purified sodium channel from rat skeletal muscle sarcolemma (*A*) and rabbit skeletal muscle T-tubule membranes (*B*). The rat channel was purified by immunoaffinity chromatography on a column of immobilized monoclonal antibodies, while the rabbit channel was isolated with traditional biochemical techniques. The subunit composition appears identical, with a 260-kilodalton α subunit and two closely spaced bands in the β region of $\sim$37 and 39 kilodaltons.

simple extraction under high salt conditions (13). In our experience, a subunit comparable in size to these β components is always present in purified preparations of sodium channel from either rat or rabbit skeletal muscle. The appearance of the β subunit, however, varies; in some cases, a single diffuse band is seen centered at about 38,000 daltons, while in other cases this band can be resolved into two separate components of approximately 37,000 and 39,000 daltons (Fig. 7). Both components are reduced in intensity when gels are run under denaturing but nonreducing conditions.

The nature of these small components is open to interpretation. They could represent true subunits of the sodium channel that are present in mammalian tissue but not in eel electroplax, for example. Alternatively, they could be tightly associated components of the cytoskeleton, shared regulatory subunits, or even proteolytic fragments generated as an artifact during channel purification.

We have used monoclonal antibodies raised against the purified rat sodium channel large subunit to address the relationship of the 38-kilodalton component to the 260-kilodalton component. These antibodies are immunospecific for the rat skeletal muscle sodium channel (12). They are capable of immunoprecipitating specific [^{3}H]-STX binding activity from crude solubilized skeletal muscle or sarcolemma even after covalent attachment to Sepharose. Each of the antibodies reacts with the purified sodium channel in solid-phase radioimmunoassay and is able to identify the solubilized channel among a mixture of crude sarcolemma membrane proteins following molecular-sieve chromatography. The three monoclonal antibodies used for this study all recognize the 260-kilodalton α subunit of the sodium channel on Western blots prepared from fresh skeletal muscle homogenized directly in NP-40 in the presence of protease inhibitors. None of these antibodies showed reactivity to components in the 38,000–40,000-molecular-weight range.

Our general approach has been to use these antibodies to purify their antigen directly from fresh skeletal muscle under nondenaturing conditions and without the need for intervening isolation of membranes or other prolonged biochemical procedures that would raise the risk of channel proteolysis due to endogenous protease activity. Fresh skeletal muscle was homogenized in NP-40 in the presence of a panel of protease inhibitors and a crude glycoprotein fraction containing the channel immediately prepared by lectin affinity chromatography. We then used three monoclonal antibodies covalently coupled to Sepharose as an immunoaffinity resin to isolate their common antigen from this fraction. To further assist in distinguishing specific channel components from nonspecifically retained contaminants, the antibody columns were eluted with linear gradients of thiocynate. Proteins eluted from the column were identified by ^{125}I labeling after denaturation in SDS (11).

An elution profile for such a column is shown in Fig. 8*a*. The 260-kilodalton protein appeared as a symmetrical peak in the elution profile between 0.5 and 1.5 molar thiocynate. In addition, two low-molecular-weight components of ~37 and 39 kilodaltons codistributed with the large glycoprotein in the elution profile. Several other minor contaminants either were seen in both control and specific antibody

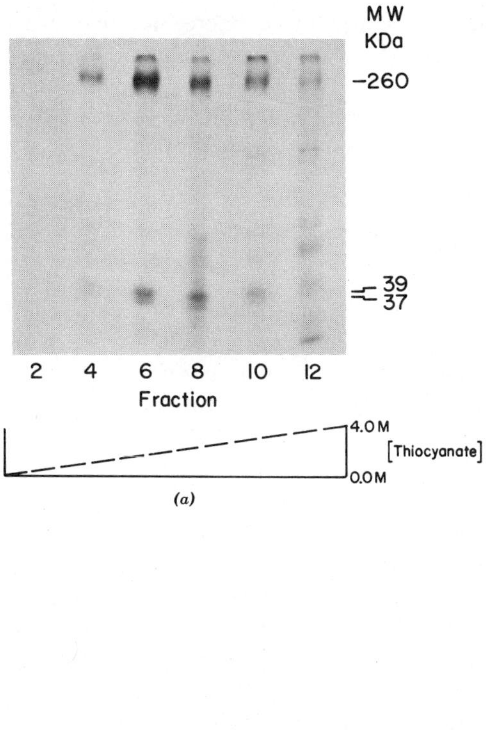

Figure 8. The β subunit(s) copurify with the 260-kilodalton α subunit on an immunoaffinity column in which monoclonal antibodies that recognize only the α subunit are covalently attached to the support matrix. (*a*) When such a column is eluted with a linear thiocyanate gradient, the β subunit(s) coelute with the α peak while the other minor contaminants are clearly separated. (*b*) In four separate immunoaffinity purifications, the same subunit composition was obtained.

column profiles or peaked elsewhere in the elution profile, indicating that they were not a part of the principal antigenic complex.

We obtained the same results in four comparable gradient immunoaffinity purification experiments (Fig. 8*b*). On some gels, a single diffuse band at 38 kilodaltons was seen and two separate bands in this region were not resolved, but in each case one or two components in the 37–40-kilodalton range coeluted with the 260-kilodalton protein.

On Western blots, the monoclonal antibodies used for these experiments recognize epitopes only on the large α subunit. The material in the β region coeluting with the α subunit from the immunoaffinity column must be retained through rigorous washing of these columns because of their association with the large 260-kilodalton subunit. The constant association of one or more components in the 37–40 kilodalton region with the 260-kilodalton protein in the material isolated from fresh muscle with our monoclonal antibodies strongly supports the contention that these small components are true subunits of the sodium channel. Furthermore, the ratio of the 260- to the 37–40-kilodalton components remained constant across the peak in each experiment, again supporting the contention that the α and β subunits are present in constant proportion in the native sodium channel complex. Within a given experiment, this ratio remained constant independent of whether one or two bands were distinguished in the β region.

4.2 Analysis of α and β Subunits

Using the rabbit T-tubular membrane as a protein source, preparative separation of the α and β subunits can be carried out under denaturing conditions. We have analyzed the isolated subunits for amino acid composition, total carbohydrate content and carbohydrate composition, and core protein size following enzymatic deglycosylation. Furthermore, the availability of these isolated subunits has allowed us to calibrate our gel systems and address questions of subunit stoichiometry in the rat and rabbit skeletal muscle system.

The 260,000- and 38,000-molecular-weight subunits of the rabbit sodium channel were separated by preparative chromatography in 0.1% SDS and 2 m*M* dithiothreitol on a 0.9 × 90-cm column of Sepharose Cl-6B after first reducing and denaturing the isolated channel in 2% SDS, 10% β mercaptoethanol. We obtained identical results when protein sulfydryl groups were reduced and alkylated prior to chromatography. With either approach, clear separation of subunits was obtained and pooled peaks that contain α and β subunits could be rechromatographed reproducibly.

Total amino acid analysis of the isolated subunits contains no surprises. The amino acid content does not deviate significantly from that reported for many other soluble and membrane proteins and in the case of the α subunit is comparable to that previously reported for the eel sodium channel.

4.3 Carbohydrate Analysis and Enzymatic Deglycosylation

We have carried out total carbohydrate analysis on both the α and β subunits of the rabbit T-tubule channel using high-resolution, gas–liquid chromatographic techniques (35), and the results of these studies are summarized in Table 3. The α subunit of the rabbit T-tubule channel is approximately 26% by weight carbohydrate. Of this carbohydrate, the highest percentage represents sialic acid and *n*-acetylhexosamines, with lesser amounts contributed by mannose and galactose. The β subunit is also heavily glycosylated, containing approximately 29% carbohydrate by weight in the rabbit T-tubule channel. The dominant sugars are again *n*-acetylhexosamines and sialic acid, and the relative contribution of sugars is similar to that obtained for the large subunit.

For comparison, Grishin et al. (16) reported that the rat brain α subunit is 20.5% carbohydrate while the β_1 and β_2 subunits are 28.9 and 31.2% carbohydrate by weight, respectively. The 260-kilodalton protein from eel electroplax is reported to be 28.5% carbohydrate by weight (31). In each case the carbohydrate composition is comparable to that detailed in Table 3 for the rabbit subunits.

TABLE 3. RABBIT SODIUM CHANNEL—CARBOHYDRATE ANALYSIS

Compound	Weight % Total Sugars	Weight % Total Glycoprotein
	260-kilodalton Subunit [a]	
N-Acetylneuraminic acid	44.1	11.7
N-Acetylhexosamines	21.4	5.7
Galactose	17.6	4.6
Mannose	15.3	4.1
Fucose	1.5	0.4
	100%	26.5%
	38-kilodalton Subunit	
N-Aceylneuraminic acid	33.5	9.8
N-Acetylhexosamines	61.7	18.1
Galactose	2.9	0.9
Mannose	2.0	0.6
Fucose	Trace	Trace
	100%	32.1%

[a]260-kilodalton subunit = 26.5% carbohydrate, 73.5% protein.
[b]38-kilodalton subunit = 29.4% carbohydrate, 70.6% protein.

Treatment of the isolated rabbit β subunit with neuraminidase reduces its apparent migration on SDS–PAGE from 38 kilodaltons to approximately 35 kilodaltons but still results in a diffuse appearance to the band. Subsequent treatment with endoglycosidase F, however, produces a sharp band at 26,500 daltons, which remains unchanged with further prolonged incubation with this enzyme (Fig. 9). Pretreatment with neuraminidase is not required for effective cleavage of the core sugar residues with endo-F in our system. Assuming a core protein molecular weight of 26.5 kilodaltons and a carbohydrate content of 29%, the molecular weight of the total glycoprotein should be approximately 37.3 kilodaltons, in reasonable accord with the apparent molecular weight of the unmodified subunit on SDS–PAGE.

4.4 Stoichiometry of the Sodium Channel Subunits

While the β component is present in all our immunoaffinity purified preparations, the resolution of the band into a separate 37- and 39-kilodalton component is not a constant finding. We find comparable behavior for the small subunit of the rabbit T-tubular sodium channel purified by traditional means. A major consideration, therefore, becomes the true stoichiometry between the 260,000-molecular-weight subunit and the β components in the native channel from skeletal muscle.

Although the large:small subunit ratio appears constant, the subunit stoichiometry

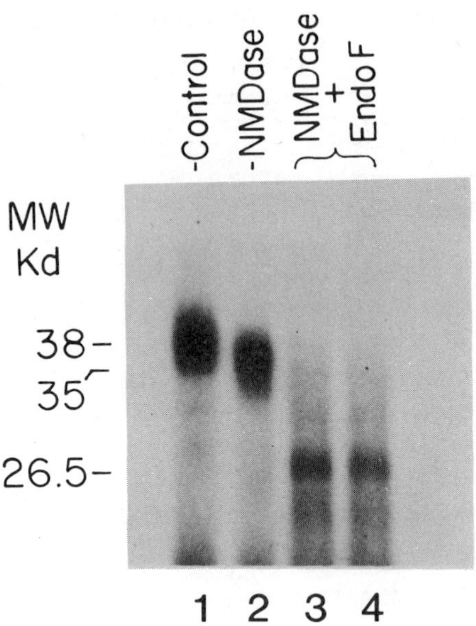

Figure 9. The isolated β subunit from the rabbit T-tubular channel can be enzymatically deglycosylated to a core protein of 26,500 molecular weight. Lane 1: isolated β subunit prior to enzymatic treatment exhibits an apparent molecular weight of 38,000. Lane 2: treatment with neuramindase increases the mobility of this band slightly. Lanes 3 and 4: subsequent treatment with endoglycosidase F produces a single sharp band at 26,500.

cannot be inferred directly from the labeling intensity of these bands since the relative efficiency of labeling for the two proteins is not the same. To calibrate the relative labeling efficiency, isolated 260- and 38-kilodalton subunits of the rabbit T-tubule sodium channel were standardized for total protein by complete amino acid analysis and known amounts subsequently mixed, labeled, electrophoresed, and quantitated in a manner identical to that used for the column fractions in the immunoaffinity purification (35). Calibration curves developed in this way indicated a radioactive labeling efficiency of 0.23 on an absolute protein weight-to-weight basis for the 260-kilodalton component as compared to the 38-kilodalton rabbit subunit.

The procedure of Shing and Ruoho (39) that we used for identifying proteins on SDS–PAGE will label both the *N*-terminal amino group and epsilon amino group of lysine residues, although the pH chosen should favor the former reaction. Since no reaction would be expected with the carbohydrate residues we have assumed that the relative molar labeling efficiency for the subunits can be estimated from values for the molecular weight of the core deglycosylated protein in each case. We have determined a value of 26,500 for the core protein of the rabbit β subunit (35). A value of 208,300 was calculated for the core protein of the eel channel 260-kilodalton glycoprotein based on its sequence as determined from the cloned message, while

Messner and Catterall (30) report an estimate of 220,000 for the core protein of rat brain α based on SDS–PAGE.

Using our value for the rabbit beta core protein and the upper and lower estimates of the alpha core in eel and rat brain, a molar labeling ratio between 1.8 and 1.9 can be calculated for α/β from our calibration data for the muscle system. Thus if one α and one β subunit are present in equimolar amounts in the rat sodium channel, the predicted labeling ratio of the SDS-PAGE autoradiogram should fall between 1.8 and 1.9. If a second β subunit of approximately the same molecular weight is assumed to be present, such that the channel stoichiometry is 1:2 α:β, an intensity ratio of the high- and low-molecular weight regions of 0.95 would be predicted.

We have quantitated the α:β intensity ratio in a total of 15 gel fractions taken from elution profiles in three separate immunoaffinity purifications of the rat sarcolemmal sodium channel (Table 4). The average measured ratio of 260:37–40-kilodalton intensity was 1.96 ± 0.45. Thus our data support a stoichiometry of one 260-kilodalton α subunit to only one subunit in the 37–40 kilodalton range for the rat sarcolemmal channel. The data for rat skeletal muscle are not consistent with a molecular model in which each 260-kilodalton subunit is associated with both a 37- and 38-kilodalton subunit as appears to be the case in the rat brain.

We have subsequently extended these observations to include preparations of purified rabbit T-tubule sodium channel isolated with traditional biochemical techniques. In three separate preparations the relative intensity of the 260- to the 37–40-kilodalton regions in the purified material was scanned. An average value of 1.70 ± 0.12 was obtained for these preparations, again consistent with a theoretical

TABLE 4. SKELETAL MUSCLE SODIUM CHANNEL SUBUNIT STOICHIOMETRY

Preparation	Fractions Analyzed	Measured 260/37–40	Theoretical 1:1	Theoretical 1:2
		Immunoaffinity-Purified Rat Channel (Gradient Elution)		
A	*5*	*1.86 ± 0.65*	*1.8–1.9*	*~0.95*
B	*4*	*2.19 ± 0.19*		
C	*6*	*1.89 ± 0.39*		
Average	(*N* = 15)	1.96 ± 0.45	1.8–1.9	~0.95
		Traditionally Purified Rabbit Channel		
A	1	1.61	1.8–1.9	~0.95
B	1	1.83		
C	1	1.66		
Average	(*N* = 3)	1.70 ± 0.12	1.8–1.9	~0.95

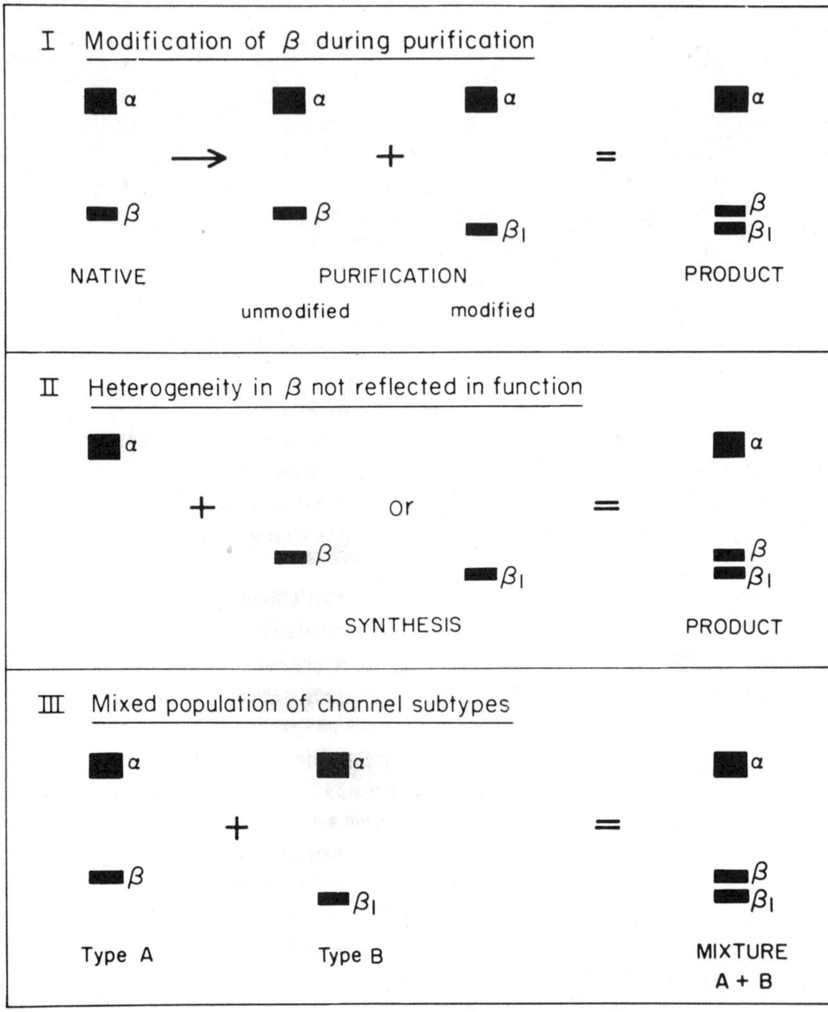

Figure 10. The appearance of two β bands on SDS-PAGE of purified channels in spite of an apparent stoichiometry of the α:β subunits of 1:1 could arise from a variety of causes (I) A single β subunit could be partially modified during channel purification by an endogenous protease or glycosidase to yield a mixture of two isoforms in the final product. (II) Native sodium channels could be constructed with β subunits that exhibit heterogeneity in glycosylation, giving rise to the appearance of several β bands in the pure channel protein. (III) Sodium channel subtypes might be distinguished by different β subunits. Purification of a mixture of subtypes could give rise to multiple β bands, although the α:β stoichiometry remains 1:1 in each channel molecule.

ratio of one α to one β subunit in the rabbit muscle channel rather than a stoichiometry of 1:2.

How can an apparent 1:1 α:β stoichiometry be reconciled with the variable appearance of two bands in the β region of the SDS–PAGE profiles of the purified muscle sodium channel? Several interpretations are possible (Fig. 10). One possibility that cannot be excluded is that the small subunit(s) have been partially dissociated from the α during channel purification, reducing the molar contribution of the subunit(s) to the final gels. It is also possible that a single class of subunits present in the native channel is susceptible to chemical modification during channel purification that leads to the variable presence of two closely spaced bands in this region. The action of an endogenous glycosidase, for example, could produce such an effect. A third possibility is that heterogeneity normally exists in this small subunit such that large subunits are associated with one or the other form of the small subunit in their native state to form channels that are functionally indistinguishable. Again the appearance of the purified material on SDS–PAGE would actually reflect a mixture of two channel forms with slightly differing β subunits.

Alternatively, the two bands in the β region may each be associated with a different subtype of sodium channel. For example, there is abundant evidence for TTX-sensitive and TTX-insensitive subtypes of sodium channels in denervated skeletal muscle and similar evidence of subtypes within developing and mature normal muscle. These subtypes may have 260-kilodalton α subunits that are indistinguishable to our antibodies and identical on SDS–PAGE but small subunits that differ slightly in their chemical composition. In this case, the appearance on SDS–PAGE of three subunits would be the result of the copurification of two very closely related channel populations, each having one large and only one small subunit.

5. SODIUM CHANNEL SUBTYPES

Small β subunits with apparent molecular weights in the 37–40 kilodalton range appear by a variety of criteria to be integral components of the sodium channels from mammalian brain and skeletal muscle. However, the stoichiometry of the small subunits may not be the same even in these two related sources. The recent successful reconstitution of the eel sodium channel that may contain only the 260-kilodalton protein further underscores the fact that the small subunits may not be required for channel function and that the structural basis for the ion channel and voltage sensing mechanisms, as well as the binding sites for many of the major neurotoxins, is contained within the α subunit. This does not necessarily decrease the importance of the β subunits, however. Issues such as long-term regulation of channel function, definition of functional channel subtypes, and differential localization of the channel within the plane of the membrane remain to be addressed. It may be in one of these contexts that the role of the small channel components is found.

The concept of sodium channel subtypes is not a new one. It has been recognized for several decades that denervation of skeletal muscle results in the appearance of sodium channels that are resistant to block by TTX and STX and are slightly different in kinetic properties from the normal skeletal muscle channel (20). More recently, the appearance of sodium channel subtypes has been described in developing skeletal muscle (9,18,38); here, the earliest stages of myoblast are characterized by the presence of TTX/STX-resistant sodium channels, while the myotubes that form later in development contain functional channels that are TTX sensitive. There is evidence that other, nonfunctional forms of the channel may also exist during development (14).

Recent work with polyclonal and monoclonal antibodies in skeletal muscle indicates that subtypes of the sodium channel may be more ubiquitous than previously suspected even in adult innervated muscle (17). In our present panel of monoclonal antibodies, for example, there are antibodies that recognize sodium channels in the sarcolemma but not the T-tubule system while others show specificity for channels located in the T system itself. Still other antibodies distinguish between channels in fast and slow fibers within the same muscle.

In binding studies with solubilized skeletal muscle channels, some antibodies show clear competition for common or overlapping sites, while others show no interaction, as expected for multiple epitopes on the same channel (10). A third group of antibodies, however, show partial interaction with other sites; the percent of maximum binding interaction is independent of the concentration of either antibody. The nature of these interactions again suggest the presence of subtypes of channels, some of which contain the binding site for both antibodies while others contain a binding site for only one.

The structural differences between these channel subtypes remain to be elucidated. While it is possible that subtypes may represent α subunits encoded by different although related genes or differing posttranslational modifications of a common α core protein, it is also possible that these various channel types are distinguished by the presence of differing β subunits. A major challenge for future work in this field is to determine the functional role of the 37–40-kilodalton peptides in relation to the obvious major role of the 260-kilodalton protein in channel function.

6. SUMMARY

Sodium channel proteins have now been isolated from a number of nerve and muscle preparations. All are characterized by the presence of a very large glycoprotein subunit of ~260 kilodaltons which may contain the structural features required for voltage-dependent channel gating and cation selectivity. These purified proteins have been reconstituted into vesicle systems and planar bilayers and demonstrate the ensemble and single-channel behavior characteristic of the native sodium channel.

Although the sodium channel from eel appears to consist of only the 260-kilodalton protein, the channels from rat brain and rat or rabbit skeletal muscle contain one or more smaller subunits of 37–39 kilodaltons. In skeletal muscle, a 38-kilodalton subunit is present in both conventionally purified channel and channel isolated with immunoaffinity techniques. The stoichiometry of the large (α) and the small (β) subunits appears to be 1:1 in skeletal muscle but 1:2 in rat brain. The role of the small subunits in the normal functioning of the sodium channel remains to be defined.

ACKNOWLEDGMENTS

Work in the authors' laboratory was supported by NIH grants NS-18013 and NS-08075 and a grant from the Muscular Dystrophy Association. The assistance of Nancy Goodman in the preparation of the manuscript is gratefully acknowledged.

REFERENCES

1. Agnew, W. S., Levinson, S. R., Brabson, J. S., and Raftery, M. A. (1978). *Proc. Natl. Acad. Sci.* (USA), **75**, 2606–2610.
2. Agnew, W. S. and Raftery, M. A. (1979). *Biochemistry*, **18**, 1912–1919.
3. Angelides, K. J., Nutter, T. J., Elmer, L. W., and Kempner, E. S. (1985). *J. Biol. Chem.*, **260**, 3431–3439.
4. Barchi, R. L. (1984). *Trends Biochem. Sci.*, **9**, 358–361.
5. Barchi, R. L., Cohen, S. A., and Murphy, L. E. (1980). *Proc. Natl. Acad. Sci.* (USA), **77**, 1306–1310.
6. Barchi, R. L. and Murphy, L. E. (1981). *Neurochem.*, **36**, 2097–2100.
7. Barhanin, R., Pauron, D., Lombet, A., Norman, R., Vijverberg, J., Giglio, J., and Lazdunski, M. (1983). *EMBO J.* **2**, 915–920.
8. Barhanin, J., Schmid, A., Lombet, A., Wheeler, K. P., and Lazdunski, M. (1983). *J. Biol. Chem.*, **258**, 700–702.
9. Baumgold, J., Parent, J. B., and Spector, I. (1983). *J. Neurosci.*, **3**. 995–1003.
10. Casadei, J. and Barchi, R. (in press) *J. Neurochem.*
11. Casadei, J., Gordon, R., and Barchi, R. (1986). *J. Biol. Chem.*, **261**, 4318–4323.
12. Casadei, J., Gordon, R., Lampson, L., Schotland, D., and Barchi, R. L.(1984). *Proc. Natl. Acad. Sci.* (USA), **81**, 6227–6231.
13. Cohen, S. A. and Barchi, R. L. (1981). *Biochem. Biophys. Acta*, **645**, 253–261.
14. Frelin, C., Vijverberg, H. P. M., Romey, G., Vigne, P., and Lazdunski, M. (1984). *Pflügers Arch.*, **402**, 121–128.
15. Furman, R. E., Tanaka, J. C., Mueller, P., and Barchi, R. L. (1986). *Proc. Natl. Acad. Sci.* (USA), **83**, 488–492.
16. Grishin, E. V., Kovalenko, V. A., Pashkov, V. N., and Shamotienko, O. G. (1984). *Membrane Biophys.* (USSR), **1**, 858–867.
17. Haimovich, B., Bonilla, E., Casadei, J., and Barchi, R. (1984). *J. Neurosci.*, **4**, 2259–2268.
18. Haimovich, B., Tanaka, J. C., and Barchi, R. L. (1986). *J. Neurochem*, **47**, 1148–1153.
19. Hanke, W., Boheim, G., Barhanin, J., Pauron, D., and Lazdunski, M. (1984). *EMBO J.*, **3**, 509–515.

20. Harris, J. B. and Thesleff, S. (1971). *Acta Physiol. Scand.*, **83**, 382–388.
21. Hartshorne, R. P. and Catterall, W. A. (1981). *Proc. Natl. Acad. Sci.* (USA) **78**, 4620–4624.
22. Hartshorne, R. P. and Catterall, W. A. (1984). *J. Biol. Chem.*, **259**, 1667–1675.
23. Hartshorne, R. P., Coppersmith, J., and Catterall, W. A. (1980). *J. Biol. Chem.*, **255**, 10572–10575.
24. Hartshorne, R., Keller, B., Talvenheimo, J., Catterall, W., and Montal, M. (1985). *Proc. Natl. Acad. Sci.* (USA), **82**, 240–244.
25. Hodgkin, A. L. and Huxley, A. F. (1952). *J. Physiol.* (London), **117**, 500–544.
26. Huang, L. M., Moran, M., and Ehrenstein, G. (1982). *Proc. Natl. Acad. Sci.* (USA), **79**, 2082–2085.
27. Kraner, S. D., Tanaka, J. C., and Barchi, R. L. (1985). *J. Biol. Chem.*, **260**, 6341–6347.
28. Levinson, S. R. and Ellory, J. (1973). *Nature New Bio.*, **245**, 122–123.
29. Lombet, A. and Lazdunski, M. (1984). *Eur. J. Biochem.*, **141**, 651–660.
30. Messner, D. J. and Catterall, W. A. (1985). *J. Biol. Chem.*, **260**, 10597–10604.
31. Miller, J. A., Agnew, W. S., and Levinson, S. R. (1983). *Biochemistry*, **22**, 462–470.
32. Noda, M., Shimizu, S., Tanabe, T., Takai, T., Kayano, T., Ideda, T., Takahashi, H., Nakayama, H., Kanaoka, Y., Minamino, N., Kangawa, K., Matsuo, H., Raftery, M. A., Hirose, T., Inayama, S., Hayashida, H., Miyata, T., and Numa, S. (1984). *Nature*, **312**, 121–127.
33. Norman, R. I., Schmid, A., Lombet, A., Barhanin, J., and Lazdunski, M. (1983). *Proc. Natl. Acad. Sci.* (USA), **80**, 4164–4168.
34. Richie, J. M. and Rogart, R. (1977). *Rev. Physiol. Biochem. Pharmacol.*, **79**, 2–50.
35. Roberts, R. and Barchi, R. (in press). *J. Biol. Chem.*
36. Rosenberg, R. L., Tomiko, S. A., and Agnew, W. S. (1984). *Proc. Natl. Acad. Sci.* (USA), **81**, 1239–1243.
37. Rosenberg, R., Tomiko, S., and Agnew, W. (1984). *Proc. Natl. Acad. Sci.* (USA), **81**, 5594–5598.
38. Sherman, S. J., Lawrence, J. C., Messner, D. J., Jacoby, K., and Catterall, W. A. (1983). *J. Biol. Chem.*, **258**, 2488–2495.
39. Shing, Y. and Ruoho, A. (1983). *Anal. Biochem.*, **110**, 171–174.
40. Tamkun, M., Talvenheimo, J., and Catterall, W. (1984). *J. Biol. Chem.*, **259**, 1679–1688.
41. Tanaka, J., Eccleston, J., and Barchi, R. (1983). *J. Biol. Chem.*, **258**, 7519–7526.
42. Weigele, J. B. and Barchi, R. L. (1982). *Proc. Natl. Acad. Sci.* (USA), **79**, 578–589.

CHAPTER 9

STUDIES ON SODIUM CHANNELS RECONSTITUTED IN LIPID BILAYERS: INFERENCES ABOUT MOLECULAR MECHANISMS DERIVED FROM SINGLE-CHANNEL RECORDINGS

M. MONTAL
R. HARTSHORNE
B. KELLER
Departments of Biology and Physics
University of California at San Diego
La Jolla, California

Dr. Montal's present address: Department of Neurosciences, Roche Institute of Molecular Biology, Nutley, New Jersey.
Dr. Hartshorne's present address: Department of Pharmacology, Oregon Health Sciences University, Portland, Oregon.
Dr. Keller's present address: Max Planck Institüt für Biophysikalische Chemie, Göttingen, West Germany.

Two characteristics fundamental in function of the nervous system are the excitability of its elements and the communication of excitability among its elements. These key features are based on the activity of ion channels, that is, membrane channel proteins. Our aim is to understand how a channel protein works. Our approach to this problem is based on the premise that understanding the mechanism of action of channel proteins depends on the convergence of knowledge about the structure of the protein with a detailed characterization of its function at the molecular level. As a first step in this endeavor, the channel protein is purified from the native membrane and thereafter is reconstituted back into an experimental membrane that readily allows the characterization of its function at the level of single channels. Thus the study of the molecular physiology of electrical excitability and signal transduction is reduced to investigations on the molecular properties of purified sodium channels or acetylcholine receptors (AChRs) reconstituted in planar lipid bilayers. The nicotinic AChR from *Torpedo californica* and the voltage-sensitive sodium channel from *Electrophorus electricus* and rat brain provide an opportunity to pursue this goal because more is known of these proteins than about any other membrane channel protein of the nervous system. In this chapter, the discussion is restricted to the voltage-sensitive sodium channel from rat brain since our investigations on the AChR have been extensively reviewed elsewhere (6,54,55,63,64). Nevertheless, the basic concepts and techniques outlined for the sodium channel are generally applicable to the AChR and may become of widespread use in the reconstitution of other purified membrane channel proteins into planar lipid bilayers (for review, see ref. 63).

1. FROM THE NERVE ACTION POTENTIAL TO THE PURIFIED SODIUM CHANNEL PROTEIN

The voltage-sensitive sodium channel is an integral membrane protein that spans the hydrophobic core of the membrane lipid bilayer and provides the permeability pathway for the inward sodium current to flow across the neuronal membrane during the rising phase of an action potential (42). Neurotoxins that interact specifically with the channel protein have proved to be extremely useful as molecular tools for identifying the channel electrophysiologically, serving as markers during the purification of the channel protein, and defining specific functional–pharmacological assays in reconstitution studies. Several distinct neurotoxin binding sites on the sodium channel have been characterized (for reviews, see refs. 13,15, and 56). Site 1 binds tetrodotoxin (TTX) and saxitoxin (STX) and consequently blocks sodium flux. Both STX and TTX are water-soluble, heterocyclic guanidines that bind to the extracellular surface of the channel protein (69) and, therefore, are valuable tools for determining the orientation of the channel protein. Site 2 binds the lipid-soluble alkaloid toxins batrachotoxin (BTX) and veratridine, which shift the voltage dependence of sodium channel activation to more negative membrane potentials and remove inactivation. The combination of these two effects causes the sodium channel to be persistently activated over a wide voltage range (4,70). The action of BTX is

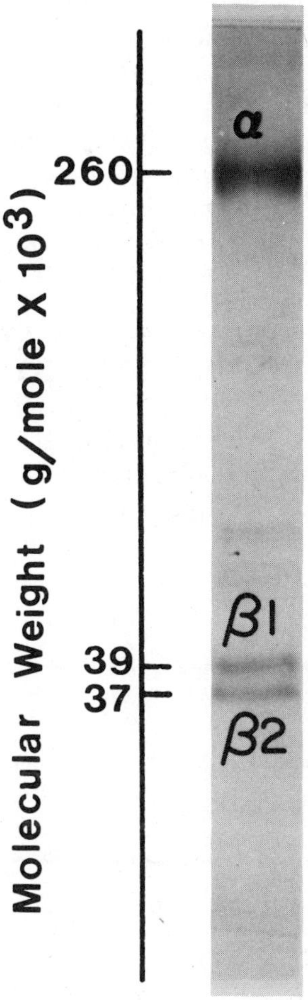

Figure 1. Subunit composition of purifed sodium channels from rat brain. Purified sodium channels were dissociated by boiling in 2% sodium dodecyl sulfate and 20 m*M* 2-mercaptoethanol and then analyzed by electrophoresis in polyacrylamide gel (58). Silver staining (73) shows the three subunits of the purified protein, α, β_1, and β_2.

inhibited by TTX in a noncompetitive fashion (3,50). Site 3 binds the basic polypeptide scorpion toxin from the North African scorpion *Leiurus quinquestriatus* (Lqtx) and leads to an inhibition of the sodium channel inactivation (15,51).

The sodium channel protein has been purified from *Electrophorus electricus* (2,60,68), rat muscle (9,11), chicken heart (57), and rat brain (34,35). The sodium channel protein from all these sources was solubilized from membranes with a variety of nonionic detergents and purified by conventional chromatographic means leading to purifications of up to about 1000-fold. All of these preparations contain a large glycopeptide with an apparent $M_r \sim 260{,}000$ (2,14,35,57). The sodium channel protein purified from mammalian sources has two smaller peptide components. In the sodium channel purified from rat brain, the large subunit is called α and the two smaller subunits β_1 ($M_r \sim 39{,}000$) and β_2 ($M_r \sim 37{,}000$). Subunit β_2 is linked to the α subunit by a disulfide bond (38,58). The β subunits are not proteolytic products of a larger precursor because the amino termini of all three subunits is blocked (see ref. 29; also Messner and Catterall, unpublished data). The protein components of the sodium channel purified from rat brain are illustrated in Fig. 1 (58). Recently, the primary structure of the *Electrophorus electricus* sodium channel was deduced from cDNA sequences by Numa and collaborators (72).

2. RECONSTITUTION OF THE VOLTAGE-SENSITIVE SODIUM CHANNEL IN LIPID BILAYERS

The functional reconstitution of the sodium channel has been an important goal since the beginning of the biochemical studies of the sodium channel more than a decade ago (12,40), but it has only recently been achieved (33,36,37,75,76,78,79,80,83). Reconstitution established that all the functional elements of the sodium channel are

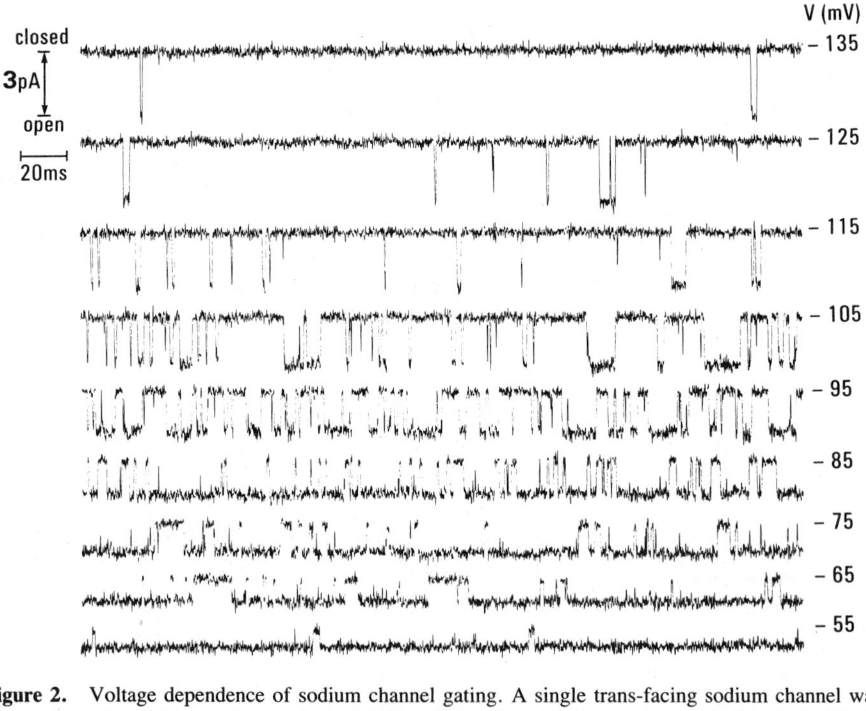

Figure 2. Voltage dependence of sodium channel gating. A single trans-facing sodium channel was incorporated into diphytanoyl phosphatidylcholine bilayer in a 70-μm aperture bathed in 0.5 *M* NaCl medium I cis and 0.2 *M* NaCl medium I plus 1 μ*M* BTX *trans*. The current was recorded under voltage clamp conditions, while the voltage was changed in 10-mV steps lasting for 1 min from −135 to −55 mV. The current records were filtered at 1 kHz, converted to digital form with a sampling frequency of 5 kHz (for details, see ref. 54), and plotted at reduced speed on a Gould 2200-S chart recorder. (From ref. 37.)

inherent to the purified protein (for reviews, see refs. 1, 10, and 39). The immediate goals are to identify the function of the β_1 and β_2 subunits of mammalian sodium channels, which subunits or part of subunits are involved in opening and closing the channel, forming its structure and involved in the control of the gating mechanism by the membrane potential. Furthermore, the sodium channel can be modified biochemically (20,21) or manipulated genetically (72) prior to or after reassembly, and the effects of such treatments on sodium channel function can then be evaluated in the reconstituted membrane. Thus the reconstituted sodium channel allows investigations of the structure–function relationships within the channel molecule and may contribute valuable information toward an understanding of its mechanism of action.

3. SODIUM CHANNELS IN LIPID VESICLES

The sodium channel protein purified from rat brain was reconstituted into phospholipid vesicles. The reconstituted vesicles exhibit a TTX-sensitive and veratridine-stimulated sodium uptake. The K_i value for the TTX inhibition is in the nanomolar range, in agreement with that observed in native membranes. The veratridine-activated sodium channels in reconstituted vesicles expressed cation selectivity with the apparent sequence $Na^+ > Rb^+ > Cs^+$ (1.0:0.25:0.10; see ref. 79). These results indicate that the purified sodium channel protein from rat brain contains the components necessary for the neurotoxin-activated and -selective ionic permeability characteristic of the sodium channel (16).

4. SODIUM CHANNELS IN PLANAR LIPID BILAYERS

The ultimate test for reconstitution is to reproduce in detail a functional sodium channel. This requires measurement of the properties of single sodium channels and determination of the voltage-dependent activation and inactivation of the channel. Planar lipid bilayers are ideally suited for recording the currents flowing through individual sodium channels with high sensitivity and time resolution. For this purpose, the purified protein reconstituted into vesicles (1,3) is transferred into planar lipid bilayers. The strategy used is that of membrane fusion. Basically, it consists of making the reconstituted vesicles interact with preformed planar bilayers (66,67) while an osmotic gradient across the bilayer is imposed. Fusion is believed to occur by the osmotic swelling of the vesicles in contact with the planar membrane, with subsequent rupture of vesicular and planar membranes in the region of contact (17,18,22,28,53,59). The functional integrity of the purified channel is assessed in the reconstituted vesicles prior to the fusion step, thus providing the necessary confidence control that the starting material for the planar bilayer studies is indeed functionally intact. A detailed description of the incorporation of sodium channels into planar lipid bilayers has appeared elsewhere (39).

4.1 Voltage Dependence of Channel Opening

Single-channel recordings from the purified sodium channel reconstituted in planar lipid bilayers provide an assay that resolves, in real time, transitions proceeding in a single molecule. Knowledge of the kinetic, pharmacological, and ion-conduction characteristics of the purified sodium channel is a prerequisite for the analysis of how modifications in its structure affect its functional properties on a molecular level. Figure 2 shows single sodium channel currents activated by 1 μM BTX and recorded from a lipid bilayer at the indicated voltages. Negative (hyperpolarizing) voltages favor channel closing, while positive (depolarizing) voltages favor channel opening. Figure 3 shows the fraction of time that the channel spends in the open state as a function of the applied voltage. For applied voltages between -100 and -50 mV, the probability of channel opening is strongly voltage dependent. For applied voltages more positive than 50 mV, the channel remains open 95% of the time.

The function displayed in Fig. 3 is a fundamental property of the system: it can be calculated by integration of conductance histograms (37,[□-symbol]), or independently from probability density analysis (19) of dwell times in the open state of the channel (o-symbol see refs. 47, 48; also below). The apparent gating charge is determined by fitting the fraction of time in the open state to a Boltzmann function. The dashed line is the Boltzmann function governed by

$$f_o = \left\{1 + \exp\left[\frac{q(V - V_{50})}{RT}\right]\right\}^{-1} \tag{1}$$

where f_o is the fraction of open time, q the apparent gating charge, and V_{50} the voltage at which $f_o = 0.5$. The dotted and solid lines correspond to different models of channel gating and are discussed in the following paragraphs. For purified sodium channels, $V_{50} = -91$ mV $\pm$ 17 mV (SD, $n = 22$) and $q = 3.8 \pm 0.3$ (SEM, $n = 5$) electronic charges per channel (37). These two parameters characteristic of the voltage dependence of sodium channel activation in BTX, V_{50} and q, are similar to those found in frog node of Ranvier ($V_{50} = -90$ mV $\pm$ 10 mV, $q = 3.5$; ref. 50), NG-108-15 neuroblastoma cells ($V_{50} = -85$ mV, $q = 4$–5; ref. 46), and rat brain (V_{50} average -90 mV, $q = 4$–6; ref. 28). The agreement, therefore, indicates that the voltage control mechanism is present and functionally intact in the purified sodium channel protein.

4.2 Tetrodotoxin Block

An exquisitely sensitive and specific test for identification of an ion channel in a planar lipid bilayer as a sodium channel is its blockade by TTX or STX. Hartshorne et al. (37) demonstrated that, indeed, the reconstituted brain sodium channel was blocked by TTX and that the TTX block was reversible and dependent on orientation and voltage. Figure 4 illustrates the concentration dependence of TTX block of purified sodium channels at -50 mV. The K_i for TTX block is 8.3 nM. This value is

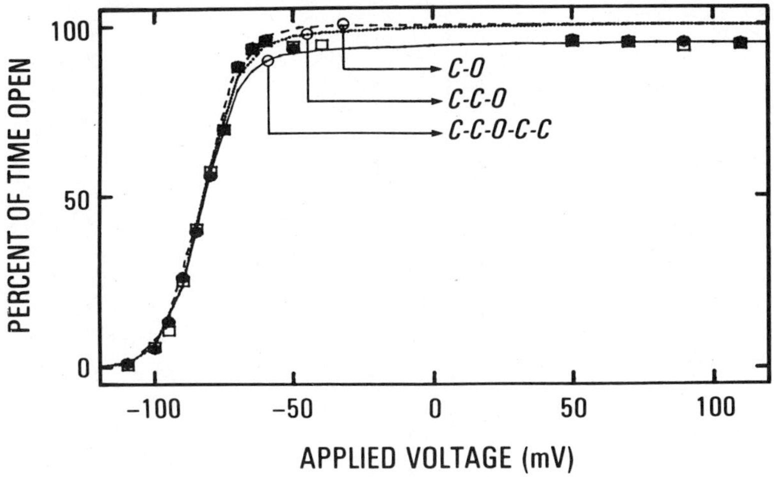

Figure 3. Probability of channel opening. The fraction of time that the channel spends in the open state f_O is strongly voltage dependent for applied voltages -100 mV $\leqslant V \leqslant -50$ mV. For positive applied voltages, the channel spends 95% of the time in the open state. Open squares were obtained by integrating conductance histograms as described in ref. 37. Filled circles were calculated independently from the distribution of open times according to

$$f_O = N \tau_O / T$$

where N represents the number of transitions, τ_O the channel open time, and T the recording time. The three lines correspond to different kinetic models of sodium channel gating and were calculated from the rate constants obtained for the best fit of the data to the models. A model with one closed (C) and one open (O) state, the C–O model; a model with two closed states and one open state, the C–C–O scheme; and a model with four closed states and a single open state, the C–C–O–C–C model. For models with exponentially voltage dependent transition rates, only the five state C–C–O–C–C model can fit the data for negative and positive applied voltages. (From refs 47 and 48.)

comparable to that obtained under similar conditions from native rat brain sodium channels (K_i = 7-16 nM; French and Krueger, personal communication). The TTX sensitivity of the reconstituted channel is unequivocable evidence that the single-channel currents observed are in fact sodium channels.

4.3 Ionic Selectivity

The purified sodium channel is more permeable to Na^+ than K^+ or Rb^+, with permeability ratios of 0.14 and 0.05, respectively (37). The permeability ratios are direct measurements of the single-channel conductance ratios in symmetrical 0.5 M solutions of chloride salts of the cations. The measured single-channel conductances are: γ_{Na^+} = 25.2 pS $\pm$ 0.5 pS (SEM, n = 10), γ_{K^+} = 3.5 pS $\pm$ 0.2 pS (SEM, n = 3), and γ_{Rb^+} = 1.2 pS $\pm$ 0.1 pS (SEM, n = 3). The single-channel conductance for Na^+ is similar to that recorded for native sodium channels incorporated in planar lipid

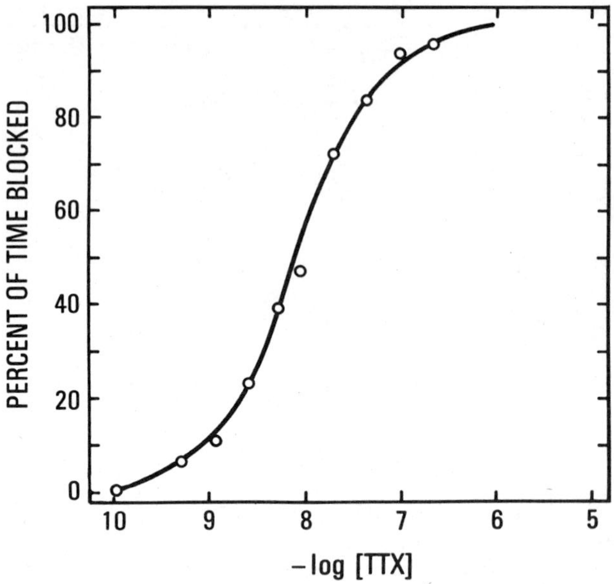

Figure 4. The concentration dependence of TTX block of purified sodium channels activated by BTX at −50 mV. Current records were examined to determine the percentage of time the channels were blocked as a function of [TTX] cis. Prior examination of the current records in the absence of TTX cis indicated that closures lasting longer than 1 sec were rare. Therefore, only channel closures lasting longer than 1 sec were attributed to TTX block. To ensure that representative samples of data were examined, 45-min records were analyzed for each [TTX] < 4 nM; 30-min records were analyzed for 4 n$M <$ [TTX] < 20 nM; and 10-min records were analyzed for [TTX] > 20 nM. The curve was drawn by plotting $\left\{\frac{100\,[\mathrm{TTX}]}{K_\mathrm{I}}\right\} + [\mathrm{TTX}]$ versus $-\log$ [TTX] for $K_\mathrm{i} = 8.3$ nM. (From ref 37.)

bilayers and activated by BTX (53,62,84). The agreement holds for the permeability ratios derived from flux measurements for BTX-activated sodium channels in neuroblastoma N-18 cells ($P_{K^+} = 0.17$ and $P_{Rb^+} = 0.08$; ref. 44) and for BTX-activated sodium channels purified from rat muscle ($P_{K^+} = 0.14$ and $P_{Rb^+} = 0.02$; ref. 80). It is clear, therefore, that the ion selectivity filter is functionally preserved in the purified channel protein.

4.4 Channel Gating Kinetics

Precise information about the number of open and closed states in which the sodium channel can exist, the pathways to enter into or exit from such states, and the rates of transition between the various states is essential for a rigorous understanding of the sodium channel at the molecular level.

Number of Channel States

Single-channel recordings are especially amenable for detailed statisticaly analysis. The major analytical tool is the computation of distributions of dwell times (lifetimes) in the open and closed states of the channel (19,55). A detailed analysis of the channel lifetimes obtained from recordings similar to those shown in Fig. 2 revealed an interesting feature of the BTX-modified sodium channel: the open-state lifetime distributions are well fitted by a single exponential; by contrast, the closed dwell time histograms are well fitted by a single exponential for voltages more negative than −75 mV and require the sum of two exponentials for voltages more positive than −75 mV (47,48). The results of the probability density analysis indicate the occurrence of a single open state and several distinct closed states in the voltage V range -120 mV $\leqslant V \leqslant +120$ mV.

Dwell Time Histograms and Transition Rates

From the time constants of the probability density analysis of dwell times in the open state and the closed state of the channel, the corresponding opening and closing rates are readily calculated (opening rate = closed time^{-1} (τ_c^{-1}); closing rate = open time^{-1} (τ_o^{-1})). Figure 5*a* illustrates the voltage dependence of the opening rate. The fast opening rate is exponentially voltage dependent for negative applied voltages with a slope of 13.5 ± 0.9 mV/*e*-fold change. In contrast, the slow opening rate, apparent only at applied voltages more positive than −75 mV, is prolonged by depolarization and decreases exponentially with a slope of −20.2 ± 2.3 mV/*e*-fold change (n = 5). For positive applied voltages, the closed dwell time histograms are well described by the sum of two exponentials. At V = 100 mV, the closed time constants are: τ_{c_1} = 0.63 ± 0.07 msec and τ_{c_2} = 244.2 ± 1.2 msec (n = 3). The corresponding opening rates are practically voltage independent, increasing an *e*-fold for voltage changes of more than 300 mV (47,48).

By contrast, the open dwell time histograms are well fitted by a single exponential for all applied voltages investigated. Depolarization greatly prolongs the open time constant from τ_o = 1.4 ms ± 0.3 ms at V = −100 mV to τ_o = 28.4 msec ± 4.5 msec at V = −60 mV. The corresponding closing rate τ_o^{-1} is exponentially voltage dependent with a slope of −13.6 ± 0.6 mV/*e*-fold change (Fig. 5*b*). For positive applied voltages, however, the closing rate is only moderately voltage dependent (71 ± 26 mV/*e*-fold change).

Autocorrelation Analysis

Correlation functions can be used to obtain information about the complexity of interconnections between the open and closed states (26,55). For reconstituted sodium channels, there is no correlation in the durations of successive opening or closing events (47,48). Autocorrelation analysis, therefore, shows that there is a single transition pathway connecting the open state with the closed states. This result is compatible with any kinetic scheme that postulates a single open state (55).

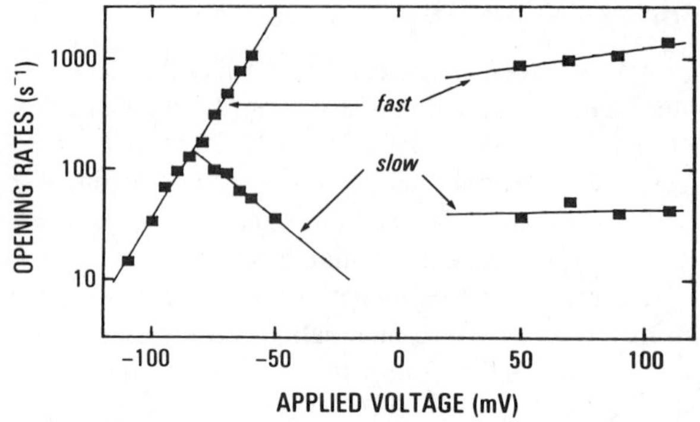

(a)

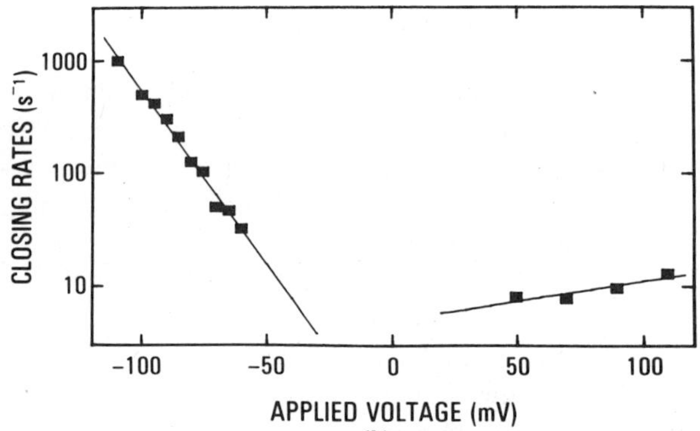

(b)

Figure 5. Voltage dependence of opening (*a*) and closing rates (*b*). (*a*) Current records similar to those in Fig. 2 lasting 30–60 sec at each voltage were digitized and the opening rates at each applied voltage determined by probability density analysis (19,24,25,54). To prevent noise from interfering with the determination of channel transitions, only records for applied voltages $V < -50$ mV or $V > 50$ mV were analyzed. For applied voltages $V < -80$ mV, the data were well fitted by a single exponential. For more positive applied voltages, the hypothesis of a single opening rate was rejected at a $p = .05$ significance level. For all voltages the data were well fitted by the sum of two exponentials. For negative applied voltages, the opening rates were found to be strongly voltage dependent; for positive applied voltages, they were virtually voltage independent. Straight lines were obtained by fitting different exponential functions to the four opening rates according to a least-squares fit. (*b*) Two exponentially voltage-dependent closing rates fitted the data for all applied voltages investigated. For negative applied voltages, the rate was strongly voltage dependent; for positive applied voltages, the rate was practically voltage independent. All fits were done by using the RS/1 software package of BBN Research Systems (Bolt Beranek and Newman, Inc., Cambridge, MA) on a VAX 11/750 computer (Digital Equipment Corporation, Marlboro, MA). (From refs. 47 and 48.)

In short, the purified brain sodium channel, modified by BTX, exhibits channel gating kinetics consistent with the occurrence of one open and several distinct closed states. At negative applied voltages, channel gating is well described by a kinetic scheme with one open and two closed states (Fig. 3): opening and closing rates are voltage dependent (Fig. 5), as previously described in mouse neuroblastoma cells

(46). At positive voltages, channel gating is adequately described by a kinetic scheme with one open and two closed states (Fig. 3): the opening and closing rates are independent of voltage (Fig. 5). The minimum kinetic scheme that is consistent with all the experimental data is the five-state model (47,48):

$$C_1 \underset{\delta}{\overset{\gamma}{\rightleftharpoons}} C_2 \underset{\beta}{\overset{\alpha}{\rightleftharpoons}} O \underset{\beta'}{\overset{\alpha'}{\leftrightharpoons}} C_3 \underset{\delta'}{\overset{\gamma'}{\leftrightharpoons}} C_4 \qquad (2)$$

5. MOLECULAR MECHANISMS THAT CAN GENERATE THE OBSERVED CHANNEL KINETICS

5.1 Inferences Derived from Primary Structure of Sodium Channel Protein

For the entire studied range of applied voltages, BTX-modified sodium channel gating can be explained by a C–C–O model with saturating transition rates or with a C–C–O–C–C model with exponentially voltage dependent transition rates (47,48).

Various mechanisms may give rise to the exponential voltage dependence of transition rates between several closed states and the open state (5,7,27,42,71). In any of them, essentially, the energy difference between channel closed states and the open state corresponds to the energy of a displaced charge. To simplify the discussion, we consider first a model that postulates charges buried in the hydrophobic interior of the membrane moving in response to an applied electric field. Such charge displacements would necessarily be conformational changes of the sodium channel protein. The location of protein charges inside the lipid bilayer would be energetically favored via the formation of ion pairs (30,82). We assume that such ion pairs are stabilized at hyperpolarizing membrane potentials (with V more negative than -100 mV). Depolarization would alter their equilibrium distribution, allowing such pairs to break and re-form with different groups. The voltage independence of the rate constants at $V \geqslant 0$ would imply that the ion pairs formed are effectively neutralized and, hence, sense no field; this is formally expressed as saturating rate constants.

These considerations are inferred from the recently derived primary structure of the *Electrophorus electricus* sodium channel (72) and the postulated models for the arrangement and folding of the polypeptide chain within the bilayer (31,32,52,72). All models consist of several transmembrane structures, most likely α-helical. The sensitivity to an applied electric field and, consequently, voltage-controlled gating, arises from the electric field-induced disruption of multiple ion pairs buried within the hydrophobic, intramembranous protein interior (31). The occurrence of multiple ion pairs (31,32,72) may account for the multiplicity of closed states (5,7,8,27,42,43,46–48). The voltage independence of the rate constant at positive voltages (47,48) may be an expression of the stable neutralization of such ion pairs.

5.2 Batrachotoxin Stabilizes the Open State of the Sodium Channel

Consider the schematic hypothetical tertiary structure for the sodium channel recently inferred from a computer-aided analysis of the *Electrophorus electricus* channel primary structure (31) (Fig. 6*B*). The structure is pseudoradially symmetrical. At its center, four amphipathic transmembrane helices (one α helix contributed by each one of the four repeated homologous domains) meet with their hydrophilic surfaces facing inward to form a putative ion channel. The central channel created by the boundary of four helices would be about 4.2 Å across its narrowest dimension. Note that the effective cutoff pore size for the sodium channel is approximately 3 × 5 Å (for review, see ref. 41). In each of the four homologous regions, the s_d helix is indicated at the center of a clustered structure of high dipole moment. The s_d helix contains positively charged arginine residues that repeat with a period of 3. The dipolar cluster is organized by the interaction between s_d, s_c, s_g, and s_f, all of which contain a number of acidic or amidic residues. Similar buried ion pairs have been observed crystallographically in myoglobin (e.g., Arg CD3, Asp E3; ref. 49). In this model, s_a, s_b, s_e, s_h, and s_f in each region are illustrated to occur at the interfaces between homologous regions, providing favorable and extensive hydrophobic boundaries.

This model suggests that the voltage control of the sodium channel gating mechanism resides in the sensitivity to applied electric fields of the four dipolar clusters each formed by s_d, s_g, and s_c. Hyperpolarization would stabilize the ion pairs, whereas depolarization would disturb their equilibrium distribution, promoting the dissociation of such pairs and reassociation among different groups. Such a rearrangement of helices may lead to channel opening (activation) (81). Suppose, now, that such a new state is unstable and decays to another state associated with a channel closure (inactivation). In other words, Na^+ channel activation is a transient, unstable state that relaxes to a more stable, inactivated state. Batrachotoxin effectively stabilizes the open state of the channel. It follows from these assumptions that BTX, a lipid-soluble toxin, may intercalate between the helices of the dipolar cluster and alter their packing arrangement. This, in turn, may change the specificity of ion-pair formation, resulting in tighter linkages and, consequently, in the stabilization of the open state. The rearrangements produced by the intercalation of BTX between helices may also be reflected in the reduction of the single-channel conductance (46,74) and the ionic selectivity (65) measured in BTX-modified sodium channels. The postulated rearrangement of the helices in the cluster could cause the widening of the central pore by a fraction of an angstrom, a change sufficient to account for the observed effects. The property of sodium channels that is not perturbed by BTX is its sensitivity to TTX. The action of BTX must be restricted to the intramembranous domain of the protein and does not appear to propagate at a distance sufficient to affect the exposed TTX binding site.

These suggestions arise from the availability of models for the sodium channel protein and an assay for the channel activity of the purified protein. Therefore, they are amenable to experimental validation or refutation.

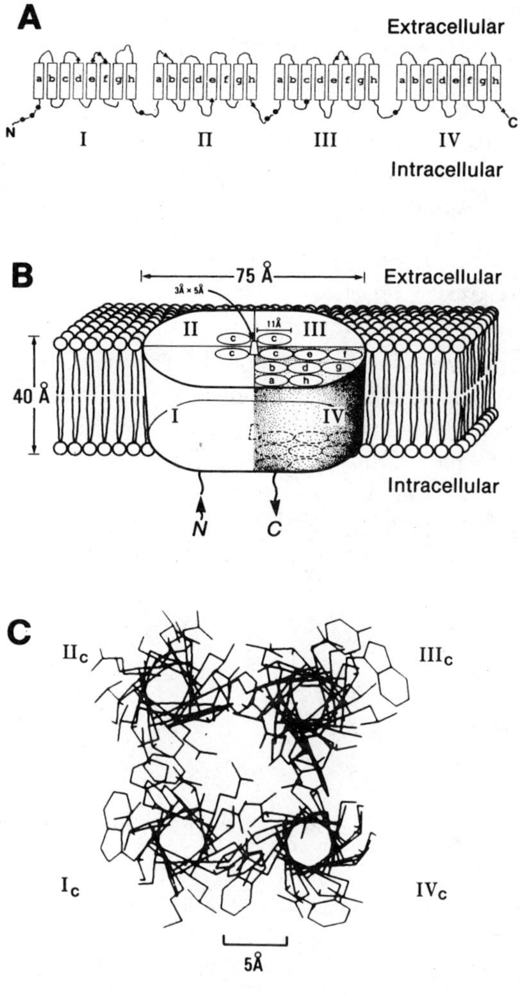

Figure 6. Hypothetical structural model of the sodium channel. (*a*) One-dimensional representation of the cytoplasmic, transmembrane, and extracellular domains. Four homologous regions, each consisting of eight membrane-spanning helical segments, are shown. Adjacent segments traverse the membrane in antiparallel orientation. Potential phosphorylation sites are shown as (⊕), potential glycosylation sites as (△). (*b*) Schematic tertiary structure model of the channel, showing the four homology regions organized in a pseudoradially symmetrical array. Note that the structure is close-packed. Carbohydrate moiety is omitted. (*c*) Orthographic projection of a possible ion pore formed by amphipathic helices s_c oriented in parallel. Distance scale indicates the spacing between Asp-188 (I_c) and Asp-634 (II_c). (From ref. 31.)

6. CONCLUDING REMARKS

The molecular physiology of electrical excitability has been reduced to investigations on the molecular properties of purified sodium channels reconstituted in lipid bilayers. This level of analysis should begin to contribute strategic information on how the structure of the channel protein produces its function. This goal appears to be more at reach than ever before with the explosive application of molecular genetics to modify the structure channel proteins (61). Concurrently, site-directed antibodies and chemical modifications of known sites in the protein combined with single-channel recordings of the reconstituted channel should increase our knowledge of how voltage controls gating of the sodium channel.

ACKNOWLEDGMENTS

We are indebted to W. A. Catterall and J. Talvenheimo for providing the reconstituted vesicles and their continuous interest; to D. Messner who supplied the gel shown in Fig. 1; to J. Daly for a kind gift of BTX; to R. Greenblatt and Y. Blatt for their collaboration in the studies shown in Fig. 6; to Tony Lee for computer programming; and to Ms. Laura Castrejon for her devoted assistance. This research was supported by grants from the National Institutes of Health (EY-02084), the Department of the Army Medical Research (17-82-C2221), and the National Multiple Sclerosis Society.

REFERENCES

1. Agnew, W. A. (1984). *Ann. Rev. Physiol.*, **46**, 517–530.
2. Agnew, W. S., Levinson, S. R., Brabson, J. S., and Raftery, M. A. (1978). *Proc. Natl. Acad. Sci.* (USA), **75**, 2606–2610.
3. Albuquerque, E. X. and Daly, J. W. (1976). In Cuatrecasas, P., Ed., *Receptors and Recognition*, Series B, Vol. 1, pp. 299–338.
4. Albuquerque, E. X., Daly, J. W., and Witkop, B. (1971). *Science*, **172**, 995–1002.
5. Aldrich, R. W., Corey, D. P., and Stevens, C. F. (1983). *Nature*, **306**, 436–441.
6. Anholt, R., Lindstrom, J., and Montal, M. (1985). In Martonosi, A., Ed., *The Enzymes of Biological Membranes*, 2nd ed., Vol. 3, Plenum, New York, Chapter 36, pp. 335–401.
7. Armstrong, C. M. (1981). *Physiol. Rev.*, **61**, 644–683.
8. Armstrong, C. M. and Bezanilla, F. (1977). *J. Gen. Physiol.*, **70**, 567–590.
9. Barchi, R. L. (1983). *J. Neurochem.*, **40**, 1377–1385.
10. Barchi, R. L. (1984). *Trends Biochem. Sci.*, **9**, 358–361.
11. Barchi, R. L., Cohen, S. A., and Murphy, L. E. (1980). *Proc. Natl. Acad. Sci.* (USA), **77**, 1306–1310.
12. Benzer, T. I., and Raftery, M. A. (1973). *Biochem. Biophys. Res. Commun.*, **51**, 939–944.
13. Bidard, J. N., Vijverberg, H. P. M., Frelin, C., Chungue, E., Legrand, A.–M., Bagnis, R., and Lazdunski, M. (1984). *J. Biol. Chem.*, **259**, 8353–8357.
14. Casadei, J. M., Gordon, R. D., Lampson, L. A., Schotland, D. L., and Barchi, R. L. (1984). *Proc. Natl. Acad. Sci.* (USA), **81**, 6227–6231.
15. Catterall, W. A. (1980). *Ann. Rev. Pharmacol. Toxicol.*, **20**, 15–43.

16. Catterall, W. A., Messner, D. J., Hartshorne, R. P., Feller, D. J., Talvenheimo, J. A., Keller, B., and Montal, M. (1984). In *Proceedings of The Ninth International Congress on Pharmacology*, MacMillan, London, pp. 219–229.
17. Cohen, F. S., Akabas, M. H., and Finkelstein, A. (1982). *Science*, **217**, 458–460.
18. Cohen, F. S., Akabas, M. H., Zimmerberg, J., and Finkelstein, A. (1984). *J. Cell Biol.*, **98**, 1054–1062.
19. Colquhoun, D. and Hawkes, A. G. (1981). *Proc. Roy Soc. Lond. B, Biol. Sci.*, **211**, 205–235.
20. Costa, M. R. C., Casnellio, J. C., and Catterall, W. A. (1982). *J. Biol. Chem.*, **257**, 7918–7921.
21. Costa, M. R. C. and Catterall, W. A. (1984). *J. Biol. Chem.*, **259**, 8210–8216.
22. Drachev, L. A., Jasaitis, A. A., Koulen, A. D., Kondrashin, H. A., Liberman, E. A., Hayrecek, I. B., Ostroumov, S. A., Semenov, A. Y., and Skulachev, V. P. (1974). *Nature*, **249**, 321–323.
23. Feller, D., Talvenheimo, J. A., and Catterall, W. A. (1984). *Soc. Neurosci. Abstr.*, **10**, 863.
24. Feller, W. (1971). *An Introduction to Probability Theory and Its Applications*, Vol. II, Wiley, New York, p. 669.
25. Fletcher, R. A. (1971). *A Modified Marquardt Subroutine for Nonlinear Least Squares*, AERE report R6799.
26. Fredkin, D. R., Montal, M., and Rice, J. A. (1985). In LeCam, L. M., and Olshen, R. A., Eds., *Proceeding of the Berkeley Conference in honor of Jerzy Neyman and Jack Kiefer,* Wadsworth, Belmont, CA., **1**, 269–289.
27. French, R. J., and Horn, R. (1983). *Ann. Rev. Biophys. Bioeng.*, **12**, 319–356.
28. French, R. J., Worley, J. F., III, and Krueger, B. K. (1984). *Biophys. J.*, **45**, 301–310.
29. Grishin, E. V., Kovalenko, V. A., Pashkov, Y. N., and Shamotienko, O. G. (1984). *Membrane Biochem.* (USSR), **1**, 858–867.
30. Gemant, A. (1962). *Ions in Hydrocarbons*, Wiley, New York, p. 261.
31. Greenblatt, R. E., Blatt, Y., and Montal, M. (1985). *FEBS Lett.*, **193**, 125–134.
32. Guy, H. R. and Seetharamulu, P. (in press). *Proc. Natl. Acad. Sci.* (USA).
33. Hanke, W., Boheim, G., Barhanin, J., Pauron, D., and Lazdunski, M. (1984). *EMBO J.*, **3**, 509–515.
34. Hartshorne, R. P. and Catterall, W. A. (1981). *Proc. Natl. Acad. Sci.* (USA), **78**, 4620–4624.
35. Hartshorne, R. P. and Catterall, W. A. (1984). *J. Biol. Chem.*, **259**, 1667–1675.
36. Hartshorne, R. P., Keller, B. U., Talvenheimo, J. A., Catterall, W. A., and Montal, M. (1984). *Soc. Neurosci. Abstr.*, **10**, Part 2, 864.
37. Hartshorne, R. P., Keller, B. U., Talvenheimo, J. A., Catterall, W. A., and Montal, M. (1985). *Proc. Natl. Acad. Sci.* (USA), **82**, 240–244.
38. Hartshorne, R. P., Messner, D. J., Coppersmith, J. C., and Catterall, W. A. (1982). *J. Biol. Chem.*, **257**, 13888–13891.
39. Hartshorne, R., Tamkun, M., and Montal, M. (1986). In Miller, C., Ed., *Ion Channel Reconstitution*, Plenum, New York, pp. 337–362.
40. Henderson, R. and Wang, J. H. (1972). *Biochemistry,* **11**, 4565–4569.
41. Hille, B. (1984). *Ionic Channels of Excitable Membranes*, Sinauer, Sunderland, MA, p. 426.
42. Hodgkin, A. L. and Huxley, A. F. (1952). *J. Physiol.* (London), **117**, 500–544.
43. Horn, R. and Vandenberg, C. A. (1984). *J. Gen. Physiol.*, **84**, 505–534.
44. Huang, L. M., Catterall, W. A., and Ehrenstein, G. (1979). *J. Gen. Physiol.*, **73**, 839–854.
45. Huang, L. M., Moran, N., and Ehrenstein, G. (1982). *Proc. Natl. Acad. Sci.* (USA), **79**, 2082–2085.

46. Huang, L. M., Moran, N., and Ehrenstein, G. (1984). *Biophys. J.*, **45**, 313–322.
47. Keller, B. U., Hartshorne, R. P., Talvenheimo, J. A., Catterall, W. A., and Montal, M. (1985). *Biophys. J.*, **47**, 439 (abstr.).
48. Keller, B. U., Hartshorne, R. P., Talvenheimo, J. A., Catterall, W. A., and Montal, M. (1986). *J. Gen. Physiol.* **88**, 1–23.
49. Kendrew, J. C., Watson, H. C., Strandberg, B. E., Dickerson, R. E., Phillips, D. C., and Shore, V. C. (1961). *Nature*, **190**, 666–670.
50. Khodorov, B. I. and Revenko, S. V. (1979). *Neuroscience*, **4**, 1315–1330.
51. Koppenhöfer, E. and Schmidt, H. (1968). *Pflügers Arch.*, **303**, 150–161.
52. Kosower, E. M. (1985). *FEBS Lett.*, **182**, 234–242.
53. Krueger, B. K., Worley, J. F., III and French, R. J. (1983). *Nature* (London), **303**, 172–175.
54. Labarca, P., Lindstrom, J., and Montal, M. (1984). *J. Gen. Physiol.*, **83**, 473–496.
55. Labarca, P., Rice, J., Fredkin, D., and Montal, M. (1985). *Biophys. J.*, **47**, 469–478.
56. Lazdunski, M. and Renaud, J. F. (1982). *Ann. Rev. Physiol.*, **44**, 463–473.
57. Lombet, A. and Lazdunski, M. (1984). *Eur. J. Biochem.*, **141**, 651–666.
58. Messner, D. J. and Catterall, W. A. (1985). *J. Biol. Chem.*, **260**, 10,597–10,604.
59. Miller, C. and Racker, E. (1976). *J. Membrane Biol.*, **30**, 283–300.
60. Miller, J. A., Agnew, W. S., and Levinson, S. R. (1983). *Biochemistry*, **22**, 462–470.
61. Mishina, M., Tobimatsu, T., Imoto, K., Tanaka, K., Fjuita, Y., Fukuda, K., Kurasaki, M., Takahashi, H., Morimoto, Y., Inayama, S., Takahashi, T., Kuno, M., and Numa, S. (1985) *Nature*, **313**, 364–369.
62. Moczydlowski, E., Hall, S., Garber, S., Strichartz, G., and Miller, C. (1984). *J. Gen. Physiol.*, **84**, 687–784.
63. Montal, M. (in press). In Ragan, C. I., and Cherry, R., Eds., *Techniques for Analysis of Membrane Proteins*, Chapman and Hall, London.
64. Montal, M., Labarca, P., Fredkin, D. R., Suarez-Isla, B. A., and Lindstrom, J. (1984). *Biophys. J.*, **45**, 165–174.
65. Mozhaeva, G. N., Naamov, A. P., Negulyaev, Y. A., and Nosyreva, E. D. (1977). *Biochim. Biophys. Acta*, **466**, 461–473.
66. Mueller, P., Rudin, D. O., (1969). In Passow, M., and Stampfli, R., Eds., *Laboratory Techniques of Membrane Biophysics*, Springer-Verlag, Berlin, pp. 141–156.
67. Mueller, P., Rudin, D. O., Tien, H. T., and Wescott, W. C. (1963). *J. Phys. Chem.*, **67**, 534–535.
68. Nakayama, H., Withey, R., and Raftery, M. A. (1982). *Proc. Natl. Acad. Sci.* (USA), **79**, 7575–7579.
69. Narahashi, T., Anderson, N. C., and Moore, J. W. (1966). *Science*, **153**, 765–767.
70. Narahashi, T., Deguchi, T., and Albuquerque, E. X. (1971). *Nature*, **229**, 221–222.
71. Neher, E. and Stevens, C. F. (1979). In *The Neurosciences*. Fourth Study Program. MIT Press, Cambridge, MA, pp. 623–629.
72. Noda, M., Shimizu, S., Tanabe, T., Takai, T., Kayano, T., Ikeda, T., Takahashi, H., Nakayama, H., Kanaoka, Y., Minamino, N., Kangawa, K., Matsuo, H., Raftery, M., Hirose, T., Inayama, S., Hayashida, H., Miyata, T., and Numa, S. (1984). *Nature*, **312**, 121–127.
73. Oakley, B. R., Kirsch, D. R., and Morris, N. R. (1980). *Anal. Biochem.*, **105**, 361–363.
74. Quandt, F. N. and Narahashi, T. (1982). *Proc. Natl. Acad. Sci.* (USA), **79**, 6732–6736.
75. Rosenberg, R. L., Tomiko, S. A., and Agnew, W. S. (1984). *Proc. Natl. Acad. Sci.* (USA), **81**, 5594–5598.

76. Rosenberg, R. L., Tomiko, S. A., and Agnew, W. S. (1984). *Proc. Natl. Acad, Sci.* (USA), **81**, 1239–1243.

77. Sigworth, F. J., and Neher, E. (1980). *Nature*, **287**, 447–449.

78. Talvenheimo, J. A., Tamkun, M. M., and Catterall, W. A. (1982). *J. Biol. Chem.*, **257**, 11868–11871.

79. Tamkun, M. M., Talvenheimo, J. A., and Catterall, W. A. (1984). *J. Biol. Chem.*, **259**, 1676–1688.

80. Tanaka, J. C., Eccleston, J. F., and Barchi, R. L. (1983) *J. Biol. Chem.*, **258**, 7519–7526.

81. Unwin, P. N. T. and Ennis, P. D. (1984). *Nature*, **307**, 609–613.

82. Warshel, A. and Russell, S. T. (1984). *Quart. Rev. Biophys.*, **17**, 283–422.

83. Weigele, J. B. and Barchi, R. L. (1982). *Proc. Natl. Acad. Sci.* (USA), **79**, 3651–3655.

84. Weiss, L. B., Green, W. N., and Anderson, O. S. (1984). *Biophys. J.*, **45**, 67 (abstr.).

CHAPTER 10

MULTIPLE TYPES OF CALCIUM CHANNEL IN EXCITABLE CELLS

R. W. TSIEN
A. P. FOX
P. HESS
E. W. McCLESKEY
B. NILIUS
M. C. NOWYCKY
R. L. ROSENBERG
Department of Physiology
Yale University School of Medicine
New Haven, Connecticut

Opening of calcium channels in response to an appropriate stimulus allows calcium ions to move down their electrochemical gradient into the cytoplasm. This calcium influx not only transfers depolarizing charge into excitable cells, but also carries a specific message to be decoded by calcium receptor proteins. The signal leads to the initiation of contraction in heart and smooth muscle cells, transmitter release from nerve cell synaptic terminals, hormone secretion by gland cells, and other important cellular responses (37,77,86). In linking membrane potential changes to the delivery of a messenger substance, calcium channels perform a function that is vital and possibly unique (45,86).

The development of new methods for recording from single cells and single channels (40,53) has led to considerable progress in understanding how calcium channels function at the level of single-channel molecules (77,85,86). Patch clamp recordings have also provided strong evidence that multiple types of calcium channel coexist in a large variety of excitable cells, in agreement with earlier work in invertebrate egg cells (33,38) and mammalian central nervous system (CNS) neurons (57,58).

This chapter reviews evidence for multiple types of calcium channel and describes similarities and differences in their properties of ion permeation, gating, and pharmacology. We summarize findings of multiple types of calcium channel in neurons; gland cells; and cardiac, skeletal, and smooth muscle cells and discuss hypotheses about their functional roles (65,66,78). This is a rapidly expanding area of research: at the September 1985 symposium of the Society of General Physiologists, there were no fewer than nine abstracts relevant to the possible existence of multiple types of calcium channel.

1. THREE TYPES OF CALCIUM CHANNEL IN SENSORY NEURONS

To illustrate similarities and differences between various types of voltage-dependent calcium channel, we will focus on patch clamp recordings in dorsal root ganglion (DRG) neurons before going on to review results from other preparations. Sensory neurons from dorsal root or petrose ganglia have provided an excellent system for studying the expression of multiple types of calcium channel (14,15,17,18,71,73). There is broad agreement in defining at least two classes of calcium channel with quite different properties. One type gives "sustained," "persistent," "long-lasting" current with maintained depolarization. Its other major characteristics are as follows: (1) activation that is rapid but requires strong depolarizations (to ~0 mV in most cells), (2) very rapid deactivation following sudden repolarization, (3) availability even with relatively depolarized holding potentials. We refer to this channel type as L (long-lasting, large unitary barium conductance), but others have used terms such as "slow," "type II," and "high-voltage-activated" (HVA).

The second most commonly reported type of calcium channel is characterized by a transient time course due to fairly rapid inactivation (time constants of the order of 20–50 msec). Other properties include (1) activation at relatively weak depolarizations (as negative as −70 mV in some preparations), (2) only moderately rapid deactivation following sudden repolarization, and (3) availability even with relatively depolarized holding potentials. We refer to this channel type as T (transient, tiny barium conductance), but others have used terms such as "fast," "type I," or "low-voltage-activated" (LVA).

A third type of calcium channel, labeled N (novel, neither T nor L, intermediate barium conductance) has been found in chick DRG neurons by our group (73) and is described in the paragraphs that follow. Although other groups have not noticed or drawn attention to N-type current as a separate entity, it can be seen in a number of earlier recordings (see Section 4).

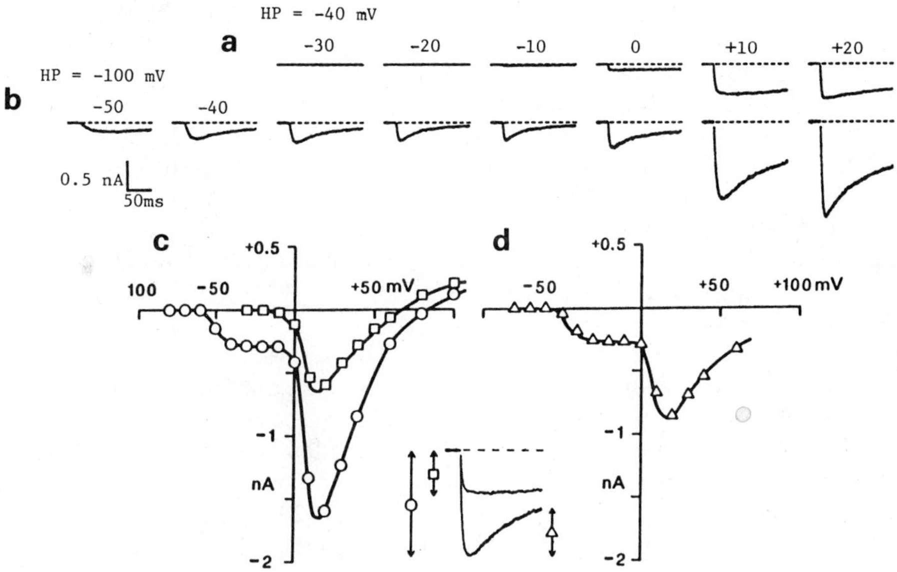

Figure 1. Three components of calcium channel current in whole-cell recordings from a chick DRG neuron with 10 m*M* extracellular calcium. (*a,b*) Current records evoked from HP = −40 (*a*) and HP = −100 mV (*b*). Test potential indicated above. Linear leak and capacitance currents have been subtracted, with 1 msec blanked at both the ON and OFF of the depolarizing pulse. (*c*) Plot of peak current versus test potential from HP = −40 (squares) and HP = −100 mV (circles). (*d*) Plot of magnitude of decaying current versus test potential from HP = −100 mV. The current relaxation is the late current subtracted from the peak current. Inset repeats records obtained at +10 mV and shows how peak and relaxing current amplitudes were measured. The internal solution (pipette) contained (in m*M*): 100 CsCl, 10 Cs-EGTA, 5 $MgCl_2$, 40 HEPES, 2 ATP, 0.25 cAMP, pH 7.3. The external solution was exchanged after formation of a gigaseal from a standard Tyrode to one containing (m*M*) 10 $CaCl_2$, 135 TEACl, 10 HEPES, and 200 n*M* TTX, pH 7.3. (From ref. 73.)

2. KINETIC SEPARATION OF T, N, AND L-TYPE CALCIUM CURRENT IN WHOLE-CELL RECORDINGS

Figure 1 illustrates the characteristic kinetic features of the three components of calcium channel current in a whole-cell recording with Ca^{2+} (10 m*M*) as the external charge carrier. Ionic conditions were chosen to minimize contamination by other currents: sodium currents were eliminated by external solutions containing TTX and TEA in place of sodium; potassium currents were inhibited by the TEA and by the use of cesium as the dominant internal cation. Different components of calcium channel current were evoked by applying depolarizing test pulses to various levels from different holding potentials.

The long-lasting (L) current was studied in isolation with depolarizations from holding potential (HP) = −40 mV (row *a*). Strong pulses (to −10, +10, +20 mV) are required to activate any inward current; the peak current–voltage relation (Fig. 1*c*) is typical for a single current component. The L current shows little inactivation over the course of the ~100-msec pulse; longer test pulses reveal a slow but essentially complete inactivation ($t_{1/2}$ of hundreds of milliseconds).

The inactivating T- and N-type calcium currents are seen if the membrane potential is held at relatively negative levels (Fig. 1*b*). Weak pulses from HP = −100 mV evoke a rapidly inactivating calcium current (component T). This is seen as a prominent shoulder at negative test potentials in a plot of the amplitude of the current decay versus potential (Fig. 1*d*). The T current first becomes noticeable at −60 mV and is nearly constant in amplitude between −40 and −10 mV, consistent with a very negative range of activation.

The N-type calcium current begins to contribute to inactivating calcium current with strong depolarizations from HP = −100 mV (e.g., traces at +10 or +20 mV in Fig. 1*b*). It is distinguished from component T in the plot of decaying current amplitude versus test potential (Fig. 1*d*). Beyond voltages where the amplitude of decaying T current reaches a plateau (between −30 and −10 mV), the decay amplitude shows a secondary rise, reaching a peak at +20 mV. Like T, but unlike L, component N inactivates relatively rapidly and requires strongly negative holding potentials for complete removal of inactivation. Like L, but unlike T, N current requires strong depolarizations for activation.

3. UNITARY ACTIVITY OF T, N, AND L-TYPE Ca CHANNELS

Patch clamp recordings from cell-attached patches on DRG cell bodies showed three types of unitary calcium channel activity that correspond to the three components of calcium current found in whole-cell recordings (Fig. 2). The T, N, and L-type calcium channels are distinguished at the single-channel level by several criteria:

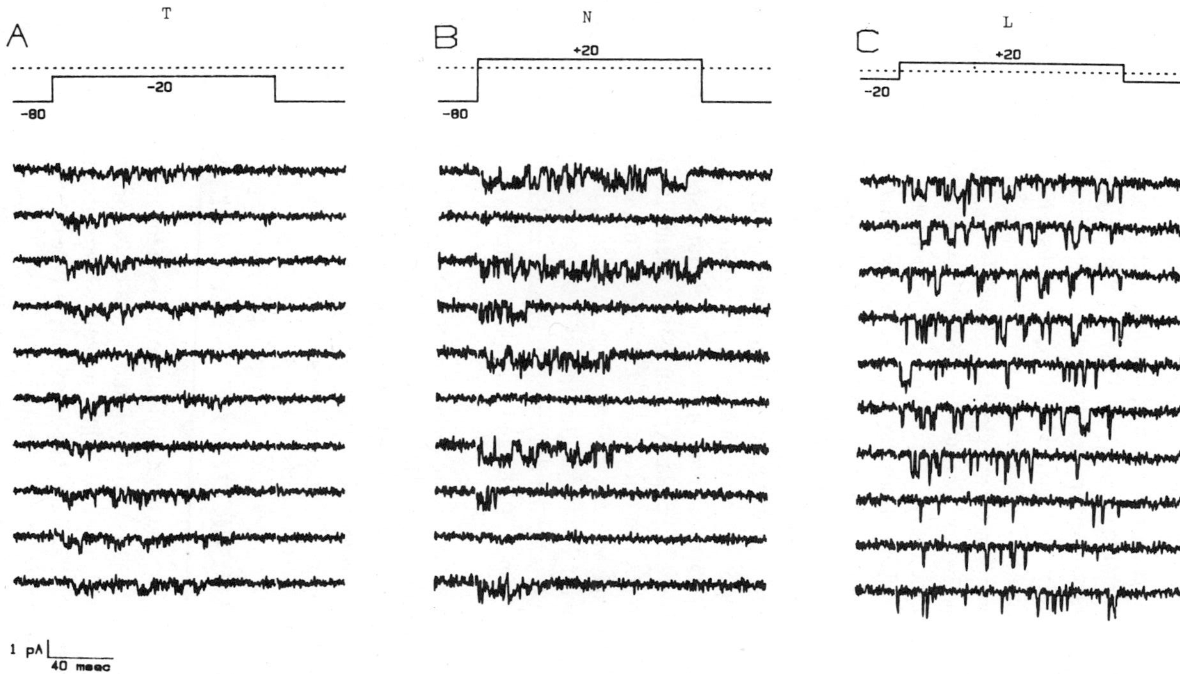

Figure 2. Three types of unitary calcium channel activity seen in cell-attached patch recordings with barium as the charge carrier: (*A*) T-type channel activity, cell N85A; (*B*) N-type channel, cell N75H; (*C*) L-type channel, cell N67J. Patch pipettes contained: 110 m*M* $BaCl_2$, 10 m*M* HEPES, 200 n*M* TTX, pH 7.4. To put patch membrane potential on an absolute scale, cell resting potential was zeroed with an external solution containing (in m*M*): 140 K-aspartate, 10 K-EGTA, 10 HEPES, 1 $MgCl_2$, pH 7.4. Current signals were filtered at 1 kHz and sampled at 5 kHz. Within each panel traces are consecutive. Values of single-channel amplitude and slope conductances given in the text were obtained from amplitude histograms. The unitary current amplitude of the T-type channel at −20 mV was given by the average of amplitudes from three different patches exhibiting long, well-resolved openings. We have seen each of the three channel types in isolation and in all possible combinations. The number of channels under the patch pipette varied considerably from one channel to dozens of channels. (From ref. 73.)

slope conductance with barium (110 m*M*) as the charge carrier, voltage dependence of open probability, and the pattern and degree of inactivation.

Figure 2*A* shows unitary current activity of a channel that we associate with component T. Inactivation of this channel is complete at holding potentials positive to −40 mV and is fully removed only at holding potentials more negative than −80 mV. Activation becomes detectable positive to −50 mV. With test depolarizations to −20 or −10 mV, the averaged current decays with a time course very similar to component T in whole-cell recordings. The T-type channel has a small slope conductance (about 8 pS with 110 m*M* barium).

Figure 2*C* illustrates unitary channel activity that corresponds to the L component. It is relatively resistant to inactivation: activity is seen at HP = −20 mV (Fig. 2*C*) or even at 0 mV. Activation of the L-type channel usually becomes significant at −10 mV, although openings can sometimes be detected as negative as − 30 mV. Averaged currents obtained from many individual traces do not relax at any test potential (see Fig. 5, bottom left). The channel is characterized by a large slope conductance in 110 m*M* Ba (25 pS).

Figure 2*B* illustrates a third type of unitary activity that we call "N" because its properties account for component N in whole-cell recordings. N-type calcium channels differ from L- and T-type calcium channels in several respects. Their slope conductance with 110 m*M* Ba is 13 pS, larger than T-type channels and smaller than L-type channels. The differences in unitary current size are highly significant: for example, at −20 mV, average values (±SEM) were 0.62 ± 0.03 pA for T (five patches), 1.22 ± 0.03 pA for N (10 patches), and 2.07 ± 0.09 pA for L (six patches). At +20 mV, the test potential illustrated in Fig. 2*B* and 2*C*, the difference between average values for N and L (0.82 and 1.12 pA) is smaller in absolute terms but is still considerable.

Each channel type shows a steep increase in the probability of channel openness with membrane depolarization. Activation of T, N, and L channels occurs over ranges of potential that are spaced ~15 mV apart. This makes it possible to see activity of T-type channels at potentials too negative to activate the larger N-type events, and likewise with N- and L-type channels.

4. SPECIAL PROPERTIES OF N-TYPE CHANNELS

The N-type calcium channels are the most difficult to study at the level of whole-cell recordings because there is no pulse protocol that activates N current exclusive of both T and L currents. This difficulty may help to explain why most other groups have not focused on N currents; with the benefit of hindsight, N-type current can be seen in whole-cell recordings from DRG neurons made by other investigators (adult guinea pig DRG neurons, ref. 48; chick DRG neurons, refs. 17 and 29).

Fortunately, N-type channels can be characterized in isolation in cell-attached patch recordings, either from single-channel patches (Fig. 2*B*) or in patches on hot

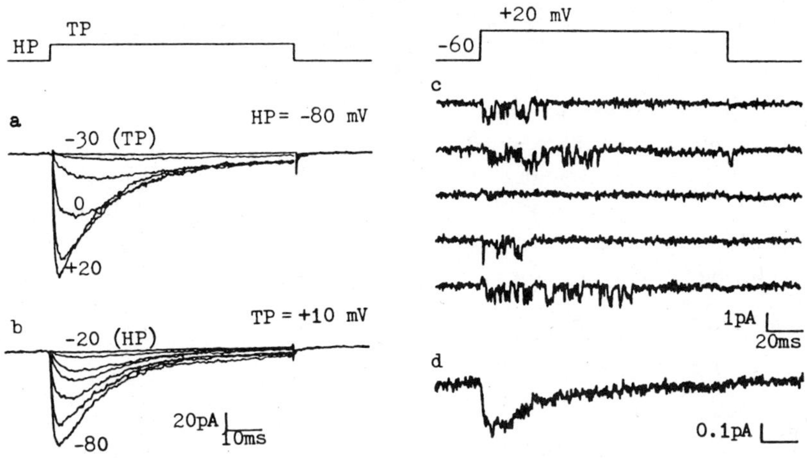

Figure 3. Gating properties of N-type calcium channels. (*a,b*) Currents recorded from a cell-attached patch containing hundreds of N-type channels and no detectable T- or L-type channel activity. (*a*) Activation studied with test depolarizations from HP = −80 mV. Test potentials spaced by 10 mV as indicated. (*b*) Inactivation studied with test depolarizations to +10 mV from various holding potentials as indicated. (*c,d*) Cell-attached patch containing only one N-type channel. (*c*) Unitary current records evoked by test depolarization indicated above; (*d*) averaged record. Patch 84G.

spots of N-type activity, containing hundreds of N-type calcium channels and no detectable T- or L-type activity (73). Figures 3*a* and 3*b* show kinetic analysis from this kind of recording. The peak inward current increases over the range between −30 and +20 mV, in reasonable agreement with recordings from whole cells (Fig. 1*d*) and patches with only one or a few N-type channels (e.g., Fig. 3*a*). The rise and fall of the N-type channel current both became faster as the test pulse depolarization was increased (Fig. 3*a*); at strong depolarizations, inactivation was largely complete within ~100 msec. Steady-state inactivation was studied by varying the holding potential (Fig. 4, middle). Inactivation was roughly 50% at HP = −60 mV and virtually complete at HP = −20 mV, a potential at which L-type calcium channels remain available (Fig. 2*C*).

Single-channel recordings provide additional kinetic criteria for distinguishing N-type from T-type calcium channels. At the level of single channels inactivation of N takes the form of a bunching of openings near the beginning of the depolarization and an absence of openings near the end (Fig. 3*b*); hence the averaged current record decays almost completely by the end of a 136-msec test pulse (Fig. 3*d*). A single N-type channel will reopen many times before inactivating; the latency-to-first opening is not the rate-limiting step in governing the time course of average current decay. This is not the case for T-type calcium channels; single T channels generally burst once during a depolarization, and inactivation is rate-limited by the latency-to-first opening, very much like sodium channels in certain cells (3).

5. PHARMACOLOGICAL DIFFERENCES BETWEEN T, N, AND L-TYPE Ca CHANNELS

Table 1 compares the various types of calcium in their kinetics, unitary conductance, and sensitivity to pharmacological agents. Having defined T-, N-, and L-type calcium channels by the voltage dependence of their activation, inactivation, and open-channel properties, we find that the three channel types show sharp differences in response to dihydropyridines, cadmium ions, and the peptide toxin ω-conotoxin GVIA (hereafter abbreviated ω-CgTX VIA).

5.1 Differential Effects of Bay K 8644 and Other Dihydropyridines

The dihydropyridine calcium agonist Bay K 8644 strongly increases the activity of L-type calcium channels in DRG neurons by promoting a pattern of gating characterized by long openings and brief closings (72). This action is very much the same as the effect of Bay K 8644 on L-type channels in other cell types such as heart (42,51) or smooth muscle (34,47). As Fig. 4 illustrates, the enhancement of averaged L-type channel currents in DRG neurons is specific to this channel type; Bay K 8644 produced no change in the current carried by T- or N-type channels (73). Nifedipine, a dihydropyridine calcium antagonist, shows similar specificity for L-type calcium

TABLE 1. PROPERTIES OF THREE TYPES OF CALCIUM CURRENT IN CHICK DRG NEURONS

	T	N	L
Activation range (10 m*M* Ca)	Positive to −70 mV	Positive to −10 mV(?)	Positive to −10 mV
Inactivation range	−100 to −60 mV	>−100 to −40 mV	−60 to −10 mV
Relaxation rate (0 mV, 10 m*M* Ca or 10 m*M* Ba)	Moderate (τ ~20–50 msec)	Moderate (τ ~50 msec)	Very slow ($\tau > 500$ msec)
Single-channel conductance	8 pS	13 pS	25 pS
Single-channel kinetics	Late opening, brief burst, inactivation	Long burst, inactivation.	Hardly any inactivation
Cadmium block	Resistant	Sensitive	Sensitive
Nickel block	Sensitive	Less sensitive	Less sensitive
ω-CgTx VI block	Weak, reversible	Persistent	Persistent
Enhancement by Bay K 8644, inhibition by nifedipine	No	No	Yes

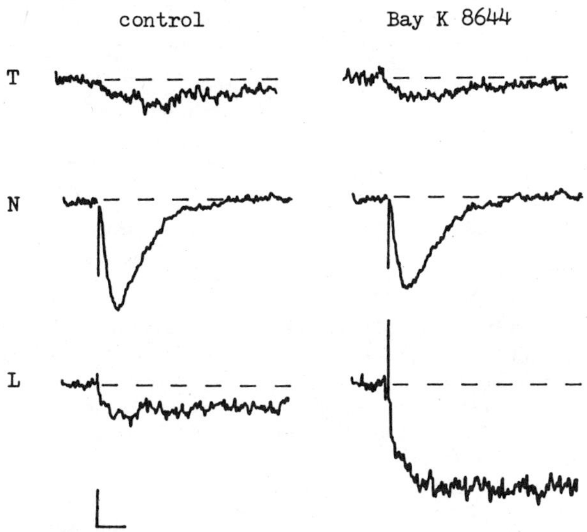

Figure 4. Single-channel current averages recorded before and after the addition of Bay K 8644. All recording conditions identical to those described in Fig. 2 legend. Averaged currents for three channel types before (left) and after exposure to 5 μ*M* Bay K 8644 (right). Top: T-type activity, voltage step from −80 to −30 mV. Patch 86D. Middle: N-type activity, voltage step from −80 to +10 mV. Patch 84K. Bottom: L-type activity, voltage step from −40 to +10 mV. Patch 89F. Vertical calibration bar gives 1 pA for middle panel and 0.25 for top and bottom panels. Horizontal calibration bar gives 20 msec for top and bottom panels and 10 msec for middle panel. (From ref. 73.)

channels in DRG neurons. Micromolar levels of nifedipine strongly reduce L currents without appreciably affecting T or N current (A. P. Fox, E. W. McCleskey, and R. W. Tsien, unpublished data).

5.2 Differential Effects of Cadmium

All three types of calcium channels are blocked by the inorganic blocker cadmium at relatively high concentrations (>200 μ*M*). Differences among the channel types were seen at lower doses (20–50 μ*M*). For example, 50 μ*M* cadmium almost completely abolishes N and L current but leaves about 50% of T (ref. 73; see also refs. 14, 21, 88).

5.3 Differential Effects of ω-CgTX VIA

Another agent with selective effects on the various types of calcium channel is the toxin ω-conotoxin GVIA (ω-CgTX VIA). This 27-amino acid peptide was derived from the venom of the fish-eating marine snail *Conus geographus*, a rich source of peptide toxins with diverse effects on ion channels (74). The peptide toxin ω-CgTX VIA inhibits transmitter release at a variety of vertebrate synapses, including the frog

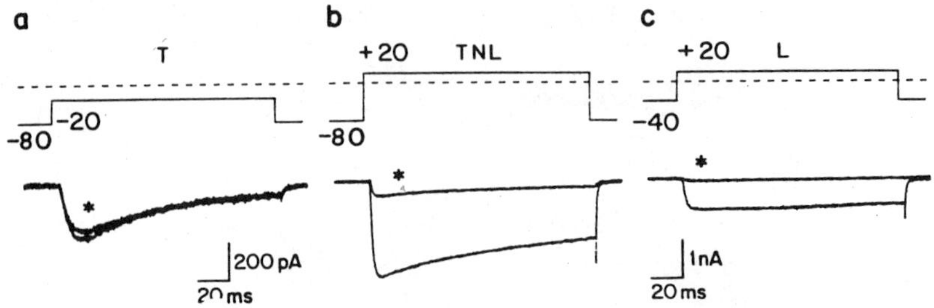

Figure 5. ω-CgTX VIA distinguishes different calcium channel types in DRG neurons. Whole-cell voltage-clamp recordings with linear leak and capacitive currents subtracted. Traces marked with * indicate data taken 3–5 min after a brief application of toxin. (*a–c*) Voltage protocols and current traces showing T current (*a*) and T, N, and L currents (*b*) and L current (*c*). Note that panel *a* has a vertical scale bar different from those in panels *b* and *c*. External solution contained (in m*M*): 10 $CaCl_2$, 135 TEA-Cl, 10 HEPES (titrated to pH 7.3 with TEA-OH), 1 μ*M* TTX. Pipette solution contained 100 CsCl, 10 EGTA, 5 $MgCl_2$, 2 NaATP, 40 HEPES (titrated to pH 7.3 with TEA–OH). Toxin pipette: 10 μ*M* ω-CgTX VIA plus 1 mg/ml cytochrome *c* in external solution. Cell D01D. (From ref. 62.)

neuromuscular junction (50). The toxin abolishes calcium-dependent but not sodium-dependent action potentials in chick DRG neurons (50), an action that results from blockade of calcium currents, not potassium currents (29).

Figure 5 illustrates the effects of ω-CgTX VIA on the three types of calcium channel in whole-cell recordings from DRG cells (62). Voltage protocols evoked pure T current (*a*), pure L current (*c*), or a combination of T, N, and L current. Each panel shows superimposed records taken before and following a brief application of toxin solution from a puffer pipette (*). Panels *b* and *c* show a persistent block of N and L current that lasts at least an hour. In contrast, T current was only weakly and transiently reduced in the presence of toxin. Following brief puffer application (*), T current shows essentially complete recovery. Thus ω-CgTX VIA blocks N- and L-type channels in preference to T-type channels (62).

The toxin seems to interact with an external binding site. A cytoplasmic second messenger mechanism or internal site of action is unlikely since channels isolated in cell-attached patches were protected from block by toxin applied to the rest of the cell; block of unitary calcium channel activity was seen with external ω-CgTX VIA application to outside-out patches (62).

The results obtained with pharmacological agents provide additional criteria for distinguishing the various types of calcium channels. Once again, the properties of N-type channels are in some sense intermediate between T- and L-type channels. Although N-type channels are similar to T-type channels, they are unlike L-type channels in their lack of responsiveness to Bay K 8644 and nifedipine; N-type channels are like L-type channels but unlike T-type channels in their responsiveness to cadmium and ω-CgTX VIA. So far, we know of no single agent that blocks T- and L-type calcium channels but spares N.

On the basis of available evidence, we believe that T, N, and L channels are distinct entities, differing in properties of voltage-dependent gating, permeability, and pharmacological sensitivity. It seems unlikely, if not impossible, that the channel "types" are merely different expressions of the behavior of one kind of molecule; if this were the case, the interval between interconversions must be considerably slower than the ~30-min duration of many single-channel recordings from DRG neurons.

6. MULTIPLE TYPES OF CALCIUM CHANNEL IN OTHER EXCITABLE CELLS

We have emphasized results in DRG neurons for the purposes of illustrating the diversity of calcium channels, but many of the same findings hold true for other excitable cells. One general finding in many preparations is the coexistence of long-lasting and rapidly inactivating calcium currents. Figure 6 illustrates results from a rat olfactory neuron (*A*), a frog sympathetic neuron (*B*), cardiac (*D*) and smooth (*E*) muscle cells, and a GH4 cultered pituitary cell (*E*). In each case,

TABLE 2. CLASSIFICATION OF MULTIPLE CALCIUM CHANNEL TYPES IN VERTEBRATE CELLS

<table>
<tr><th></th><th></th><th colspan="3">Calcium Channel Type</th><th></th></tr>
<tr><th></th><th></th><th colspan="2">Fast, Inactivating</th><th>Slow, Persistent</th><th></th></tr>
<tr><th>Tissue</th><th>Representative Preparation</th><th>LVA T</th><th>HVA N</th><th>HVA L</th><th>Reference</th></tr>
<tr><td>Neuronal</td><td>Chick, rat DRG</td><td>+</td><td>+</td><td>+</td><td>73[a]</td></tr>
<tr><td></td><td>Frog sympathetic</td><td>−</td><td>+</td><td>+</td><td>56</td></tr>
<tr><td></td><td>Neuroblastoma</td><td>+</td><td>−</td><td>+</td><td>88</td></tr>
<tr><td></td><td>Rat olfactory</td><td colspan="2">+</td><td>+</td><td>16</td></tr>
<tr><td></td><td>Rat hippocampal</td><td colspan="2">+</td><td>+</td><td>39</td></tr>
<tr><td>Heart</td><td>Dog, frog atrial</td><td>+</td><td>−</td><td>+</td><td>8</td></tr>
<tr><td></td><td>Guinea pig ventricular</td><td>+</td><td>−</td><td>+</td><td>67,70</td></tr>
<tr><td></td><td>Frog ventricular</td><td>−</td><td>−</td><td>+</td><td>62</td></tr>
<tr><td>Smooth muscle</td><td>Rat mesenteric artery</td><td>+</td><td>−</td><td>+</td><td>9</td></tr>
<tr><td></td><td>Rat venous</td><td>+</td><td>−</td><td>+</td><td>83</td></tr>
<tr><td></td><td>Rabbit ear artery</td><td>+</td><td>−</td><td>+</td><td>1</td></tr>
<tr><td>Skeletal</td><td>Mouse myotubes</td><td>+</td><td>−</td><td>+</td><td>7</td></tr>
<tr><td>Endocrine</td><td>Rat myoballs</td><td>+</td><td>−</td><td>+</td><td>20a</td></tr>
<tr><td></td><td>GH_3 or GH_4</td><td>+</td><td>−</td><td>+</td><td>5,21</td></tr>
<tr><td></td><td>Adrenal chromaffin</td><td>−</td><td>−</td><td>+</td><td>30,46</td></tr>
<tr><td></td><td>Lactotroph and somatotroph</td><td>+</td><td>−</td><td>+</td><td>23</td></tr>
<tr><td>Glial</td><td>Type 2 astrocyte</td><td>+</td><td>−</td><td>−</td><td>6</td></tr>
</table>

[a]See ref. 73 for additional references cited therein.

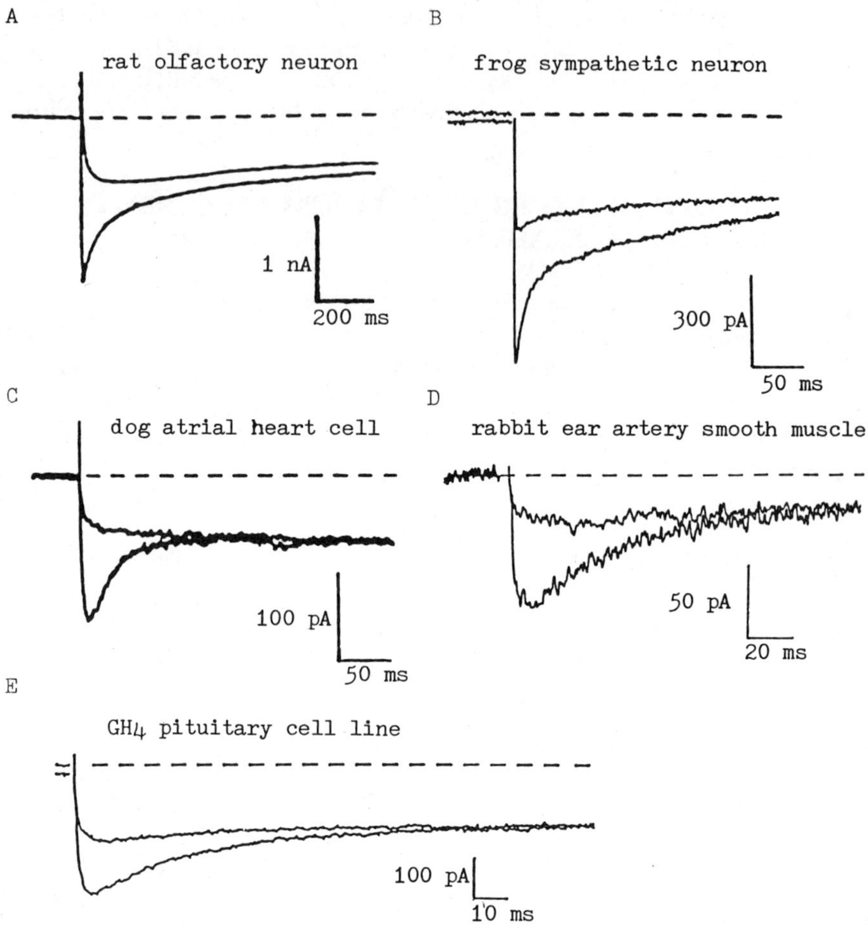

Figure 6. Coexistence of sustained and rapidly inactivating calcium current in a wide variety of excitable cells. Each panel superimposes current records evoked by depolarization from two different holding potentials (HP) to a fixed test potential (TP). The larger and more prominently decaying inward current is associated with the more negative holding potential. (*A*) Rat olfactory pyramidal neuron: TP = −20 mV, HP = −50 or −125 mV (from ref. 16). (*B*) Frog sympathetic neuron: TP = 0 mV, HP = −40 or −80 mV (from ref. 56). (*C*) Dog atrial cell: TP = 0 mV, HP = −30 or −80 mV (from ref. 8). (*D*) Rabbit ear artery smooth muscle cell: TP = +20 mV, HP = −30 or −80 mV (from ref. 1). (*E*) GH4 pituitary cell line: TP = −10 mV, HP = −35 or −90 mV (from ref. 21).

depolarizations from depolarized holding potentials evoked a sustained Ca current while depolarizations from strongly negative holding potentials gave a combination of sustained and inactivating calcium current components.

Table 2 provides an overview of the expression of the various types of calcium channel in a large number of excitable cells, including the preparations illustrated in Fig. 6. The classification is based on properties such as gating kinetics;

pharmacology; and, where known, single-channel conductance. As detailed in the following paragraphs, T, N, and L are used as broad categories with the presumption that broad similarities between channels within each class greatly outweigh detailed differences. The entries for olfactory and hippocampal neurons merely indicate observations of fast inactivating current; experimental information is insufficient for classification of the current as T or N. In some other cases, the classification must be viewed as tentative pending more experimental data.

A number of points deserve comment. Table 2 includes examples of preparations that seem to express only one component of calcium current. Whereas adrenal chromaffin cells and frog ventricular cells show sustained L-type current exclusively, type 2 astrocytes (6) display low-voltage activating, rapidly inactivating T-type calcium current, but not N or L current. The most common behavior is a combination of sustained and inactivating components, as already illustrated in Fig. 6. Although this pattern is widespread, there is wide variation in the relative magnitude of the sustained and inactivating currents, ranging from a large predominance of inactivating (presumably T) current in smooth muscle cells from veins (83) and ear artery (1), to a large predominance of L-type current in guinea pig ventricular muscle (8,67,72). The DRG neurons and frog and dog atrial cells are intermediate cases between the extremes. Channels identified with certainty as N type have so far been found only in neurons. Sympathetic neurons are interesting because they show a large amount of N-type current but little or no detectable T current (refs. 18 and 56; A. P. Fox, L. Hirning, E. W. McCleskey, R. J. Miller and R. W. Tsien, unpublished observations).

7. DISTINCTIVE FEATURES OF T-TYPE CHANNELS COMMON TO DIFFERENT CELLS

Detailed analysis of T-type calcium channel activity in heart cells has been carried out at the level of single cells (8,67) and single channels (72). There is a strong resemblance between T-type channels in cardiac cells and DRG neurons. Characteristic features are (1) the low "threshold" for activation, (2) similar slope conductance (~8 pS) with either Ca^{2+} or Ba^{2+} (110 m*M*) as charge carriers, (3) relative insensitivity to DHP calcium antagonists and Cd^{2+}, and (4) ability of activity to survive excision.

8. SIMILARITIES AND DIFFERENCES AMONG VARIOUS L-TYPE CALCIUM CHANNELS

The sustained current component has been studied at the level of single channels in a wide variety of preparations, including heart cells, DRG neurons, smooth muscle cells, A10 cells, adrenal chromaffin cells, and PC12 cells (for references, see Table 2). In all of these preparations, the calcium channels that generate the long-lasting current can be classified as L type because they share a number of important features: (1) their slope conductance with 110 m*M* Ba^{2+} as the charge carrier is large (20–25 pS); (2) channel activation is steeply voltage dependent between −20 and +30 mV; (3) inactivation is relatively slow, and its rate increases with increasing voltage; (4) the pattern of gating shifts between different gating modes, with brief openings (mode 1) as the normal form of gating but with occasional episodes of another form of gating characterized by long openings and occasional short closings (mode 2); (5) DHP calcium channel agonists such as Bay K 8644 increase L-type current by promoting mode 2 gating; and (6) excision leads to an irreversible decrease of channel activity and, correspondingly, cell dialysis leads to irreversible rundown of whole-cell current.

Although we have labeled channel types with similar gating pattern and conductance in different cells with the same letter, it is also important to point out some significant differences between L-type channels in heart and DRG neurons:

1. Cardiac L-type channels show both a voltage and calcium entry-dependent mechanism of inactivation (10,49,54,63), while a calcium entry-dependent inactivation seems less important in DRG L-type channels, given the close similarity of the kinetics of inactivation with either Ca^{2+} or Ba^{2+} as the charge carrier (71).
2. The toxin ω-CgTX blocks L-type channels in DRG neurons with high affinity but has no detectable effect on cardiac L-type channels (62).
3. Interventions leading to elevated cAMP greatly enhance cardiac L-type channel activity, whereas diffusable cAMP derivatives fail to increase DRG L-type channel activity (A. P. Fox, unpublished observations). The list of dissimilarities between "L-type" calcium channels in different preparations is likely to grow as more detailed experimental information becomes available. Meanwhile, the idea of calcium channel types will remain useful as long as general similarities outweigh detailed differences.

9. DIFFERENCES BETWEEN DHP-SENSITIVE CALCIUM CHANNELS IN SKELETAL MUSCLE AND IN OTHER PREPARATIONS

The transverse tubules of skeletal muscle are endowed with an extremely high density of dihydropyridine binding sites, of the order of 50–500 times higher than in other membrane preparations (33), and have provided the starting material for the most successful efforts at calcium channel purification (12,22,36,84a). It is important, therefore, to know whether skeletal T-tubule calcium channels are representative of dihydropyridine-sensitive L-type calcium channels in other preparations.

Significant differences have already been observed at the level of whole cell recordings. Skeletal calcium channels activate much more slowly than do L-type calcium channels in neurons, heart, or smooth muscle. Skeletal and cardiac calcium channels differ significantly in magnesium permeability (43,52,60) despite similarities in pore size (60,61).

Striking differences have also been seen at the level of unitary recordings. Although calcium channels in the T system are inaccessible to conventional patch-clamp recordings, their single-channel properties can be analyzed following incorporation of T-tubule membranes into planar bilayers (ref. 2; cf. refs. 25, 26, 68,

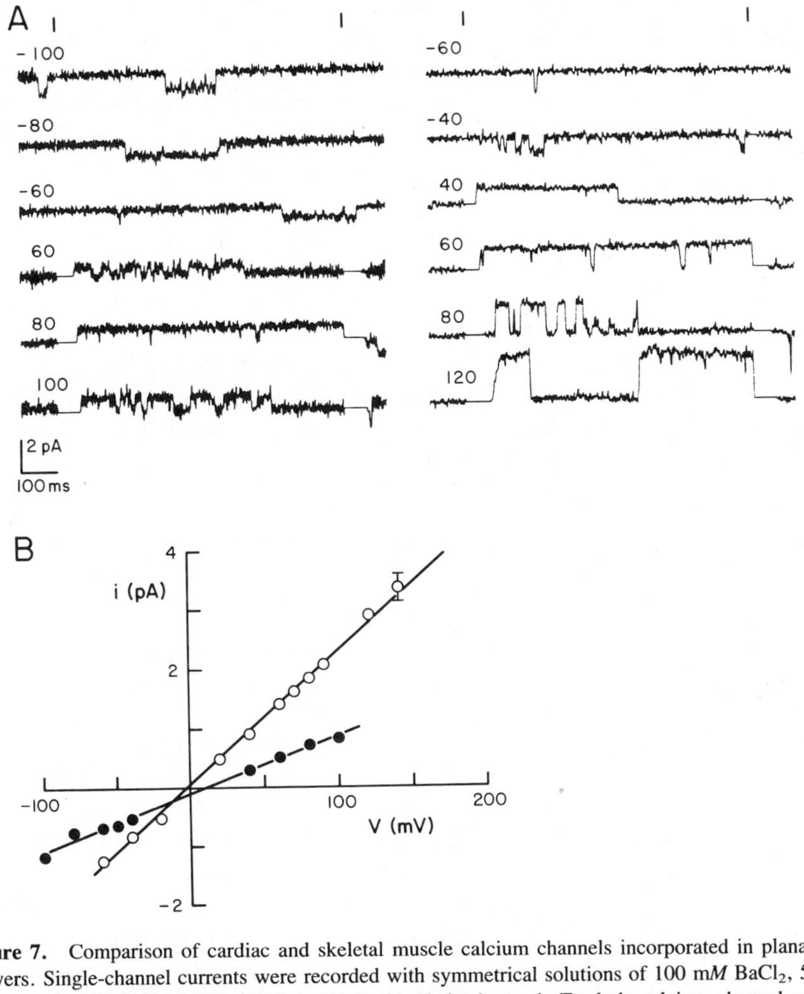

Figure 7. Comparison of cardiac and skeletal muscle calcium channels incorporated in planar lipid bilayers. Single-channel currents were recorded with symmetrical solutions of 100 m*M* $BaCl_2$, 50 m*M* NaCl, 10 m*M* HEPES–NaOH, pH 7.4. (*A*) (Left): skeletal muscle T-tubule calcium channel activity. Holding potential was −100 mV; the vertical marks indicate the start and termination of depolarizations to the test potentials shown. Bilayer R3A1. (*A*) (Right): cardiac calcium channel records. Holding potential was −80 mV. Bilayer J5H1. (*B*) Current–voltage curves of cardiac (open symbols) and skeletal (closed symbols) calcium channels. Lines represent least-squares fits. (From ref. 79.)

and 69). Like cardiac L-type channels, skeletal calcium channels respond to Bay K 8644 with an increased proportion of long openings (2). As Fig. 7 illustrates, there is a skeletal calcium channel with a significantly smaller slope conductance (10.6 ± 0.8 pS) than the cardiac L-type calcium channel (22.7 ± 1.5 pS) despite identical lipid and ionic conditions (100 mM Ba on both sides) (79; see also ref. 2). However, recordings from calcium channels purified from T-tubules and reconstituted in planar bilayers indicate a slope conductance of 20 pS in symmetrical 90 mM Ba (84a). In all experiments, the skeletal channels can activate at strongly negative potentials where there is essentially no activity of the cardiac L-type calcium channel. The skeletal calcium channels activate more slowly and have much longer mean open times, as might be expected from earlier whole-cell recordings (4,7,20a,80). These differences in single-channel properties suggest further functional and structural diversity among calcium channels that share sensitivity to dihydropyridines.

10. ARE THERE OTHER MAJOR CLASSES OF CALCIUM CHANNEL?

As recording techniques have improved, we have seen a rapid increase in the number of distinguishable channel subtypes in the case of sodium, potassium, and chloride channels (45). If recent experiments are any indication, the end is not yet in sight for calcium channels, either. Additional types of calcium channel almost certainly exist in addition to the voltage-activated calcium channels that are the main focus of this chapter.

10.1 Calcium Channels Induced by Transmitters and/or Intracellular Enzymes

There are a number of systems where one might expect to see recruitment of dormant calcium channels through the action of neurotransmitters or intracellular second messengers. This is a promising area for further study. "Receptor-operated calcium channels" have been postulated to exist in smooth muscle (11,88) but have never been identified with direct electrophysiological recording. A "transmitter-induced calcium current" was described by Pellmar and Carpenter (75) in their study of serotonin action in Aplysia neurons but has not been pursued at the level of single channels.

Calcium channels induced by protein kinase C stimulation have been found in Aplysia bag cell neurons. In these neurosecretory cells, the height of the calcium-dependent action potential is increased by injection of protein kinase C or stimulation of protein kinase C with the phorbol ester 12-O-tetradecanoylphorbol-13-acetate (TPA). Cells pretreated with TPA show twofold larger calcium currents than do control cells (24). Single-channel recordings from TPA-pretreated cells reveal a new class of calcium channel not detected in untreated cells (82). The TPA-induced calcium channel is easily distinguished because it has a slope conductance of 23 pS, considerably larger than the normally functional calcium channel (10–15 pS), although both types of calcium channel share similar voltage dependence.

10.2 Leak Calcium Channels

Thus far, most investigators have focused on calcium channels that require membrane depolarization for activation. Recent experiments have provided evidence for channels capable of mediating divalent cation entry without depolarization, with the following two principal observations:

1. Cardiac sarcolemmal membranes incorporated in planar bilayers often display unitary activity with a ~10-pS slope conductance (100 m*M* Ba). Unlike L- or T-type channels, the activity shows very weak voltage dependence, and its activity may readily be observed at bilayer potentials corresponding to cellular resting potentials (R. L. Rosenberg, P. Hess, and R. W. Tsien, unpublished observations).
2. Excised patches from Aplysia neurons show a divalent cation selective unitary conductance (~16 pS with 60 m*M* Ca^{2+} as charge carrier). The channel shows significant opening probability over a broad range of potentials, including the resting potential, and only mildly voltage-dependent gating (ref. 20; also J. A. Strong, A. P. Fox, R. W. Tsien, and L. K. Kaczmarek, unpublished observations).

It remains to be seen whether these types of activity can be detected under physiological conditions in intact cells and whether their activity is linked in some way to levels of intracellular messengers or metabolites. Meanwhile, it seems reasonable to consider these channels as possible "leak calcium channels" because their activity in resting cells would constitute a pump–leak balance when combined with calcium extrusion mechanisms such as a calcium pump or a calcium–sodium exchange.

11. CELLULAR DISTRIBUTION AND FUNCTIONS OF T-, N-, AND L-TYPE CHANNELS

Biophysical analysis of various types of calcium channel has preceded knowledge about their cell biology, but it is likely that questions about where calcium channels are located and what they do may soon be unraveled. The work of Llinas and colleagues has had a strong influence on present thinking about the localization and functional roles of the multiple types of calcium channel. Their recordings of low- and high-threshold spikes in various CNS neurons provided strong suggestions of the existence of two types of calcium channel, distinguished by their voltage range of activation (57,58). On the basis of extracellular recordings, they suggested that rapidly inactivating, LVA calcium channels were localized on the cell body, and noninactivating, HVA calcium channels were localized on dendrites. The LVA channels were thought to be active in pacemaker activity and rebound excitation following afterhyperpolarizations; the HVA channels were thought to underly dendritic action potentials and transmitter release.

The working hypothesis can be updated on the basis of more direct analysis with patch-clamp recordings. Apart from reckoning with the possibility of three rather than two distinct channel types, our present view must consider evidence that HVA channels are most certainly found on the soma, not only on dendrites. The relative proportion of these channels on cell bodies and peripheral processes awaits further study. A search for LVA (T-type) calcium channels in dendritic membranes might also be interesting in light of possible mechanisms for helping the spread of weak depolarizations from peripheral dendritic inputs and for the generation of low-threshold spikes on dendritic spines (59).

The function of N-type channels, the most recently discovered types of calcium channel, is perhaps the most mysterious. These channels often appear clustered in hot spots on the surface of the cell body, with as many as 600 channels within a patch no larger than a few square micrometers in area, a far higher density than that found for the cell body as a whole (73). Such geographic segregation is reminiscent of the clustering of putative calcium channel proteins in the synaptic zones of presynaptic terminals (65,66) and raises questions about the possible involvement of N-type channels in the control of neurosecretion. There is indirect evidence that N-type calcium channels might trigger transmitter release from frog motor terminals and rat sympathetic neurons: in both systems, neurosecretion is inhibited by ω-CgTX VIA but not by nifedipine (50,76). The other types of calcium channel may be more important for neurotransmitter release in other systems. Involvement of L-type channels is indicated by dihydropyridine-sensitive secretion in chromaffin cells (35), PC12 cells (84), and DRG neurons (76). Participation of T-type channels in the secretion of prolactin is suggested by whole-cell recordings from pituitary lactotrophs that show T current but not N or L currents (23).

12. WHAT MAKES A CALCIUM CHANNEL A CALCIUM CHANNEL?

The main focus of this chapter has been the multiplicity of calcium channels, but it is fair to ask what properties the diverse types of calcium channel have in common. The most obvious characteristic, almost a definition, is selectivity for calcium over other ions. In the case of all types of calcium channel studied thus far, reversal potential measurements indicate that the pore strongly prefers calcium over sodium, with Fatt–Ginsborg selectivity ratios of the order of 10^3 or 10^4 (54). The mechanism of ion permeation and selectivity has been studied in detail in dihydropyridine-sensitive calcium channels in cardiac and skeletal muscle. These channels behave like large-diameter, single-file pores with high-affinity binding sites for calcium ions (4,43,44,52,60).

Much less is known about ion permeation through T- or N-type calcium channels. As already mentioned, T channels differ from L channels in that they are considerably less sensitive to block by cadmium ions, and they do not appear to discriminate between calcium and barium ions. Such differences are important in the initial distinction between various channel types, but they might well be explained by relatively small structural differences among a closely related family of membrane proteins. An analogy might be made to calcium binding proteins such as troponin, calmodulin, or parvalbumin, which exhibit a variety of selectivity sequences for alkaline earth cations even though their calcium binding sites share the same basic "*E–F* hand" structure (81).

REFERENCES

1. Aaronson, P., Benham, C. B., Bolton, T. B., Hess, P., Lang, R. J., and Tsien, R. W. (in press). *J. Physiol.*
2. Affolter, H. and Coronado, R. (1985). *Biophys. J.*, **48**, 341–347.
3. Aldrich, R. W., Corey, D. P., and Stevens, C. F. (1983). *Nature*, **306**, 436–441.
4. Almers, W. and McCleskey, E. W. (1984). *J. Physiol.*, **353**, 585–608.
5. Armstrong, C. M. and Matteson, D. R. (1985). *Science*, **227**, 65–67.
6. Barres, B. A., Chun, L. L. Y., and Corey, D. P. (1985). *Soc. Neurosci. Abstr.*, **11**, 147.
7. Beam, K. G., Knudson, C. M., and Powell, J. A. (1985). *J. Gen. Physiol.*, **86**, 36a.
8. Bean, B. P. (1985). *J. Gen. Physiol.*, **86**, 1–30.
9. Bean, B. P., Sturek, M., Puga, A., and Hermsmeyer, K. (1985). *J. Gen. Physiol.*, **86**, 23a.
10. Bechem, M. and Pott, L. (1985). *Pflügers Arch.*, **404**, 10–20.
11. Bolton, T. B. (1979). *Physiol Rev.*, **59**, 606–718.
12. Borsotto, M., Barhanin, J. Norman, R. I., and Lazdunski, M. (1984). *Biochem. Biophys. Res. Commun.*, **122**, 1357–1368.
13. Bossu, J. L. and Feltz, A. (1984). *Neurosci. Lett.*, **51**, 241–246.
14. Bossu, J. L., Feltz, A., and Thomann, J. M. (in press). *Pflügers Arch.*
15. Brown, A. M., Camerer, H., Kunze, D. L., and Lux, H. D. (1982). *Nature*, **299**, 156–158.
16. Brown, D. A., Constanti, A., Docherty, R. J., Galvan, M., Gaehwiler, B., and Halliwell, J. V. (1984). *IUPHAR Proceedings.*
17. Carbone, E. and Lux, H. D. (1984). *Biophys. J.*, **46**, 413–418.
18. Carbone, E. and Lux, H. D. (1984). *Nature*, **310**, 501–502.
19. Cavalie, A., Ochi, R., Pelzer, D., and Trautwein, W. (1983). *Pflügers Arch.*, **398**, 284–297.
20. Chesnoy-Marchais, D. (1985). *J. Physiol.*, **367**, 457–488.

20a. Cognard, C., Lazdunski, M., and Romey, G. (1986). *Proc. Natl. Acad. Sci.* (USA), **83**, 517–521.

21. Cohen, C. J. and McCarthy, R. T. (1985). *Biophys. J.*, **47**, 513a.
22. Curtis, B. M. and Catterall, W. A. (1984). *Biochemistry*, **23**, 2113–2117.
23. DeRiemer, S. A. and Sakmann, B. (in press), *Exptl. Br. Res.*
24. DeRiemer, S. A., Strong, J. A., Albert, K. A., Greengard, P., and Kaczmarek L. K. (1985). *Nature*, **313**, 313–316.
25. Ehrlich, B. E., Finkelstein, A., Fortes, M., and Kung, C. (1984). *Science*, **225**, 427–428.

26. Ehrlich, B. E., Schen, C. R., Garcia, M. L., and Kaczorowski, G. J. (in press). *Proc. Natl. Acad. Sci.* (USA).
27. Fedulova, S. A., Kostyuk, P. K., and Veselovsky, N. S. (1985). *J. Physiol.*, (London), **359**, 431–446.
28. Feldman, D., Fox, A. P., McCleskey, E. W., Olivera, B. M., Tsien, R. W., and Yoshikami, D. (in press). *Nature*.
29. Feldman, D. H. and Yoshikami, D. (1985). *Soc. Neurosci, Abstr.*, 517.
30. Fenwick, E. M., Marty, A., and Neher, E. (1982). *J. Physiol.* (London), **331**, 599–635.
31. Fishman, M. C. and Spector, I. (1981). *Proc. Natl. Acad. Sci.* (USA), **78**, 5245–5249.
32. Fosset, M., Jaimovich, E., Delpont, E., and Lazdunski, M. (1983). *J. Biol. Chem.*, **258**, 6086–6092.
33. Fox, A. P. and Krasne, S. (1984). *J. Physiol.* (London), **356**, 491–505.
34. Fox, A. P., Hess, P., Lansman, J. B., Nilius, B., Nowycky, M. C., and Tsien, R. W. (1986). In Poste, G. and Crooke, S. T., Eds., *New Insights into Cell and Membrane Transport Processes*, Smith-Kline-French Symposium.
35. Garcia, A. G., Sala, F., Reig, J. A., Viniegra, S., Frias, J, Fonteriz, R., and Gandia, L. (1984). *Nature*, **308**, 69–71.
36. Glossmann, H. and Ferry, D. R. (1983). *Naunyn-Schmiedeberg's Arch. Pharmacol.*, **323**, 279–291.
37. Hagiwara, S. and Byerly, L. (1981). *Ann. Rev. Neurosci.*, **4**, 69–125.
38. Hagiwara, S., Ozawa, S., and Sand, O. (1975). *J. Gen. Physiol.*, **65**, 617–644.
39. Halliwell, J. V. (1983). *J. Physiol.*, **341**, 10P.
40. Hamill, O., Marty, A., Neher, E., Sakmann, B., and Sigworth, F. J. (1981). *Pflügers Arch.*, **391**, 85–100.
41. Hermsmeyer, K. and Sturek, M. (1985). *J. Gen. Physiol.*, **86**, 24a.
42. Hess, P., Lansman, J. B., and Tsien, R. W. (1984) *Nature*, **311**, 538–544.
43. Hess, P., Lansman, J. B., and Tsien, R. W. (in press). *J. Gen. Physiol.*
44. Hess, P. and Tsien, R. W. (1984). *Nature*, **309**, 453–456.
45. Hille, B. (1984). *Ionic Channels in Excitable Cell Membranes*, Sinauer Associates, Sunderland, MA.
46. Hoshi, T. (1985). Doctoral dissertation, Yale University.
47. Josephson, I. R., Caffrey J. M., and Brown, A. M. (in press). *Science*.
48. Kameyama, M. (1983). *J. Membrane Biol.*, **72**, 195–203.
49. Kass, R. S. and Sanguinetti, M. C. (1984). *J. Gen. Physiol.*, **84**, 705–726.
50. Kerr, L. M. and Yoshikami, D. (1984). *Nature*, **308**, 282–284.
51. Kokubun, S. and Reuter, H. (1984). *Proc. Natl. Acad. Sci.* (USA), **81**, 4824–4827.
52. Lansman, J. B., Hess, P., and Tsien, R. W. (in press). *J. Gen. Physiol.*
53. Lee, K. S., Akaike, N., and Brown, A. M. (1980). *J. Neurosci. Methods*, **2**, 51–78.
54. Lee, K. S., Marban, E., and Tsien, R. W. (1985). *J. Physiol.* (London), **364**, 395–411.
55. Lee, K. S. and Tsien, R. W. (1984). *J. Physiol.* (London), **354**, 253–272.
56. Lipscombe, D. L. (1986). Doctoral thesis, University College, London.
57. Llinas, R. and Sugimori, M. (1980). *J. Physiol.* (London), **305**, 171–195.
58. Llinas, R. and Yarom, Y. (1981). *J. Physiol.* (London), **315**, 549–567.
59. Malinow, R. and Miller, J. P. (1985). *Soc. Neurosci. Abstr.*, **11**, 856.
60. McCleskey, E. W. and Almers, W. (1985). *Proc. Natl. Acad. Sci.* (USA), **82**, 7149–7153.
61. McCleskey, E. W., Hess, P., and Tsien, R. W. (1985). *J. Gen. Physiol.*, **86**, 22a.

62. McCleskey E. W., Fox, A. P., Feldman, D., Olivera, B. M., Tsien, R. W., and Yoshikami, D. (1986). *Biophys. J.*, **49**, 431a.
63. Mentrard, D., Vassort, G., and Fischmeister, R. (1984). *J. Gen. Physiol.*, **83**, 105–131.
64. Middlemiss, D. N. and Spedding, M. (1985). *Nature*, **314**, 94–96.
65. Miller, R. J. (1985). *Trends Neurosci.*, **8**, 45–47.
66. Miller, R. J. (1986). In Triggle, D. J., and Venter, J. C., Eds., *Structure and Physiology of the Slow Inward Calcium Channel, Receptor Biochemistry and Methodology*, Vol. 7.
67. Mitra, R. and Morad, M. (1985). *J. Gen. Physiol.*, **86**, 22a.
68. Nelson, M. T., French, R. J., and Krueger, B. K. (1984). *Nature*, **308**, 77–80.
69. Nelson, M. T. (1986). *J. Gen. Physiol.*, **87**, 201–222.
70. Nilius, B., Hess, P., Lansman, J. B., and Tsien, R. W. (1985). *Nature*, **316**, 443–446.
71. Nowycky, M. C., Fox, A. P., and Tsien, R. W. (1984). *Biophys. J.*, **45**, 36a.
72. Nowycky, M. C., Fox, A. P., and Tsien, R. W. (1985). *Proc. Natl. Acad. Sci.* (USA), **82**, 2178–2182.
73. Nowycky, M. C., Fox, A. P., and Tsien, R. W. (1985). *Nature*, **316**, 440–443.
74. Olivera, B. M., Gray, W. R., Zeikus, R., McIntosh, J. M., Varga, J., Rivier, J., de Santos, V., and Cruz, L. J., (1985). *Science*, **230**, 1338–1343.
75. Pellmar, T. C. and Carpenter, D. O. (1980). *J. Neurophysiol.*, **44**, 423–439.
76. Perney, T. M., Hirning, L. D., and Miller, R. J. (1986). *Biophys. J.*, **49**, 200a.
77. Reuter, H. (1983). *Nature*, **301**, 569–574.
78. Reuter, H. (1985). *Nature*, **316**, 391.
79. Rosenberg, R. L., Hess, P., Reeves, J., Smilowitz, H., and Tsien, R. W. (in press). *Science*.
80. Sanchez, J. A. and Stefani, E. (1978). *J. Physiol.* (London), **283**, 197–209.
81. Seamon, K. B. and Kretsinger, R. H. (1983). In Spiro, T. G., Ed., *Calcium in Biology*, Wiley, New York.
82. Strong, J. A., Fox, A. P., Tsien, R. W., and Kaczmarek, L. K. (1986). *Biophys. J.*, **49**, 430a.
83. Sturek, M. and Hermsmeyer, K. (1985). *J. Gen Physiol.*, **86**, 23a.
84. Toll, L. (1982). *J. Biol. Chem.*, **257**, 13189–13192.
84a. Trautwein, W., Cavalié, A. and Pelzer, D. (in press). *Circulation Res.*
85. Trautwein, W. and Pelzer, D. (in press). In Grune & Stratton, Orlando, FL. *Cardiac Electrophysiology and Arrhythmias*, Zipes, D. P. and Jalife, J., Eds.
86. Tsien, R. W. (1983). *Ann. Rev. Physiol.*, **45**, 341–358.
87. Tsien, R. W., Bean, B. P., Hess, P., and Nowycky, M. (1983). *Cold Spring Harbor Symp. Quant. Biol.*, **48**, 201–212.
88. Tsunoo, A., Yoshii, M., and Narahashi, T. (1984). *Biophys. J.*, **47**, 433a.
89. Van Breemen, C., Aaronson, P., and Loutzenhiser, R. (1979). *Pharmacol. Rev.*, **30**, 167–208.

CHAPTER 11

VOLTAGE-DEPENDENT BINDING AND ACTION OF 1,4-DIHYDROPYRIDINE ENANTIOMERS IN INTACT CARDIAC CELLS

H. REUTER
H. PORZIG
S. KOKUBUN
B. PROD'HOM
Department of Pharmacology
University of Berne
Berne, Switzerland

The important role of calcium ions (Ca^{2+}) in the regulation of several cellular functions, like muscle contraction, secretion of hormones and neurotransmitters, or enzymatic activities, is widely recognized. One mechanism by which the cytosolic calcium concentration can be controlled is the voltage-dependent opening and closing of calcium channels in the surface membrane of cells. These channels are integral membrane proteins forming minute pores through which calcium ions can flow down their electrochemical gradient from the intercellular space into the cell. The molecular structure of these proteins determines which ions are allowed to pass through, that is, the *ion selectivity* of these channels. It also determines the voltage-dependent opening and closing kinetics, commonly called *gating*, of the channels. Recently it has become apparent that there is more than one type of calcium channels in muscle and neuronal tissues (7,26,28). The various calcium channels can be distinguished by their respective ion selectivity, by their distinct modes of gating, and by their different sensitivities to pharmacological interventions (2,30).

There is one particular type of calcium channel present in many excitable cells, including cardiac muscle, which is highly sensitive to 1,4-dihydropyridines. Several compounds of this class of drugs have gained wide therapeutic application. However, they have also been extremely useful experimentally as radioligands in binding assays for the identification of calcium channels in particulate membrane fractions, as tools for isolation and purification of these channels and, in electrophysiological experiments, as drugs that modulate the gating behavior of these channels (31). In the present study we have investigated two enantiomers of the dihydropyridine derivative Sandoz 202-791 (21) (Fig. 1) that have opposite effects on one type of calcium channel in cardiac tissue. One enantiomer [(+)-(*S*)-202-791] acts as a channel activator, while the other enantiomer [(−)-(*R*)-202-791] is a calcium channel blocker. The binding affinities of both enantiomers are strongly dependent on the membrane potential of these cells.

1. PROPERTIES OF SINGLE CALCIUM CHANNELS

Calcium channels in neuronal and muscle tissues have been extensively analyzed by the patch-clamp method over the last few years. With this method a very small area of membrane is electrically isolated from its surrounding (17). Such a membrane patch

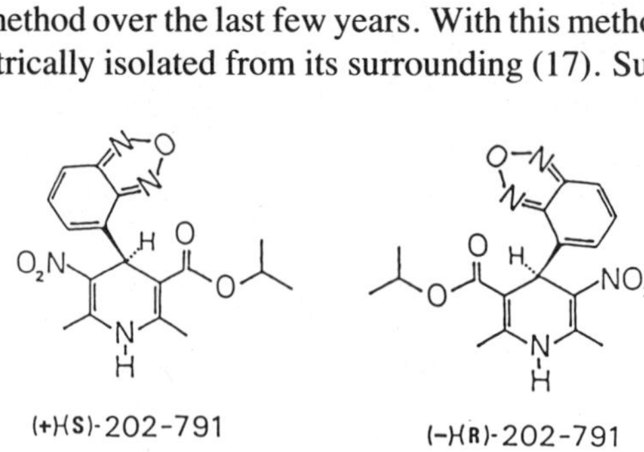

Figure 1. Structures of the enantiomers of Sandoz 202–791 (see ref. 21).

may contain one or more functional calcium channels that can open and close if the potential across the membrane is changed. Figure 2 shows elementary currents flowing through a single calcium channel as recorded from a cultured neonatal rat heart cell. The charge carriers were Ba^{2+} ions. Depolarizing step pulses of 70- or 80-mV amplitude were applied from a holding potential identical to the resting potential (RP)~ -70 mV. Calcium channel openings often occur in bursts separated by longer intervals of closures (mode 1 gating) (19). Quite rarely very long openings can also be observed (mode 2 gating: (19); see second and third current traces with $\Delta V = 70$ mV in Fig. 2). Sometimes the channel does not open at all during repetitive depolarization [mode 0 gating, (19); see traces labeled $p = .00$ in Fig. 2]. The overall probability p of the channel to be in the open state increases with increasing depolarization. With $p = 1.0$ the channel would be open all the time throughout the voltage step, while with $p = 0$ it would remain shut. The open-state probability can be calculated from $I = Npi$, where I is the time-averaged mean current flowing through an open channel during depolarizing pulses, i is the magnitude of the single channel current that decreases with increasing depolarizing step sizes because the driving force for the inward moving Ba^{2+} ions decreases (Fig. 3), and N is the number of functioning channels in the small membrane patch ($N = 1$ in Figs. 2 and

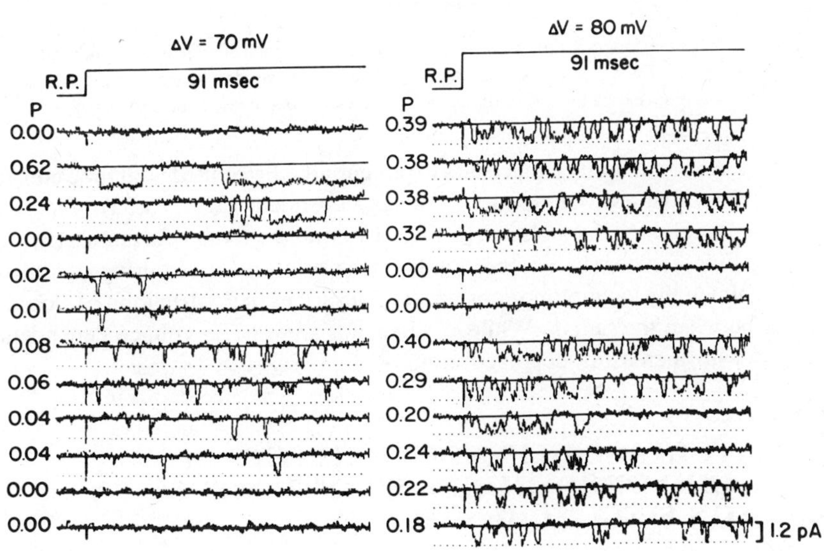

Figure 2. Elementary calcium channel currents recorded from a cell-attached membrane patch in a cultured cardiac cell (day 2). The pipette was filled with 96 m*M*/liter $BaCl_2$; thus the current carrier was Ba^{2+}. Depolarizing clamp steps (ΔV) of 70 mV (left) or 80 mV (right) were applied from a holding potential identical to the resting potential (~ -70 mV). Pulse duration was 91 msec. Twelve consecutive single-channel current records are shown for each pulse potential. The numbers on the left-hand side of each current record indicates the open-state probability p of the channel during that pulse. The heavy lines indicate the closed state of the channel; the dotted lines, the open-state current.

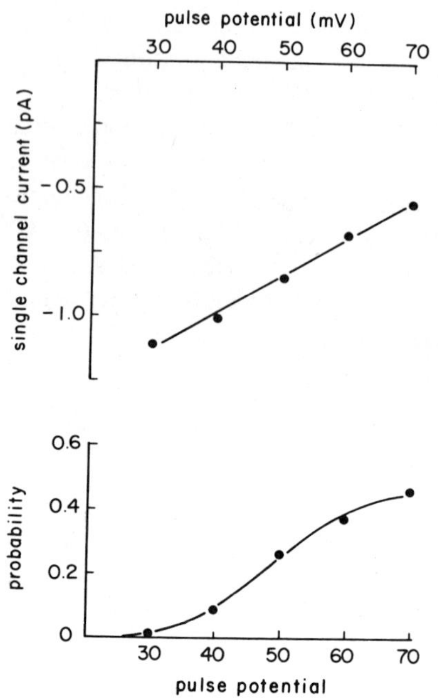

Figure 3. Voltage dependencies of open-channel current (upper part) and open-state probabilities of the channel (lower part). Each point is the average of at least 100 single-channel current traces at each potential. The straight line of the *i/V* relationship indicates a slope conductance of 18 pS.

3). Open times of calcium channels are exponentially distributed with time constants, ($\tau_o \sim 1$ msec) that are only slightly voltage dependent (5,11,16,32). Closed time histograms of calcium channels can be fitted by double-exponential curves (5,11,16). The short time constant τ_{c_1}, corresponds to the brief closures during the bursts of openings, while the longer time constant, τ_{c_2}, describes the distribution of closed time intervals between the bursts. While τ_{c_1} does not seem to be voltage dependent, τ_{c_2} decreases with increasing depolarization.

2. VOLTAGE-DEPENDENT EFFECTS OF DIHYDROPYRIDINE ENANTIOMERS ON SINGLE CHANNELS

1,4-Dihydropyridines are highly lipid soluble compounds that seem to have access to individual calcium channels through the lipid phase of the membrane (23). There are chemically closely related 1,4-dihydropyridines with opposite effects on calcium channels. The main group that also has therapeutic applications (nifedipine, nitrendipine, nimodipine, nisoldipine) reduces calcium currents (blockers), while the

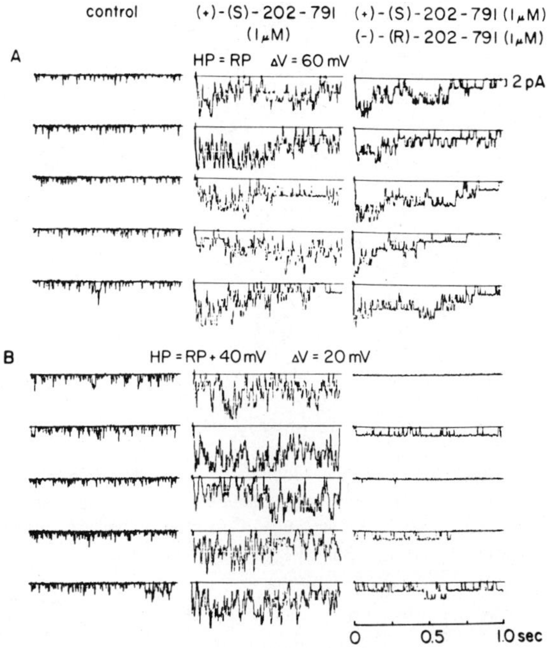

Figure 4. Single-channel current traces recorded from an adult rat heart cell at two different holding potentials in the absence and presence of (+)- and (−)-enantiomers of the drug 202-791; Ba^{2+} ions were the charge carriers through open channels. There were at least six calcium channels in the membrane patch. Single-channel currents in the controls are poorly resolved because of the slow time resolution. Increased channel activity in the presence of the drug is due to marked prolongation of the times the channel spends in the open state. (*A*) Holding potential was equal to the resting potential, and voltage clamp pulses of 60 mV amplitude lasted 1 sec. (*B*) The holding potential was adjusted 40 mV positive to the resting potential, and pulse potentials of 20 mV amplitude were applied. Note the sharp decrease of channel openings at this holding potential with both enantiomers present in the bathing solution.

other group (CGP 28392, Bay K 8644) increases these currents (activators) (22). Most interestingly, the two enantiomers of the compound Sandoz 202-791 (Fig. 1) have also opposite effects on calcium channels (21,24). Figure 4 shows a result obtained with these two enantiomers. The membrane potential of the patch was held at the resting potential (Fig. 4*A*), or 40-mV positive to the resting potential (Fig. 4*B*). The depolarizing voltage steps eliciting calcium channel activity were 60 and 20 mV, respectively. The calcium channel activity was greatly increased shortly after application of (+)-(*S*)-202-791. The increase in activity is due to a considerable prolongation of the mean open times of the channels by the drug. This result shows that the (+)-enantiomer acts as a calcium channel activator by promoting mode 2 (19) behavior of the channel. When the (−)-enantiomer was also added, there was a pronounced speeding in the rate of decay of single calcium channel activity during the

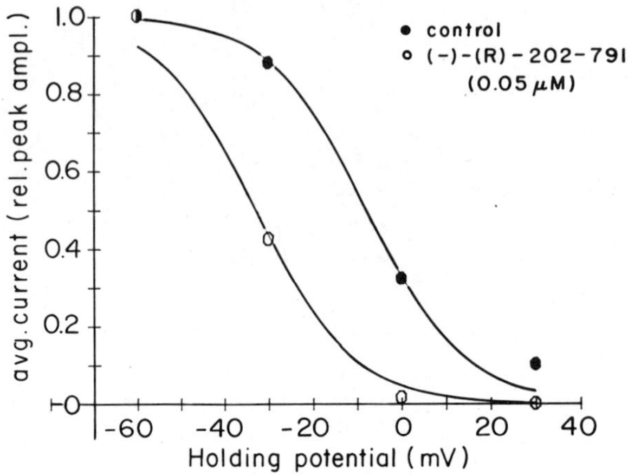

Figure 5. Effect of (−)-(*R*)-202-791 (channel blocker) on the voltage dependence of averaged calcium channel currents. Normalized current scale; absolute peak currents (I_{max}) at holding potential −60 mV were 2.8 pA in the control and 1.6 pA in the presence of the drug. Pulse potential was always +40 mV. Solid lines were drawn according to $(1 + \exp[(V_m - V_{0.5})/K])^{-1}$ with $V_{0.5} = -8$ mV, $K = 12$ mV (control) and $V_{0.5} = -34$ mV, $K = 12$ mV (drug).

pulse and an increase in the number of failures of the channels to open. This indicates that the (−)-enantiomer acts rather as a calcium channel blocker mainly by promoting mode 0 (19) of calcium channel activity.

The blocking effect was strongly dependent on membrane potential (Fig. 4*B*). When the holding potential was depolarized, the blocking potency of the

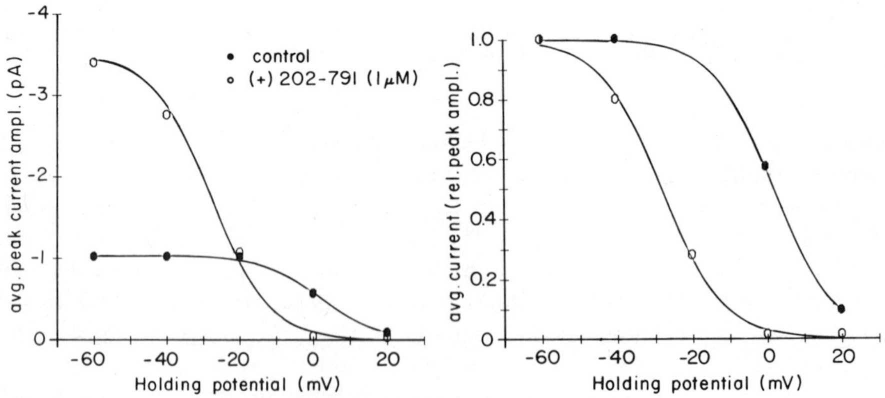

Figure 6. Effect of (+)-(*S*)-202-791 (channel activator) on the voltage dependence of averaged calcium channel currents. Left-hand side: absolute current scale; right-hand side: normalized current scale. Note that the channel activating effect turns over into a blocking effect at a holding potential of −20 mV. Solid lines fitting the normalized data points were drawn according to $(1 + \exp[(V_m - V_{0.5})/K])^{-1}$ with $V_{0.5} =$ 0 mV, $K = 7$ mV (control) and $V_{0.5} = -32$ mV, $K = 7$ mV (drug).

(−)-enantiomer was greatly increased. This is shown more systematically in Fig. 5, where the relative peak amplitudes of averaged calcium channel currents *I*, have been plotted as a function of the holding potential. In the absence of the drug, *I* decreases as the holding potential is made more positive. This voltage-dependent calcium current inactivation is due to a decrease of the overall opening probability of the channel when the holding potential is depolarized. The calcium channel blocking enantiomer (−)-202-791 causes a shift of the inactivation curve toward more negative holding potentials. This shift is consistent with a 29-fold increase in the binding affinity of the blocking drug and agrees qualitatively with similar results obtained with other dihydropyridine compounds (1,33). The (+)-enantiomer that increases calcium channel activity also shows a considerable voltage dependence of action. However, at negative membrane potentials, the average current amplitude is greatly increased (Fig. 6, left-hand side). With increasing depolarization of the holding potential, the average current amplitude decreased, and at potentials positive to −20 mV there was an inhibition of the calcium current; thus the activating activity of (+)-202-791 turns over into a blocking effect (Fig. 6). Normalization of the curves on the left-hand side of Fig. 6 illustrates a shift of the relative peak amplitude of the average currents along the voltage axis similar to that obtained with the (−)-enantiomer (Fig. 6, right-hand side).

In conclusion, then, the electrophysiological evidence with the two enantiomers of the Sandoz compound 202-791 clearly shows an interference with calcium channel gating. The (−)-enantiomer predominantly favors the inactivated (mode 0) state of the channel, thereby reducing the calcium current, while the (+)-enantiomer primarily stabilizes the prolonged open (mode 2) state of the channel, which causes enhancement of the current. Both enantiomers seem to have the capability of producing both effects, but one of them prevails.

The 1,4-dihydropyridine compounds are not only useful tools in studying the gating of calcium channels, but also in assessing the ion selectivity of those channels. The activator compounds promoting long open states of the channel facilitated the analysis of the sequence of inorganic cations that can pass through the open channel or that cause blockade of the channel (18,20). The channels preferentially pass barium, strontium, and calcium, while they usually are much less (1/1000) permeable for the monovalent cations potassium and sodium. Other divalent cations (Mg, Cd, Co) cause blockade of the channels.

3. VOLTAGE-DEPENDENT BINDING OF DIHYDROPYRIDINE ENANTIOMERS STUDIED WITH RADIOLIGAND

^{3}H-Nitrendipine was the first useful radioligand for biochemical characterization of calcium channels (3,15). Meanwhile, various other radioactive compounds of the dihydropyridine class have been used for labeling calcium channels in membrane fragments from various tissues (22). Radiolabeled dihydropyridines bind with high affinity to these membrane preparations (K_D 10^{-10}–10^{-9} M). In the majority of studies a single class of high-affinity binding sites has been found. Different compounds of the dihydropyridine group displace each other in a competitive way, whereas other calcium channel blockers (verapamil, diltiazem, etc.) seem to bind at different, but possibly interacting sites (13).

A discrepancy between high binding affinity and low blocking potency of calcium channel blockers of the 1,4-dihydropyridine type has long been a puzzle. In cardiac muscle the concentration required for half-maximal inhibition of calcium currents was usually 10–1000 times higher than expected from the K_D value obtained from binding studies in membrane fragments (25). The demonstration in cardiac cells that the potency of dihydropyridine action increases greatly as the membrane is depolarized provides a solution to this puzzle (1,33). In membrane fragments calcium channels are no longer functional, but rather in an inactivated state (11). Electrophysiological studies have provided indirect evidence that the binding affinity to inactivated channels is much higher than to channels that are available to open (1,19,33). Therefore, it would be expected that binding affinities of radioactively labeled 1,4-dihydropyridine ligands in intact cardiac cells depend on the membrane potential. This has been shown to be indeed the case (31). Figure 7 shows such an experiment where the dihydropyridine enantiomer [^{3}H]-(+)-PN-200-110 has been bound to calcium channels in cultured cardiac cells with two different membrane potentials. The compound PN-200-110 is chemically closely related to 202-791 and

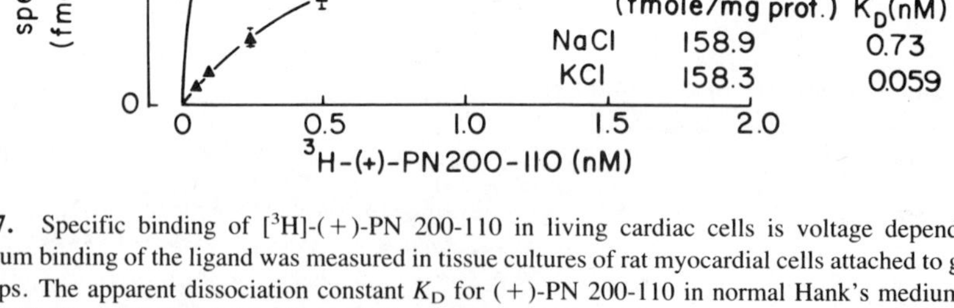

Figure 7. Specific binding of [^{3}H]-(+)-PN 200-110 in living cardiac cells is voltage dependent. Equilibrium binding of the ligand was measured in tissue cultures of rat myocardial cells attached to glass cover slips. The apparent dissociation constant K_D for (+)-PN 200-110 in normal Hank's medium (in mM): 137 NaCl, 5.4 KCl, 0.8 Mg SO_4, 4.2 Na HCO_3, 0.34 Na_2HPO_4, 0.44 KH_2PO_4, 20 HEPES, 5.5 glucose) was 0.73 ± 0.13 (▲), in potassium-rich medium (137 KCl, 5.4 NaCl, other components as above) the K_D was 0.059 ± 0.005 (●). The maximal binding capacities B_{max} were not significantly different under both conditions. The K_D values were estimated from computerized nonlinear least-square fits to the experimental data. For the fitting procedure, a single homogenous population of binding sites was assumed for both conditions.

acts as a calcium channel blocker. In calcium-free Hank's solution with 137 mM NaCl and 5.4 mM KCl the K_D value of the drug is approximately 12 times higher than in depolarized cells, where all the sodium chloride in the Hank's solution has been replaced by KCl. The membrane potentials under those conditions were -38 ± 4 mV and about 0 mV, respectively. This shows that binding 1,4-dihydropyridines to calcium channels is voltage dependent and agrees, at least qualitatively, with the corresponding electrophysiological experiments. The quantitative discrepancy may be due to differences in membrane potential under the various experimental conditions and/or in the voltage dependence of the binding constants of individual dihydropyridines.

We have also used [^{3}H]-(+)-PN 200-110 as a radiolabel to characterize the binding properties of the two enantiomers of the compound Sandoz 202-791. In cultured cells from rat hearts [^{3}H]-(+)-PN 200-110 was applied at concentrations of 0.7 nM in NaCl-containing Hank's solution or 0.06 nM in KCl–Hank's solution. This corresponds to the respective K_D values of the radioligand in normal and depolarized cultured cells (see Fig. 7). When the radioligand was added together with increasing concentrations of the (−)-enantiomer or (+)-enantiomer of the compound 202-791, there was competition in binding between the radioligand and the enantiomers in NaCl–solution as well as in KCl–solution. Figure 8 shows such competitive displacement curves under equilibrium conditions. The antagonistic (−)-enantiomer displaced the radioligand in a one-to-one fashion in both solutions. However, the respective K_D values were very different in the two solutions. The calculated K_D value for the (−)-enantiomer in the NaCl–solution was 1.26×10^{-7} M, while in a KCl–solution it was 2.7×10^{-9} M, indicating a 46.7-fold increase in the binding affinity of this enantiomer when the cells were depolarized by approximately 40 mV in the KCl–solution. The agonistic (+)-enantiomer also showed considerable voltage dependence of its binding. However, in NaCl–solution the steepness of the

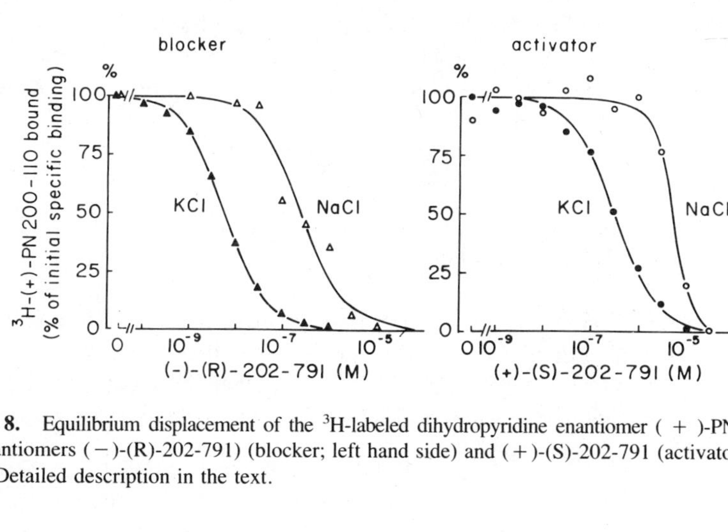

Figure 8. Equilibrium displacement of the ^{3}H-labeled dihydropyridine enantiomer (+)-PN 200-110 by the enantiomers (−)-(R)-202-791) (blocker; left hand side) and (+)-(S)-202-791 (activator; right hand side). Detailed description in the text.

displacement curve was much greater (Hill coefficient 1.9) than in KCl–solution (Hill coefficient 0.9). This indicates multiple binding sites for the activator enantiomer and a single site for the blocker. In the depolarized cells, the affinity of the (+)-enantiomer was much greater than in cells incubated in NaCl–solution and the radioligand was displaced in a one-to-one fashion. This agrees with the electrophysiological experiments, which show that calcium channel activation by (+)-202-791 turns over into a blocking effect if the cell is depolarized (Fig. 6).

The demonstration of voltage-dependent binding of the ligands and the electrophysiological results provide strong evidence that these drugs are bound at or near the site of the channel that is responsible for its voltage-dependent gating.

4. THE 1,4-DIHYDROPYRIDINE BINDING SITE

Recently various techniques, including target size analysis (12,27,34), photoaffinity labeling (6), and protein purification techniques, have been used to obtain first results on the molecular structure of the dihydropyridine binding site. The dihydropyridine–receptor complex of transverse tubular membranes of skeletal muscle has been solubilized and partially purified (4,8,14). Target size analysis of the 1,4-dihydropyridine receptor *in situ* is consistent with relative molecular weights M_r of 185,000–210,000 in skeletal muscle and brain membranes (12,27). The biochemical analysis suggests that this complex is a glycoprotein consisting of three subunits (α,β,γ) with apparent molecular weights of 130–142, 32–50, and 33 kilodaltons (4,8). Calcium channels can be functionally modulated by a cAMP-dependent phosphorylation reaction (29). In intact T-tubule membranes from skeletal muscle the γ subunit of the dihydropyridine binding site is phosphorylated by the catalytic subunit of cAMP-dependent protein kinase (9,10) and may play a physiological role in functional modulation of the channel. The next step will be to reincorporate the subunits into artificial lipid bilayers to determine whether they can be made functional and resemble the recordings from intact calcium channels. First attempts in this direction have already been successful (9). Future work will require cloning of the genes and sequencing of the channel protein, and the final goal will be to obtain a complete three-dimensional molecular structure of the calcium channel which can be correlated with its functional properties.

ACKNOWLEDGMENTS

The authors thank Charlotte Becker for unfailing experimental help with the binding experiments. They also thank Dr. R. Hof, Sandoz AG, Basel, for providing them with the enantiomers of the compound 202-791. The financial support by the Swiss National Science Foundation (grants 3.154-0.81 and 3.059-0.84) is gratefully acknowledged.

REFERENCES

1. Bean, B. P. (1984). *Proc. Natl. Acad. Sci.* (USA), **81**, 6388–6392.
2. Bean, B. P. (1985). *J. Gen. Physiol.*, **86**, 1–30.
3. Bellemann, P., Ferry, D., Lübbecke, F., and Glossmann, H. (1981). *Arzneimittelforschung*, **31**, 2064–2067.
4. Borsotto, M., Barhanin, J., Norman, R. I., and Lazdunski, M. (1984). *Biochem. Biophys. Res. Commun.*, **122**, 1357–1366.
5. Cachelin, A. B., de Peyer, J. E., Kokubun, S., and Reuter, H. (1983). *Nature*, **304**, 462–464.
6. Campbell, K. P., Lipshutz, G. M., and Denney, G. H. (1984). *J. Biol. Chem.*, **259**, 5384–5387.
7. Carbone, E. and Lux, H. D. (1984). *Nature*, **310**, 501–502.
8. Curtis, B. M. and Catterall, W. A. (1983). *J. Biol. Chem.*, **258**, 7280–7283.
9. Curtis, B. M. and Catterall, W. A. (1985). *J. Gen. Physiol.*, **86**, 13a.
10. Curtis, B. M. and Catterall, W. A. (1985). *Proc. Natl. Acad. Sci.* (USA), **82**, 2528–2532.
11. Fenwick, E. M., Marty, A., and Neher, E. (1982). *J. Physiol.* (London), **331**, 599–635.
12. Ferry, D. R., Goll, A., and Glossmann, H. (1983). *Naunyn-Schmiedeberg's Arch. Pharmacol.*, **323**, 292–297.
13. Gengo, P. J., Luchowski, E., Rampe, D. E., Rutledge, A., Triggle, A. M., Triggle, D. J., and Janis, R. A. (1983). *Cold Spring Harbor Symp. Quant. Biol.*, **48**, 279–286.
14. Glossmann, H. and Ferry, D. R. (1983). *Naunyn-Schmiedeberg's Arch. Pharmacol.*, **323**, 279–291.
15. Glossmann, H., Ferry, D. R., Lübbecke, F., Mewes, R., and Hofmann, F. (1982). *Trends Pharmacol. Sci.*, **3**, 431–437.
16. Hagiwara, S. and Ohmori, H. (1983). *J. Physiol.* (London), **336**, 649–661.
17. Hamill, O. P., Marty, A., Neher, E., Sakmann, B., and Sigworth, F. J. (1981). *Pflügers Arch.*, **391**, 85–100.
18. Hess, P., Lansman, J. B., and Tsien, R. W. (1984). *J. Physiol.* (London), **358**, 61P.
19. Hess, P., Lansman, J. B., and Tsien, R. W. (1984). *Nature*, **311**, 538–544.
20. Hess, P. and Tsien, R. W. (1984). *Nature*, **309**, 453–456.
21. Hof, R. P., Rüegg, U. T., Hof, A., and Vogel, A. (1985). *J. Cardiovasc. Pharmacol.*, **7**, 689–693.
22. Janis, R. A. and Triggle, D. J. (1984). *Drug Dev. Res.*, **4**, 257–274.
23. Kokubun, S. and Reuter, H. (1984). *Proc. Natl. Acad. Sci.* (USA), **81**, 4824–4827.
24. Kongsamut, S., Kamp, T. J., Miller, R. J., and Sanguinetti, M. C. (1985). *Biochem. Biophys. Res. Commun.*, **130**, 141–148.
25. Lee, K. S. and Tsien, R. W. (1983). *Nature*, **302**, 790–794.
26. Nilius, B., Hess, P., Lansman, J. B., and Tsien, R. W. (1985). *Nature*, **316**, 443–446.
27. Norman, R. I., Borsotto, M., Fosset, M., Lazdunski, M., and Ellory, J. C. (1983). *Biochem. Biophys. Res. Commun.*, **111**, 878–883.
28. Nowycky, M. C., Fox, A. P., and Tsien, R. W. (1985). *Nature*, **316**, 440–443.
29. Reuter, H. (1983). *Nature*, **301**, 569–574.
30. Reuter, H. (1985). *Nature*, **316**, 391.
31. Reuter, H., Porzig, H., Kokubun, S., and Prod'hom, B. (1985). *Trends NeuroSci.* **8**, 396–400.
32. Reuter, H., Stevens, C. F., Tsien, R. W., and Yellen, G. (1982). *Nature*, **297**, 501–504.
33. Sanguinetti, M. C. and Kass, R. S. (1984). *Circ. Res.*, **55**, 336–348.
34. Venter, J. C., Fraser, C. M., Schaber, J. S., Jung, C. Y., Bolger, G., and Triggle, D. J. (1983). *J. Biol. Chem.*, **258**, 9344–9348.

CHAPTER 12

MOLECULAR PROPERTIES OF VOLTAGE-SENSITIVE CALCIUM CHANNELS

WILLIAM A. CATTERALL
BENSON M. CURTIS
University of Washington
Department of Pharmacology
Seattle, Washington

In muscle tissues, voltage-sensitive calcium channels mediate calcium influx during cellular depolarization and play an important role in excitation–contraction coupling (reviewed in refs. 25 and 39). In neurons, they produce action potentials in dendrites (32,47) and couple changes in membrane potential at nerve terminals to the release of neurotransmitter (6,28). Multiple classes of calcium channels have been distinguished in neurons (4,12,37) and cardiac muscle cells (35). This chapter focuses on calcium channels that are blocked by dihydropyridine calcium antagonists and are termed ''L-type channels'' by Tsien and colleagues (ref. 37 and Chapter 10, this volume). These are the most prominent calcium channels in smooth, cardiac, and skeletal muscle and are also present in neurons and neurosecretory cells.

Dihydropyridine-sensitive calcium channels are nonconducting at the resting membrane potential. They are activated by membrane depolarization with half-maximal probability of activation at -20 to $+10$ mV in various tissues (51). Calcium channels in some tissues are slowly inactivated on prolonged membrane depolarization, whereas in others inactivation is partial or absent. Inactivation results from a decrease in the probability of channel activation. Two different mechanisms have been described: inactivation produced by membrane depolarization per se and inactivation due to increased intracellular calcium concentration (51). Calcium channels in secretory cells and skeletal muscle seem not to inactivate significantly, while those in cardiac muscle exhibit both voltage- and calcium-dependent inactivation mechanisms.

Under physiological conditions, calcium channels are highly selective yet transport up to 3×10^6 ions/sec (51). Both Ca^{2+} and Ba^{2+} are efficiently transported, while other divalent cations are poor permeants and block Ca^{2+} and Ba^{2+} conductance (25). Surprisingly, in the absence of Ca^{2+}, calcium channels are much less selective and will transport a variety of monovalent cations effectively (1,3,26,33). Thus Ca^{2+} regulates channel selectivity. A model involving two high-affinity Ca^{2+} binding sites within the ion-conducting pore located close enough for strong intersite repulsive interactions accommodates these findings. When neither site is occupied by Ca^{2+}, monovalent cations are bound weakly and transported. When one site is occupied by Ca^{2+}, the channel is blocked. When both sites are occupied, transport of Ca^{2+} is efficient and selective and monovalent cations are excluded.

The functional properties of calcium channels are being rapidly elucidated by electrophysiological methods. Although study of the molecular properties of this physiologically important ion channel is of interest, only recently have high-affinity chemical probes for the calcium channel become available.

UNIVERSITY LIBRARY COLLEGE

1. IDENTIFICATION AND PURIFICATION OF CALCIUM CHANNELS

1.1 Organic Calcium Antagonists as Probes of Calcium Channel Structure and Function

Calcium channels are inhibited by three different classes of organic antagonists: the dihydropyridines, including nifedipine and nitrendipine, the phenylalkylamines and diphenylalkylamines, including verapamil, and the thiobenzazepine diltiazem (49). Certain ^{3}H-labeled derivatives of nitrendipine and other dihydropyridines have been prepared and their binding to high-affinity receptor sites in homogenates and membrane preparations from excitable tissues extensively examined (27). A single class of high-affinity binding sites is observed with a K_D for nitrendipine in the range of 0.1–1.0 n*M*. The structure–activity relationships for occupancy of these high-affinity binding sites correlates closely with the concentrations of various dihydropyridines required to block contraction of smooth muscle over a wide range of K_D values (7). Similar relative structure–activity relationships are observed in other tissues, but the absolute concentration required to block calcium currents often is substantially greater than that required to occupy the high-affinity nitrendipine binding sites (2,31). This apparent discrepancy has now been resolved by the finding that dihydropyridine binding and block of calcium channels is highly voltage sensitive (5,46,53). In cardiac cells, half-maximal block of calcium currents at the resting membrane potential requires 700 n*M* nitrendipine. In contrast, when cells are depolarized to -20 mV to inactivate 70% of the calcium current, only 0.8 n*M* nitrendipine is required for half-maximal block. This value is in close agreement with K_D values for high-affinity binding sites in cardiac membrane preparations that are depolarized by tissue homogenization and cell disruption (27). These results resolve an important discrepancy between previous biochemical and physiological measurements and provide additional strong support for the view that measurements of specific binding of nitrendipine and other dihydropyridines represent drug interaction with calcium channels or a closely associated regulatory component.

The binding of nitrendipine to its receptor site is modulated via an allosteric mechanism by diltiazem and various mono- and diphenylalkylamines such as verapamil (34) that interact with two separate receptor sites (22). Verapamil decreases nitrendipine binding and enhances the rate of dissociation of the nitrendipine–receptor complex, whereas diltiazem increases binding and slows the rate of dissociation of the nitrendipine–receptor complex. Thus the calcium antagonist receptor of the calcium channel contains at least three drug binding sites, one specific for dihydropyridines and one each for diltiazem and phenylalkylamines, which interact allosterically in modulating calcium channel function and have opposite effects on binding of dihydropyridine calcium channel blockers.

The discovery that the binding of dihydropyridine calcium antagonists is state dependent and favors the inactivated state of the channel suggests that they may block channel function by altering normal channel gating rather than by physically occluding the ion pore. This conclusion is illustrated more clearly by the discovery of dihydropyridine calcium channel activators, exemplified by Bay K 8644 (45). These agents bind at the same receptor site as the dihydropyridine channel blockers but cause smooth muscle contraction and stimulation of the heart (45) by favoring activation of calcium channels at more negative membrane potentials (26,30). Thus occupancy of the same receptor site by agents of closely related structure can cause either channel block or channel activation. These results provide firm evidence that the dihydropyridine receptor modulates the gating of the calcium channel as its principal mode of action.

The development of the calcium channel blockers and activators has been important for the therapy of hypertension, atrial arrhythmia, and angina pectoris. They have allowed clearer definition of the role of calcium channels in various physiological pathways. In addition, the development of radioactively labeled derivatives of high affinity and specificity now allows the use of these drugs as probes to identify and isolate the protein components of calcium channels and understand their molecular properties.

1.2 Molecular Size of the Calcium Antagonist Receptor of the Calcium Channel

The first estimates of the molecular size of the component(s) of the calcium channel that bind dihydropyridines have been derived from radiation inactivation measurements. Estimates of the target size for inactivation of [^{3}H] nitrendipine binding to receptor sites are 180,000 or 210,000 for skeletal muscle (17,36), 210,000 for brain (36), and 280,000 for guinea pig ileum (54). The apparent target size derived from radiation inactivation studies depends on the radiolabeled calcium antagonist used in the receptor binding assay and on the assay and irradiation conditions, suggesting that the calcium antagonist receptor is a complex oligomeric structure (18).

1.3 Identification of Protein Components of the Calcium Antagonist Receptor by Covalent Labeling

The dihydropyridines are amenable to structural modification at several positions and thus are potentially valuable tools for affinity chromatography and covalent labeling. An isothiocyanate analog of nitrendipine covalently labels a protease-sensitive polypeptide with an apparent molecular weight of 45,000 in guinea pig ileum membrane preparations (54) or 36,000 in skeletal muscle transverse tubules (29). Nitrendipine itself is photolabile and can be covalently linked to a polypeptide with apparent molecular weight of 32,000 in cardiac sarcolemma by high-intensity irradiation (11). A photoreactive azidobenzoyl analog of nitrendipine has been prepared by Ferry and co-workers (19). This compound labels a polypeptide with an apparent molecular weight of 145,000 in skeletal muscle transverse tubule membranes. In each case, the covalent labeling reactions described were prevented by incubation with excess unlabeled calcium antagonist, suggesting that the polypeptides labeled are components of the calcium antagonist receptor in these different tissues. Either the calcium antagonist receptor from these different tissues has different polypeptide components, or, as argued below, the receptor is an oligomer of subunits of different molecular weight that are differentially labeled by the probes used.

1.4 Characterization of the Detergent-Solubilized Calcium Antagonist Receptor

Since it is likely to be an intrinsic membrane protein, the first essential step in purification and biochemical characterization of the calcium antagonist receptor is solubilization from an appropriate membrane source and characterization of the solubilized protein. After a survey of several detergents, we concluded that digitonin is the most effective detergent for solubilization of a specific [^{3}H]nitrendipine–receptor complex from brain and skeletal muscle transverse tubule membranes (13,14). Up to 40% of the receptor–ligand complex is solubilized. The dissociation of bound [^{3}H]nitrendipine from the complex is accelerated by verapamil and slowed by diltiazem through allosteric interactions between the nitrendipine binding site and the binding sites for those ligands as previously observed in intact membranes. These results show that the three different binding sites for calcium antagonist drugs remain associated as a complex after detergent solubilization of the calcium antagonist receptor and provide further support for the conclusion that a specific receptor complex has been solubilized under conditions that allow retention of the functional allosteric regulation of dihydropyridine binding.

Sedimentation of the solubilized [^{3}H]nitrendipine–receptor–digitonin complex through sucrose gradients gives a single peak of specifically bound nitrendipine with a sedimentation coefficient of 19*S* to 20*S* (13). Comparison of the sedimentation behavior of the solubilized complex from brain, heart, and skeletal muscle indicates that they have identical size. These results provide support for the view that the calcium antagonist receptor in different tissues is quite similar.

Many plasma membrane proteins are glycosylated during their synthesis and transport to the cell surface. The solubilized calcium antagonist receptor from brain or skeletal muscle is specifically adsorbed to and eluted from affinity columns with immobilized wheat germ agglutinin or other lectins (8,13,14,23). Evidently, one or more of the subunits of the calcium antagonist receptor are glycoproteins.

1.5 Skeletal Muscle Transverse Tubules as a Model System for Examination of the Biochemical Properties of Calcium Channels

Skeletal muscle fibers have substantial voltage-activated calcium currents that have been measured under voltage clamp (43). These currents originate almost entirely in the transverse tubular system and are blocked by dihydropyridine calcium antagonists

(2). They have been presumed to play a role in excitation–contraction coupling, although direct evidence for such a role has not been obtained. Transverse tubule membranes can be extensively purified from skeletal muscle by a combination of differential and density-gradient centrifugation (41). Antibodies prepared against the most highly purified fractions of T-tubule membranes stain only T tubules and not sarcolemma or sarcoplasmic reticulum of intact muscle, indicating a high degree of purity of this preparation (41). Analysis of [^{3}H]nitrendipine binding to membrane fractions from skeletal muscle reveals a specific localization of the calcium antagonist receptor in the T-tubule fraction (21,24). These membranes contain a tenfold greater concentration of calcium antagonist receptor than any other membrane preparation described to date (27). Since T-tubule membranes are the most enriched source of calcium antagonist receptors and have a substantial voltage-activated calcium current that is blocked by dihydropyridines, they provide a favorable experimental preparation for examination of the molecular properties of the calcium antagonist receptor and its relationship to voltage-sensitive calcium channels. It is anticipated that the T-tubule calcium channel will resemble those of other tissues that are blocked by dihydropyridines and, therefore, that information on the molecular properties of this calcium channel will give insight into others.

1.6 Purification and Characterization of the Calcium Antagonist Receptor from Skeletal Muscle Transverse Tubules

If the calcium antagonist receptor has a molecular weight of 210,000 (36) and one dihydropyridine binding site per mole, the pure protein would have a specific binding capacity of 4760 pmol/mg. The specific binding capacity of purified T-tubule membranes is 10–50 pmol/mg (14,21,24). Thus only 100- to 500-fold purification is required to achieve a homogeneous preparation. This degree of purification can usually be obtained by conventional procedures.

We have purified the calcium antagonist receptor solubilized from T-tubule membranes by digitonin 330-fold by affinity chromatography on wheat germ agglutinin–Sepharose, ion-exchange chromatography on DEAE–Sephadex and velocity sedimentation through sucrose gradients (14). The purified preparation contained 1950 pmol of calcium antagonist receptor per milligram of protein, implying a purity of at least 42%. Analysis of the purified preparation by polyacrylamide gel electrophoresis after denaturation in sodium dodecyl sulfate (SDS) revealed three major protein bands as illustrated in Fig. 1. Their molecular weights are: α, 135,000; β, 50,000; and γ, 33,000. All three of these polypeptides quantitatively comigrate with the [^{3}H]nitrendipine–receptor complex during velocity sedimentation in sucrose gradients (14) and thus are likely to be subunits of the calcium antagonist receptor. The concentration of the β subunit varies among preparations. This seems to result from both dissociation and high protease sensitivity.

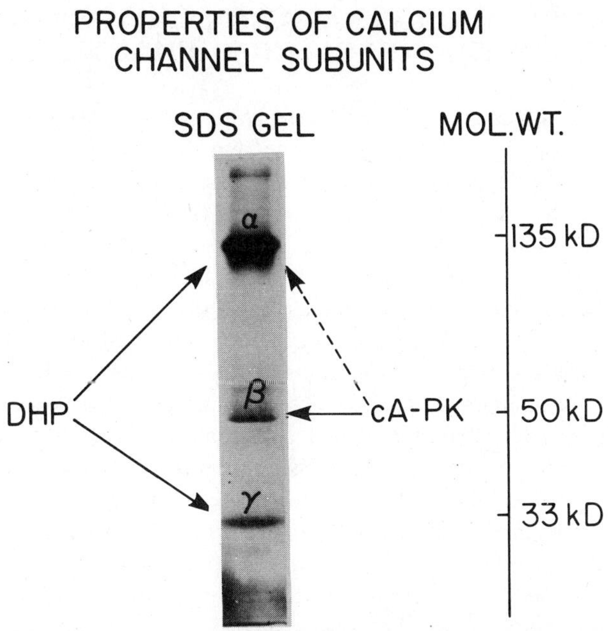

Figure 1. Properties of calcium channel subunits. The purified calcium antagonist receptor from skeletal muscle transverse tubules was purified and analyzed by gel electrophoresis in SDS after reduction and dissociation by boiling in SDS and β-mercaptoethanol. The molecular weights of the α, β, and γ subunits are indicated along with sites of incorporation of ^{32}P by reaction with cAMP-dependent protein kinase and dihydrophyridine calcium channel blockers (DHP) by photoaffinity labeling as described in the text.

It is of interest to compare the apparent molecular weights of the subunits of the purified calcium antagonist receptor from T tubules with those of the polypeptides covalently labeled by dihydropyridine derivatives (Fig. 1). The polypeptide with apparent molecular weight of 145,000 labeled by azidopine (19) has a size similar to that of the α subunit of the purified preparation. The polypeptide with apparent molecular weight of 32,000 covalently labeled by nitrendipine itself (11) has the same size as the γ subunit of the purified preparation. The polypeptides with apparent molecular weights of 45,000 in ileum and 36,000 in transverse tubules that are covalently labeled with the isothiocyanate derivative of nitrendipine are significantly larger than the γ subunit of the purified preparation, but the specificity of labeling suggests that they may be related. Thus the disparate results of covalent labeling experiments in different tissues may be reconciled by the hypothesis that the dihydropyridine binding site of calcium antagonist receptor is located near the contact regions of the α, β, and γ subunits such that different dihydropyridine ligands preferentially label different subunits. The covalent labeling results provide further evidence that these subunits are components of the calcium antagonist receptor in intact membranes.

2. PHOSPHORYLATION AND cAMP-DEPENDENT REGULATION OF CALCIUM CHANNELS

The classic work of Reuter, Tsien, and their colleagues (38,40,50) established that the positive inotropic effect of epinephrine and norepinephrine on the heart is mediated by a cAMP-dependent increase in inward calcium current. This effect is due to an increase in the number of calcium channels that are active during the cardiac action potential (10,51). These effects are mimicked by intracellular injection of the catalytic subunit of cAMP-dependent protein kinase, suggesting that protein phosphorylation mediates the effects of cAMP (9). It is uncertain whether the substrate for cAMP-dependent phosphorylation is the calcium channel itself or another regulatory component. In addition to these studies on mammalian heart, calcium channels in mammalian dorsal root ganglion (DRG) neurons and both calcium channels and calcium-activated K^+ channels in molluscan neurons are regulated by cAMP-dependent protein phosphorylation (reviewed in ref. 42). Recent results indicate that calcium channels in cultured skeletal muscle cells are also regulated in this manner (44). Thus modulation of calcium channel properties by cellular regulatory processes may be a general phenomenon.

Although regulation of calcium channels via a pathway involving cAMP and protein phosphorylation is now well established, it is not known whether the site of phosphorylation is the calcium channel itself or another intracellular protein that, in turn, regulates channel function. As a step toward resolving this question, we have investigated whether the subunits of the purified calcium antagonist receptor can serve as substrates for cAMP-dependent protein kinase in the purified state and in intact T-tubule membranes (15). When incubated with physiological concentrations of cAMP-dependent protein kinase, the α and β subunits are phosphorylated as diagrammed in Fig. 1. Up to 0.85–0.9 mole of ^{32}P can be incorporated into each subunit at rates that are consistent with a physiologically significant phosphorylation reaction. In contrast, the γ subunit is not a substrate for cAMP-dependent protein kinase, even under forcing reaction conditions (15). These results identify the α and β subunits of the calcium antagonist receptor as potential sites of regulation of the voltage-sensitive calcium channel by cAMP-dependent phosphorylation.

If these subunits are indeed physiological substrates of the enzyme, they should be phosphorylated in intact T-tubule membranes. To examine this, we incubated purified T-tubule membranes with the catalytic subunit of cAMP-dependent protein kinase and [$\gamma^{32}P$]ATP, analyzed the phosphorylated proteins by one- and two-dimensional gel electrophoresis, and compared the migration of the membrane phosphoproteins to that of the purified α and β subunits of the calcium antagonist receptor (15). These experiments showed that the β subunit is a preferred substrate for cAMP-depenent protein kinase in intact T tubules as well as in purified preparations. Migration of this protein as a series of closely spaced spots in pH-gradient electrophoresis suggested that the β subunit could be phosphorylated on up to three sites. In contrast, under the conditions of our experiements, the α subunit of the calcium antagonist receptor was not selectively phosphorylated in T-tubule membranes as assessed by two-dimensional gel electrophoresis. These results focus attention on the β subunit of the calcium antagonist receptor as a potential site of regulation of voltage-sensitive calcium channels by protein phosphorylation.

3. FUNCTIONAL PROPERTIES OF THE PURIFIED CALCIUM ANTAGONIST RECEPTOR IN PHOSPHOLIPID VESICLES

We purify the calcium antagonist receptor as a preformed complex with the calcium antagonists [^{3}H]nitrendipine or [^{3}H]PN 200–110. After solubilization, it retains allosteric interactions among the separate binding sites for verapamil, diltiazem, and dihydropyridines (13). However, other aspects of calcium channel function cannot be assessed in detergent solution. It is important, therefore, to return the purified calcium antagonist receptor to a membrane environment and determine whether the purified protein is capable of mediating voltage-dependent calcium flux.

To maximize the probability of purification of the calcium channel in an active state, the calcium antagonist receptors in T-tubule membranes were incubated with sufficient [^{3}H]PN 200–110 to label approximately 1% of the binding sites. This label was used to identify the calcium antagonist receptor during purification as in previous studies. The T-tubule membranes were then incubated in an excess of the specific calcium channel activator Bay K 8644 so that the remaining 99% of dihydropyridine sites would be occupied by this agent. The calcium antagonist receptors were then solubilized in digitonin and purified in the continued presence of Bay K 8644 by previously described procedures (14). Phosphatidylcholine for reconstitution was solubilized in the zwitterionic detergent CHAPS because it is poorly soluble in digitonin. Purified calcium antagonist receptor dispersed in digitonin was then mixed with phosphatidylcholine dispersed in CHAPS, and single-walled phospholipid vesicles were formed by removal of the detergents by molecular-sieve chromatography. Analysis of the resulting vesicle preparations by sucrose density-gradient sedimentation shows that the purified calcium antagonist receptors are quantitatively incorporated into phosphatidylcholine vesicles and 15–25% of vesicles contain at least one calcium antagonist receptor. These preparations thus provide a suitable system for analysis of the ion-transport properties of this purified protein.

Initial rates of influx of $^{45}Ca^{2+}$ or $^{133}Ba^{2+}$ into reconstituted phosphatidylcholine vesicles were measured under counter transport conditions in which the intravesicular compartment contains a high concentration of unlabeled Ca^{2+} or Ba^{2+}. These conditions greatly increase the amount of $^{45}Ca^{2+}$ or $^{133}Ba^{2+}$ uptake required to achieve isotopic equilibrium and greatly slow the approach to isotopic equilibrium as previously described for reconstituted sodium channels (48). They thus maximize any ion flux mediated by reconstituted channels. Under these conditions, calcium

TABLE 1. ION FLUX MEDIATED BY RECONSTITUTED CALCIUM CHANNELS

Activator	Inhibitor	$^{45}Ca^{2+}$ Influx (nmol/min)
None	None	8.5
None	Verapamil (100 μ*M*)	5.8
Bay K 8644 (1 μ*M*)	None	20.5
Bay K 8644 (1 μ*M*)	Verapamil (100 μ*M*)	5.7

influx into phosphatidylcholine vesicles containing reconstituted calcium antagonist receptors was two- to threefold greater than influx into protein-free phosphatidylcholine vesicles (Table 1). This increase is completely blocked by verapamil at a concentration of 100 μ*M*. If the calcium channel activator Bay K 8644 is removed from the vesicle preparation by molecular-sieve chromatography, $^{45}Ca^{2+}$ influx is markedly reduced. These results show that at least a fraction of the purified calcium antagonist receptors can function as calcium channels when incorporated into phosphatidylcholine vesicles.

The inhibition of the reconstituted calcium channels by different concentrations of organic calcium channel blockers from 10^{-9} to 10^{-4} *M* was examined. Half-maximal inhibition was observed with approximately 1.5 μ*M* verapamil, 1.0 μ*M* D600, or 0.2 μ*M* PN200–110 as summarized in Table 2. These concentrations are similar to those that give half-maximal inhibition of voltage-activated calcium currents in intact skeletal muscle fibers (Table 2), which is consistent with the conclusion that the calcium flux in reconstituted vesicles is mediated by functional purified calcium channels.

To determine whether the calcium influx stimulated by Bay K 8644 and blocked by PN200–110 and verapamil required the presence of the subunits of the calcium antagonist receptor and not other detectable proteins, purified preparations were sedimented through sucrose gradients, and each fraction was examined for bound [^{3}H]PN 200–110, polypeptide composition and the ability to mediate $^{133}Ba^{2+}$ influx when incorporated into phosphatidylcholine vesicles (16). A close quantitative

TABLE 2. INHIBITION OF RECONSTITUTED CALCIUM CHANNELS BY ORGANIC CALCIUM CHANNEL BLOCKERS

Drug	$K_{0.5}$ for Reconstituted Vesicles (μ*M*)	$K_{0.5}$ for Intact Muscle Fibers (μ*M*)
PN200-110	0.2	0.43
D600	1.0	10.0
Verapamil	1.5	—

correlation was observed between the presence of the α, β, and γ subunits of the calcium antagonist receptor and the ability to mediate barium flux. No other polypeptides were noted whose presence correlated with ion flux activity. Thus these results are also consistent with the conclusion that the purified calcium antagonist receptor is capable of mediating ion flux with the pharmacological characteristics expected of voltage-sensitive calcium channels.

Although voltage-sensitive sodium and calcium channels have roughly comparable single-channel conductances, the initial rate of $^{45}Ca^{2+}$ influx measured in our present reconstitution experiments is much slower than previously observed for sodium channels (48). We have estimated the fraction of functionally active vesicles in our reconstituted preparations by comparing the $^{45}Ca^{2+}$ influx mediated by calcium channels with that mediated by the calcium ionophore A23187 added to the same vesicles. We find that the initial rate of $^{45}Ca^{2+}$ influx mediated by A23187 is much faster than for reconstituted calcium channels and that the equilibrium level of $^{45}Ca^{2+}$ uptake is 200-fold greater. Evidently only 0.5% of the intravesicular volume in the vesicle population is accessible to functional calcium channels. Since 15–25% of the reconstituted vesicles contain purified calcium antagonist protein, we conclude that only a small fraction (2–3.3%) of the purified calcium antagonist receptors can form functional calcium channels when reconstituted by our present approach.

The finding that only a small fraction of purified calcium antagonist receptors from skeletal muscle transverse tubules can be reconstituted to give functional calcium channels is disappointing but may have a physiological basis. Schwartz et al. (46) have concluded that no more than a few percent of the high-affinity dihydropyridine binding sites in transverse tubules of intact skeletal muscle fibers can be associated with calcium channels that can be simultaneously activated during votage-clamp experiments. Our purified calcium channels may retain this inactive state during isolation and reconstitution, allowing only a small fraction to mediate $^{45}Ca^{2+}$ influx. The small fraction of active channels may contain additional protein components or have a different phosphorylation state. Alternatively, the present methods of purification and/or reconstitution may themselves damage the functional properties of calcium channels.

4. CONCLUSION

The molecular properties of dihydropyridine-sensitive calcium channels are now being elucidated by following a general strategy that was used previously in studies of voltage-sensitive sodium channels in this laboratory: identification by specific ligand binding and covalent labeling, solubilization and isolation by conventional protein purification methods, and reconstitution of channel function *in vitro*. The work reviewed here illustrates the substantial progress achieved with this approach to date. Future directions include further definition of the functional properties of the purified calcium channel complex in reconstituted phospholipid vesicles, analysis of the mechanism of regulation of the channel by protein phosphorylation, determination of the primary structure of the channel subunits, and comparison of the structural features of calcium channels with those of voltage-sensitive sodium channels in order to define common structural themes underlying the function of voltage-sensitive ion channels in general.

REFERENCES

1. Almers, W. and McCleskey, E. W. (1984). *J. Physiol.* (London), **353**, 585–608.
2. Almers, W., Fink, R., and Palade, P. T. (1981). *J. Physiol.* (London), **312**, 177–207.
3. Almers, W., McCleskey, E. W., and Palade, P. T. (1984). *J. Physiol.* (London), **353**, 565–583.
4. Armstrong, C. M. and Matteson, D. R. (1985). *Science*, **227**, 65–67.
5. Bean, B. P. (1984). *Proc. Natl. Acad. Sci.* (USA), **81**, 6388–6392.
6. Blaustein, M. P. (1975). *J. Physiol.* (London), **247**, 617–655.
7. Bolger, G. T., Gengo, P. J., Luchowski, E. M., Siegel, H., Triggle, D. J., and Janis, R. A. (1982). *Biochem. Biophys. Res. Commun.*, **104**, 1604–1609.
8. Borsotto, M., Barhanin, J., Norman, R. I., and Lazdunski, M. (1984). *Biochem. Biophys. Res. Commun.*, **122**, 1357–1366.
9. Brum, G., Flockerzi, V., Hofmann, F., Osterrieder, W., and W. Trautwein (1983). *Pflügers Arch.*, **398**, 147–154.
10. Cachelin, A. B., de Peyer, J. E., Kokubun, S., and Reuter, H. (1983). *Nature*, **304**, 462–464.
11. Campbell, K. P., Lipschutz, G. M., and Denney, G. H. (1984). *J. Biol. Chem.*, **259**, 5384–5387.
12. Carbone, E. and Lux, H. D. (1984). *Nature*, **310**, 501–502.
13. Curtis, B. M. and Catterall, W. A. (1983). *J. Biol. Chem.*, **258**, 7280–7283.
14. Curtis, B. M. and Catterall, W. A. (1984). *Biochemistry*, **23**, 2113–2118.
15. Curtis, B. M. and Catterall, W. A. (1985). *Proc. Natl. Acad. Sci.* (USA), **82**, 2528–2532.
16. Curtis, B. M. and Catterall, W. A. (1986). *Biochemistry*. **25**, 3077–3083.
17. Ferry, D. R., Goll, A., and Glossman, H. (1983). *Naunyn-Schmiedeberg's Arch. Pharmacol.*, **323**, 292–297.
18. Ferry, D. R., Goll, A., and Glossman, H. (1983). *EMBO J.*, **2**, 1729–1732.
19. Ferry, D. R., Goll, A., and Glossman, H. (1984). *FEBS Lett.*, **169**, 112–167.
20. Flockerzi, U., Mewes, R., Ruth, P., and Hofmann, F. (1983). *Eur. J. Biochem.*, **135**, 131–142.
21. Fosset, M., Jaimovich, E., Delpont, E., and Lazdunski, M. (1983). *J. Biol. Chem.*, **258**, 6086–6092.
22. Garcia, M. L., King, V. F., Arcuri, K. E., Siegl, P. K. S., Reuben, J. P., and Kaczorowski, G. J. (1985). *J. Gen. Physiol.*, **86**, 21a.
23. Glossman, H. and Ferry, D. R. (1983). *Naunyn-Schmiedeberg's Arch. Pharmacol.*, **323**, 279–291.
24. Glossman, H., Ferry, D. R., and Boschek, C. B. (1983). *Naunyn-Schmiedeberg's Arch. Pharmacol.*, **323**, 1–11.
25. Hagiwara, S. and Byerly, L. (1981). *Ann. Rev. Neurosci.*, **4**, 69–125.
26. Hess, P. and Tsien, R. W. (1984). *Nature*, **309**, 453–456.
27. Janis, R. A. and Scriabine, A. (1983). *Biochem. Pharmacol.*, **32**, 3499–3507.
28. Katz, B. and Miledi, R. (1969). *J. Physiol.* (London), **203**, 459–487.
29. Kirley, T. L. and Schwartz, A. (1984). *Biochem. Biophys. Res. Commun.*, **123**, 41–49.
30. Kokubun, S. and Reuter, H. (1984). *Proc. Natl. Acad. Sci.* (USA), **81**, 4824–4827.

31. Lee, K. S. and Tsien R. W. (1983). *Nature*, **302**, 790–794.
32. Llinas, R., Yarom, Y., and Sugimori, M. (1981). *Fed. Proc.*, **40**, 2240–2245.
33. McClesky, E. W. and Almers, W. (1985). *Proc. Natl. Acad. Sci.*, **82**, 7149–7153.
34. Murphy, K. M. M., Gould, R. J., Largent, B. L., and Snyder, S. H. (1983). *Proc. Natl. Acad. Sci.* (USA), **80**, 860–864.
35. Nilius, B., Hess, P., Lansman, J. B., and Tsien, R. W. (1985). *Nature*, **316**, 443–446.
36. Norman, R. I., Borsotto, M., Fosset, M., Lazdunski, M., and Ellory, J. C. (1983). *Biochem. Biophys. Res. Commun.*, **111**, 878–883.
37. Nowycky, M. C., Fox, A. P., and Tsien, R. W. (1985). *Nature*, **316**, 440–443.
38. Reuter, H. (1974). *J. Physiol.* (London), **242**, 429–451.
39. Reuter, H. (1979). *Ann. Rev. Physiol.*, **41**, 413–424.
40. Reuter, H. (1983). *Nature*, **301**, 569–574.
41. Rosemblat, M., Hidalgo, C., Vergara, C., and Ikemoto, N. (1981). *J. Biol. Chem.*, **256**, 8140 8148.
42. Rossie, S., and Catterall, W. A. (1986). *Fed. Proc.* **45**, 1009.
43. Sanchez, J. A. and Stefani, E. (1978). *J. Physiol.* (London), **283**, 197–209.
44. Schmid, A., Renaud, J., and Lazdunski, M. (1985). *J. Biol. Chem.*, **260**, 13041–13046.
45. Schramm, M., Thomas, G., Towart, R., and Franckowiak, G. (1983). *Nature*, **303**, 535–537.
46. Schwartz, L. M., McCleskey, E. W., and Almers, W. (1985). *Nature*, **314**, 747–751.
47. Schwartzkroin, P. A. and Slawsky, M. (1977). *Brain Res.*, **135**, 157–161.
48. Talvenheimo, J. A., Tamkun, M. M., and Catterall, W. A. (1982). *J. Biol. Chem.*, **257**, 11868–11871.
49. Triggle, D. J. (1982). In Flaim, S. F. and Zelis, R., Eds., *Calcium Blockers: Mechanisms of Action and Clinical Applications*, Urban and Schnurzenberg, Baltimore, pp. 121–134.
50. Tsien, R. W., Giles, W., and Greengard, P. (1972). *Nature New Biol.*, **240**, 181–183.
51. Tsien, R. W. (1983). *Ann. Rev. Physiol.*, **45**, 341–358.
52. Tsien, R. W., Bean, B. P., Hess, P., and Nowycky, M. (1983). *Cold Spring Harbor Symp. Quant. Biol.*, **48**, 201–212.
53. Uehara, A. and Hume, J. R. (1985). *J. Gen. Physiol.*, **85**, 621–647.
54. Venter, J. C., Fraser, C. M., Schaber, J. S., Jung, C. Y., Bolger, G., and Triggle, D. J. (1983). *J. Biol. Chem.*, **258**, 9344–9348.

CHAPTER 13

EFFECTS OF PERMEANT IONS AND BLOCKERS ON PROPERTIES OF SINGLE CALCIUM CHANNELS FROM BRAINS

MARK T. NELSON

University of Vermont College of Medicine
Department of Pharmacology
Burlington, Vermont

Calcium ions trigger or modulate a variety of physiological responses, including neurotransmitter release, muscle contraction, and cell motility (3). Despite their importance, calcium channels have not been as well characterized as sodium or acetylcholine receptor (AChR) channels, as discussed in this symposium. Calcium channels have been difficult to study for a variety of technical reasons (9). In the last 4 years, progress in the calcium channel field has increased significantly, largely as a result of improvements in single-cell isolation and culturing procedures, the discovery of high-affinity ligands for calcium channels, and the use of whole-cell and single-channel recording techniques.

One approach to the study of single ion channels has been to incorporate them into planar lipid bilayers. The planar lipid bilayer system permits direct electrical measurement of ion channels that are normally inaccessible using other techniques and allows control over the solution composition on both sides of the membrane and over the lipid composition of the membrane. A variety of "biological" channels have been incorporated into planar lipid bilayers, including the AChR channel (25), tetrodotoxin-sensitive sodium channel from brain (13) and muscle (16), the calcium-activated potassium channel from muscle (14), and the amiloride-sensitive sodium channel from epithelia (29). Functional calcium channels incorporated into planar lipid bilayers were first reported over 2 years ago, with calcium channels from rat brain synaptosomes (21,23) and *Paramecium* cilia (8) observed. Recently, calcium channels from skeletal muscle transverse tubule (1) and sarcoplasmic reticulum (28,31) have been incorporated into planar lipid bilayers. These recordings constitute the first single-channel measurements from these tissues and indicate that single calcium channels in a cell-free system can be studied. Furthermore, this technique has made it possible to study the calcium channel with a defined divalent cation gradient. In addition, these studies suggest that the dihydropyridine binding site (1) and the verapamil binding site (this chapter) can be incorporated into a bilayer, indicating that they may be intimately associated with the channel.

In this chapter, studies of a calcium channel from brain synaptosomes are presented in order to address the following general problems: (1) how divalent cations move through calcium channels, (2) how inorganic calcium channel blockers work, (3) whether the binding of a permeant ion to a site or sites in the calcium channel can affect other conformation transitions of the protein, (4) how the organic calcium channel blocker verapamil affects channel activity, and (5) whether calcium channel molecules open and close in a random and independent fashion.

1. METHODS

The median eminence of the rat brain was chosen as a source of calcium channels because it is rich in nerve terminals (17). Nerve terminals are thought to have a high density of calcium channels. Ten to fifteen median eminences were homogenized and subjected to differential centrifugation steps as described elsewhere (23,24). The membrane preparation ("synaptosomes") was sonicated and added to one side (defined as cis) of a preformed planar lipid bilayer (phosphatidylethanolamine and phosphatidylserine from bovine brain; Avanti Polar Lipids, Birmingham, AL; hole diameter 100–250 μm). The solutions on both sides of the bilayer were identical (100–250 m*M* $CaCl_2$, $BaCl_2$, $SrCl_2$, $MnCl_2$, pH = 7.0), unless otherwise noted. The solution compositions were changed by either adding a small volume of a concentrated stock or perfusing one side with a push–pull syringe system.

2. ION PERMEATION AND SELECTIVITY

Voltage-sensitive calcium channels typically permit calcium, barium, or strontium to permeate and select for these divalent cations over magnesium, sodium, potassium, and chloride ions (9). Divalent cations are thought to permeate calcium channels by interacting with specific sites in the channel (9). On the basis of macroscopic current measurements and noise analysis, a multisite, single-file model for ion permeation in the calcium channel has recently been developed (2,11). The interpretation of macroscopic current data can, however, be complicated by effects of divalent cations on surface charges (4,33) and on channel kinetics (23,30), as well as by the existence of multiple channel types (26,27). Recently, by use of patch-clamp and bilayer techniques, the permeation properties of single calcium channels have been examined, thus avoiding many of the difficulties present in the macroscopic measurements. These studies support the existence of divalent cation binding sites in the calcium channel (see Chapter 10).

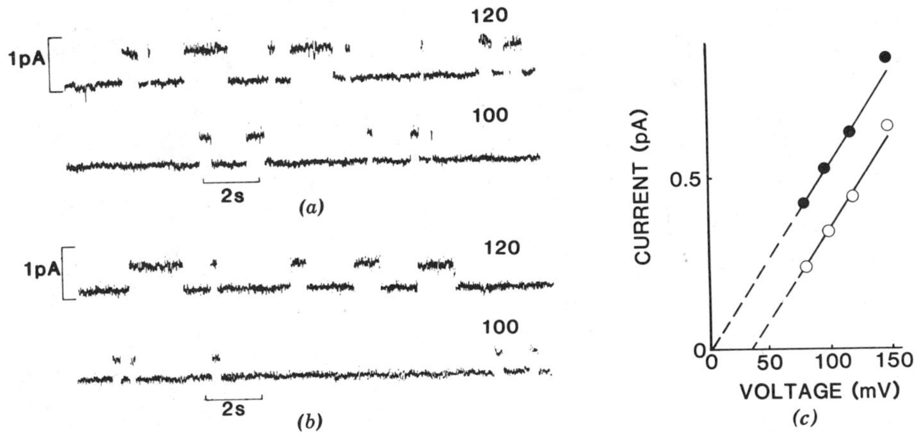

Figure 1. Single calcium channel current fluctuations: (*a*) original record at two voltages, +100 and +120 mV, with identical concentrations of divalent cations on both sides of the channel; (*b*) effect of magnesium on single-channel fluctuations; *c*) single-channel current–voltage relationship with symmetrical (filled circles) and asymmetric conditions (open circles). With the asymmetric conditions, E_{Ca} was calculated to be +55 mV and E_{Mg} = negative infinity, E_{Cl} = 0 mV.

With single calcium channels in a planar lipid bilayer, one can investigate the permeation properties of calcium channels under very simplified conditions, such as with identical concentrations of divalent cations on both sides of the channel. Figure 1 shows current fluctuations of single calcium channels incorporated into a planar lipid bilayer (with calcium as the charge carrier). Figures 1*b* and *c* illustrate data indicating that these channels select for calcium over magnesium and chloride ions. With identical concentrations of a permeant divalent cation on both sides of a membrane, the single-channel current–voltage relationship is approximately linear (23,24) and extrapolates to 0 mV (Fig. 1*c*). Reduction of the calcium concentration on the cis side and increase of the magnesium concentration caused the single current–voltage relationship to shift in the positive direction. Under these experimental conditions, the equilibrium potential E for magnesium was nominally minus infinity, E_{Cl} was 0 mV, and E_{Ca} was about +55 mV. The reversal potential for current through these channels was determined to be +35 mV (by extrapolation), indicating that these channels select for calcium over chloride and magnesium. The slope conductances (about 5.1 pS) and mean open times (about 250 msec) under both symmetrical and these asymmetric conditions were the same, suggesting that magnesium under these experimental conditions is inert to the channel. Other reversal potential experiments have demonstrated that these channels select for the permeant divalent cations over monovalent cations (23).

Another feature of calcium channels is that barium and strontium ions permeate. As seen in Fig. 2, the different divalent cations permeate at different rates, with the order of single-channel conductances as Ba > Sr (with Sr = Ca) (23). In addition, this figure illustrates that manganese can also permeate, at a lower rate (see also Fig. 3). Figure 3 also illustrates the voltage dependence of these channels; the mean open times increase *e*-fold for a 30-mV depolarization, and the mean closed times decrease *e*-fold for a 30-mV depolarization. Although manganese ions are often regarded as a calcium channel blocker, they also permeate many calcium channels (7). This apparent conflict can be resolved if ion permeation is determined by interaction of divalent cations with binding sites in the channel (11). In fact, manganese ions when added to a solution that contains a more permeant divalent cation (e.g., Ba, Ca, Sr) reduce the single conductance of the calcium channels from brain synaptosomes (24).

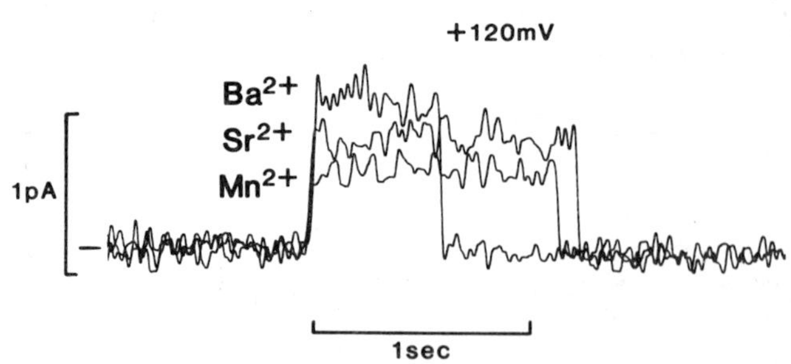

Figure 2. Comparison of the charge-carrying ability of barium, strontium, and manganese ions under identical conditions. Three representative single-channel openings were superimposed.

These results support the existence of at least one saturable divalent cation binding site in the channel.

Inorganic calcium channel blockers such as cadmium and lanthanum appear to exert their effects by competing with the permeant ions for a binding site in the permeation pathway (9). To test this proposal at the single-channel level, the effects of lanthanum ions on ion movement through single calcium channels were examined (Fig. 4). Lanthanum ions were found to be the most potent of all the ions tested, with 150 μM La reducing the single-channel current by 50% (Fig. 4). Lanthanum or cadmium ions were found to reduce the single-channel current in a manner consistent with binding of these ions to a single site. While bound to this site, other ions cannot pass through the channel. Lanthanum or cadmium ions leave this site too slowly to support a measurable current.

Although the inorganic calcium channel blockers reduce ion movement through the open channel, this does not necessarily indicate that the site(s) involved in the block is the same as the site(s) that the permeant ions interact with on their passage through the channel. In support of the idea that the blocking site is the same as the permeation site, the nature of the permeant ion affected the block by cadmium of single-channel currents, with the order of half-block constants being $K_{\mathrm{Mn}} > K_{\mathrm{Ca}} = K_{\mathrm{Sr}} > K_{\mathrm{Ba}}$. Mean blocked times for a cadmium ion or a lanthanum ion have been estimated to be about 25 μsec and 1 msec, respectively (24). A mean blocked time of 1 msec is consistent with the effects seen in Fig. 4. In the presence of lanthanum ions, the single-channel current fluctuations are clearly less rectangular and there is an increase in the frequency of nonconducting events. Under the recording conditions employed in this experiment, the time resolution was about 2–3 msec. Thus some of the nonconducting events could represent longer-lived blocking events or multiple block events. Furthermore, mean block times of 25 μsec and 1 msec indicate that the maximal cadmium or lanthanum currents would be 0.015 and 0.0004 pA, that is, below the limit of detection (about 0.05 pA).

These results are consistent with the idea that divalent cations move in single file through a calcium channel by binding to specific sites in the channel. The order of apparent affinities for this site(s) is: La > Cd > Mn > Ca > Sr > Ba >> Mg, Na, K.

3. ION DEPENDENCE OF THE OPEN TIMES

The mean open times were affected by voltage (Fig. 3B) and the nature of the permeant ion (Fig. 5). The order of closing rate constants (1/mean open times) and the single-channel conductances was the same: Ba > Ca = Sr > Mn (23,24). This parallel sequence suggests the possibility that the binding of a permeant ion to a site in the channel may affect the probability of the channel going from conducting to a nonconducting conformation. This was further supported by the observations that these effects on open time depended only on the nature of the ion moving through the channel (21) and that the nature of the permeant ion did not affect the channel mean closed time (24).

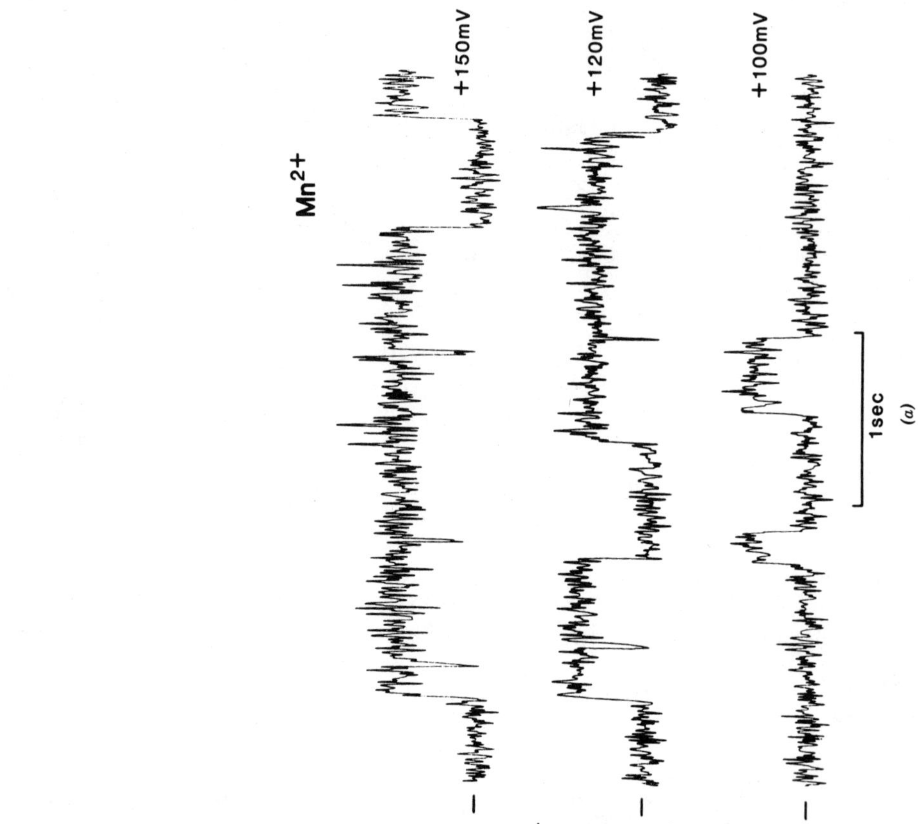
Mn2+
+150mV
+120mV
+100mV
1pA
1sec
(a)

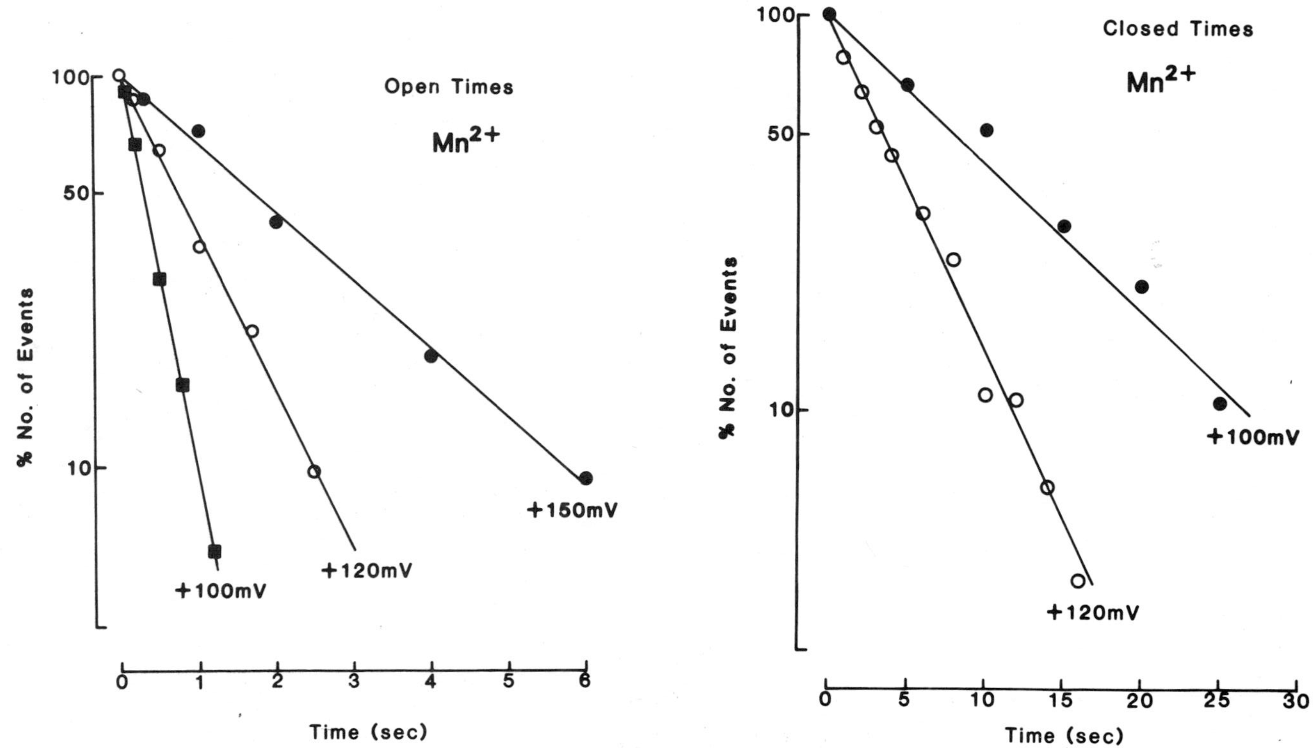

(b) (c)

Figure 3. Voltage dependence of single calcium channels from brain. (*a*) Original record at +100, +120, and +150 mV. (*b*) Effect of voltage on the open time lifetimes. The total number of events, slopes, and correlation coefficients of the linear regression lines through the semilog transformed data were at +100 mV, 96, −2.39 sec^{-1}, −0.999; at +120 mV, 221, −0.937 sec^{-1}, −0.937; at +150 mV, 43, −0.403 sec^{-1}, −0.998. (*c*) Effect of voltage on closed time distribution at +100 and +120 mV. The total number of events, slopes, and correlation coefficients of the linear regression lines through the semilog transformed data were at +100 mV, 38, −0.086 sec^{-1}, −0.992; +120 mV, 109, −0.195 sec^{-1}, −0.998. Manganese ions were the charge carrier.

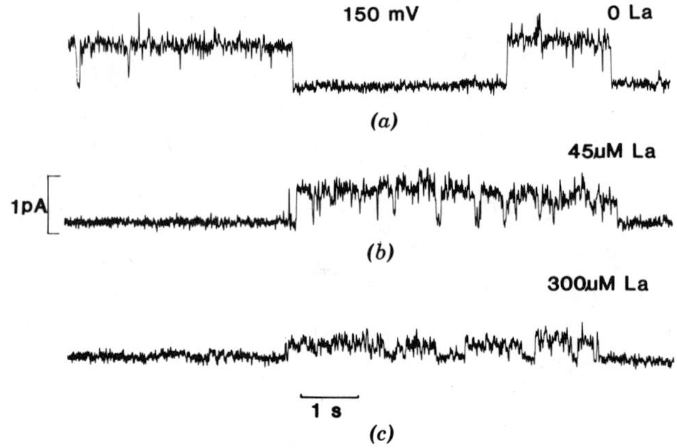

Figure 4. Block by lanthanum ions of current through single calcium channels. Lanthanum from 1 *M* stock was added to the cis side of the bilayer. Strontium ions were the charge carriers.

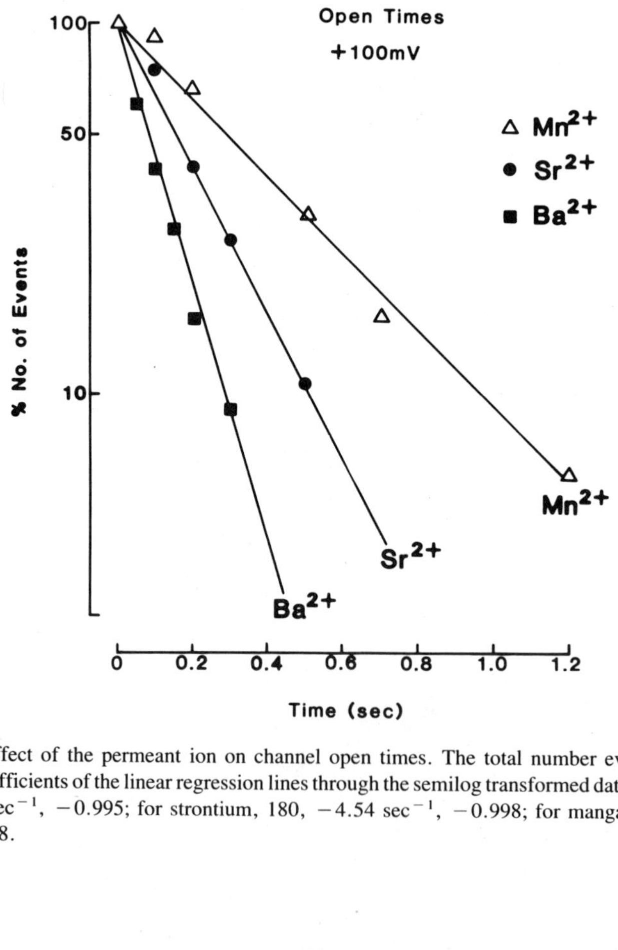

Figure 5. Effect of the permeant ion on channel open times. The total number events, slopes, and correlation coefficients of the linear regression lines through the semilog transformed data were for barium, 48, $-8.085\ sec^{-1}$, -0.995; for strontium, 180, $-4.54\ sec^{-1}$, -0.998; for manganese, 96, $-2.39\ sec^{-1}$, -0.998.

To explore this possibility further, the effects of the permeant ions on the free energies of closing and binding are compared. Because these closing rate constants k appear to be exponential functions of voltage, they can be expressed in the following manner:

$$k = A \exp -\Delta G/RT \qquad \text{(at 0 mV)} \tag{1}$$

where A is a vibration or attempt frequency and may be on the order of 10^{13} $\sec^{-1}$ (12) and ΔG is the activation free energy change for the particular transition in the pathway from conducting to nonconducting configuration of the channel. The difference between the free energies of closing G^* of two different permeant ions (at the same voltage) can then be expressed as

$$G^* = RT \ln \frac{k_{\text{Ba}}}{k_{\text{Ca}}} \tag{2}$$

Similarly, the difference between the free energies of binding of the permeant ions is

$$G^* = RT \ln \frac{K_{\text{D}_{\text{Ba}}}}{K_{\text{D}_{\text{Ca}}}} \tag{3}$$

where K_{D} is the dissociation constant constant for the permeant ions (24).

The difference in energy of binding between barium and calcium or strontium can be estimated to be $0.8RT$ and the difference in the energies of closing to be $0.69RT$ (about 80% of that of binding) (see Table 1). The parallel changes in free energies supports the idea that the binding of permeant ion to a site in the channel can affect another conformation transition of the channel, that is, the transition from the open state to the closed state. Implicit in this analysis is the assumption that the single-channel currents represent the true rates of ion movement through the open channel. Therefore, the effects of the permeant ions on the single-channel currents represent differences in binding affinities of these ions to sites in the channel.

The single exponential fits of open and closed time distributions suggest that the channel can be described by a two-state model: one open state and one closed state

TABLE 1. DIFFERENCES IN FREE ENERGY (IN *RT* UNITS)[a]

Between	Binding	Closing	G^*(closing)/G^*(binding)
(Ba) and (Ca)	−0.80	−0.64	0.80
(Ba) and (Sr)	−0.80	−0.64	0.80
(Ba) and (Mn)	−1.69	−1.39	0.82

[a]Values were determined by using Equations (2) and (3). Rate constants and dissociation constants are from ref. 24.

(see Figs. 3*b* and 3*c*). With this model both the opening and closing rate constants would be voltage dependent. The closing rate constant would also depend on the permeant ion. However, the situation is clearly not so simple. As can be seen clearly in Figs. 3, 4, and 7, the open state is punctuated by very brief often unresolved closures, suggesting the existence of a second closed state. With a three-state model for the calcium channel, an alternative explanation for the effects of the permeant ion on the single-channel conductances and mean open times presents itself. With this model (see the formula following this paragraph), the fluctuations between the open state (O) and the closed state (C_1) would be so rapid that the measured single-channel currents and open times are really averages of the time spent in the two states. Thus, the measured unitary current does not represent the true permeation rate of an ion through the open channel.

$$C_1 \leftrightarrows O \rightleftharpoons C_2$$

The effects of the permeant ion on the channel kinetics and single-channel conductances can be explained by a single effect on the rate constant from the open state (O) to closed state (C_1). Replacement of barium by manganese would increase the rate constant for the transition from the open to the closed states (C_1), thereby reducing the single-channel current and the mean open time. This model is directly analogous to the "sequential block scheme" as described by Neher and Steinbach (19) for the block of AChR channels by local anesthetics. This scheme also predicts that the total charge passing through the channel during a "conducting" period should be independent of the nature of the permeant ion (20). This is clearly not the case for the calcium channel. When barium is the charge carrier (at + 100 mV), the total charge transferred during an event lasting as long as the mean open time was about 100 femtocoulombs (fC) (mean unitary current × mean open time) and the total charge transferred with manganese as the charge carrier was 150 fC. Furthermore, such a model would suggest that the true permeation rates for barium, calcium, strontium and manganese would be just the opposite from those observed. Such a permeation sequence would be in conflict with the effects of the permeant ions on cadmium block (24). In addition, studies on other types of calcium channel support a C_1–C_2–O sequence of states (32). Higher-resolution recordings may help to resolve mode of action of the permeant ions on the kinetics of the channel.

4. ORGANIC CALCIUM CHANNEL BLOCKERS

Verapamil and methoxy derivative, D600, have been reported to reduce potassium-stimulated calcium isotopic fluxes through calcium channels in brain synaptosomes (18) and other preparations (9). Verapamil has proved particularly important in the treatment of certain types of cardiac arrhythmias and cardiovascular disorders. The effects of verapamil and D600 on single calcium channels in the bilayer were tested. In contrast to the inorganic blockers, these agents did not affect currents through the open channel. However, the addition of verapamil (25 μM) to the trans side of bilayer that contained at least five calcium channels caused the mean current through the channels to decrease from 0.270 to 0.078 pA, or by 71%. Addition of verapamil to the cis side caused a small but significant further reduction in the mean current from 0.078 to 0.041 pA. The percent reduction by verapamil of the mean single-channel currents was comparable to the inhibition of potassium-stimulated isotopic calcium influx measured by Nachshen and Blaustein (18). Verapamil did not significantly affect the single-channel conductance or the mean open time. The effect of verapamil on the mean currents can be accounted for by an increase in the mean closed time, that is, decrease in the opening rate constant.

The sensitivity of these calcium channels to verapamil suggests that the verapamil receptor is intimately associated with the channel molecule and that the receptor–channel complex can function in a cell-free system. Thus it may be reasonable to assume that a biochemically purified verapamil receptor would have a functional channel associated with it.

5. POSSIBLE INTERACTIONS OF CHANNEL MOLECULES: NONINDEPENDENCE OF CALCIUM CHANNELS

An important and difficult issue is whether calcium channels behave in a random and independent fashion. As can be seen clearly in Fig. 6*b*, there were three occurrences out of 50 bins of mean currents greater than 1.0 pA (one of these occurrences can be seen in Fig. 6*a* as three channels open at the same time). If the distribution of currents around the mean is normal or Gaussian, the probability of such occurrences would be less than 0.005. Yet 6% of the occurrences were of such magnitude. There are three obvious explanations for these results: (1) silent channels spontaneously switch on and off; (2) the channels open and close nonrandomly; (3) the channels interact with one another, that is, they are not independent. The first possibility would require a group of channels to switch on and off together within ~2 sec. Although this possibility is difficult to disprove, it would seem unlikely. The second possibility was tested by performing a runs analysis on the records shown in Figs. 6*d* and 6*f* (12). If channel activity were nonrandom, one would expect that the bins of channel activity would be clustered. This analysis failed to uncover nonrandom behavior. Furthermore, in multichannel membranes, the lifetimes of two channels that are open at the same time can be accurately predicted, assuming independence, from the mean closed times and the mean open times of channel. Although these results are consistent with the behavior of the channels in a random or independent manner, this does not prove that the channels always behave in such a fashion. For example, Fig. 7 provides evidence that the channels can behave nonindependently. In 20 experiments (>100 fluctuations) examined, events similar to the one shown in Fig. 7 were

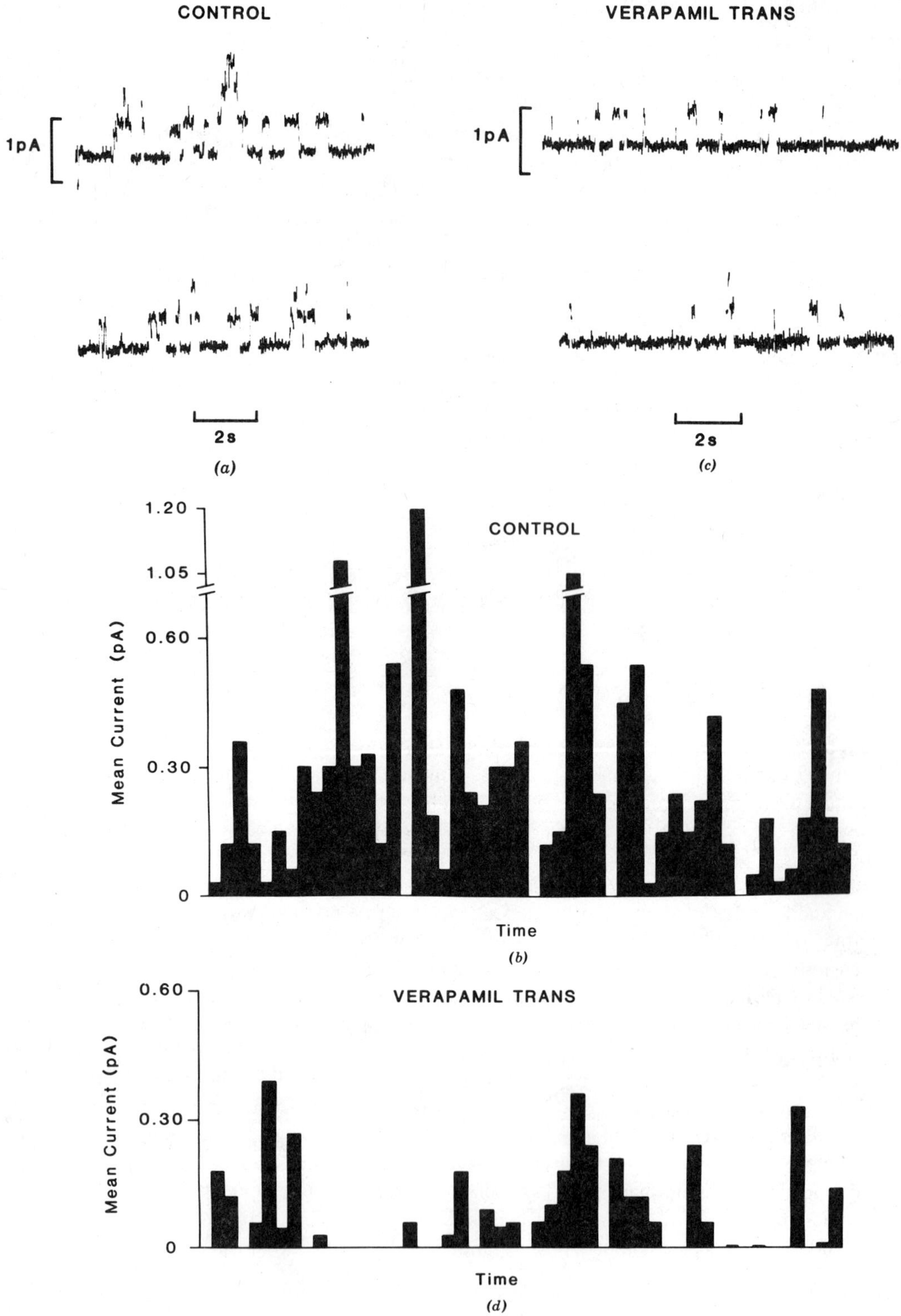
CONTROL
VERAPAMIL TRANS
1pA
1pA
2s
2s
(a)
(c)
1.20
1.05
0.60
0.30
0
Mean Current (pA)
CONTROL
Time
(b)
0.60
0.30
0
Mean Current (pA)
VERAPAMIL TRANS
Time
(d)

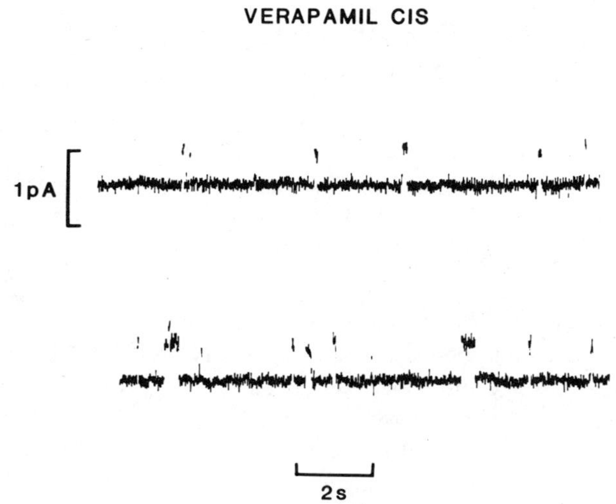

(e)

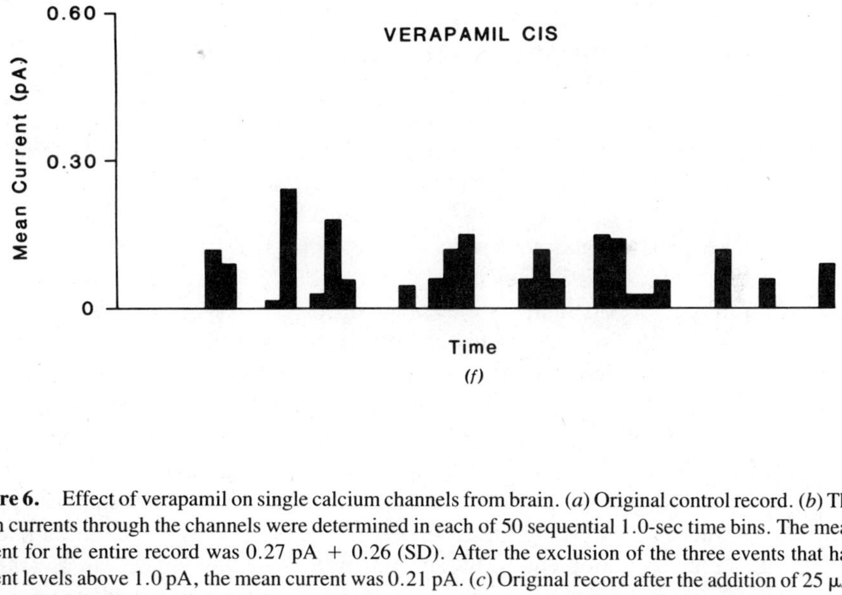

(f)

Figure 6. Effect of verapamil on single calcium channels from brain. (*a*) Original control record. (*b*) The mean currents through the channels were determined in each of 50 sequential 1.0-sec time bins. The mean current for the entire record was 0.27 pA + 0.26 (SD). After the exclusion of the three events that had current levels above 1.0 pA, the mean current was 0.21 pA. (*c*) Original record after the addition of 25 μ*M* verapamil to the trans side of the bilayer. (*d*) Effect of *trans*-verapamil on the mean currents through the channels. The mean current for the entire record was 0.08 pA + 0.11 (SD). (*e*) Original record after the addition of 25 μ*M* verapamil to the cis side of the bilayer (i.e., there was 25 μ*M* verapamil on both sides of the membrane). (*f*) Effect of trans and then cis verapamil on the mean currents through the channels. The mean current for entire record was 0.04 + 0.06 pA (SD). Strontium was the charge carrier (V_M = +100 mV). The effects of verapamil occurred without a delay.

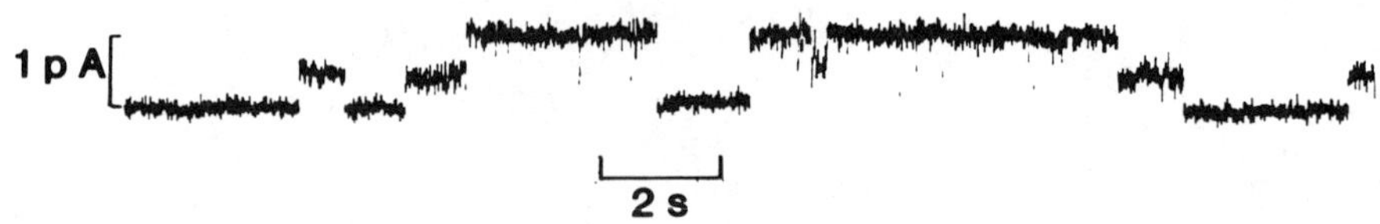

Figure 7. Apparent nonindependent behavior of single calcium channels from brain. Manganese was the charge carrier and the single conductance was about 4 pS. Excluding the apparent nonindependent events, the mean open time or mean closed time was 1.07 or 5.13 sec.

observed. These events were characterized by two channels opening and closing in synchrony and a substantial increase in the probability of channel being open. The charge that crosses through membrane during this event was over seven standard deviations from the mean charge movement before and after the event. One possible explanation for these results is that two channel molecules can interact. Although these events constituted a small percentage (<5%) of the total, they were responsible for a significant amount of ion movement. Thus the regulation of interactions between calcium channels may conceivably regulate calcium flow across membranes.

6. CONCLUSIONS AND PERSPECTIVES

Although calcium channels can come in a variety of types with different behaviors, the data presented in this chapter support the following three general characteristics of calcium channels (see also Chapter 10):

1. Ion permeation and selectivity are determined by interactions of divalent cations with specific sites in the channel (2,11). The affinities of these sites for inorganic divalent cations may differ from type to type of calcium channels (26), suggesting that there may be structural differences in binding sites from different calcium channels. However, the basic structure of binding site may indeed be conserved. For example, Troponin C has four binding sites, two with high affinity for magnesium and two with lower affinity for magnesium and specific for calcium (10). Although these pairs of sites have very different affinities for divalent cations, the basic supersecondary structure of the binding site is conserved and consists of two helices linked by an amino acid loop chain. In fact, this basic structure appears in a variety of calcium binding proteins (e.g., calmodulin, Wasserman protein, parvalbumin, S-100) (6). It will be of interest to see whether the binding sites in the calcium channel also exhibit a similar helix–loop–helix configuration.
2. Inorganic calcium channel blockers reduce macroscopic calcium currents by competing with the permeant ions for this site(s). The affinities of blocking ions for different types of calcium channels may provide a means for classifying the calcium channel binding sites (4).

3. Calcium channel modulators (e.g., verapamil, dihydropyridines, phoshorylation) do not block ion movement through the open channel. However, they affect the opening and closing of the channel, often in a complicated manner (see chapter by Hess et al.).

Furthermore, information presented in this chapter suggests that the binding of a permeant ion to a site or sites in a calcium channel can affect another conformational transition, the closing of the channel. It is possible that other functional properties of the channel such as antagonist and agonist effects may be allosterically influenced by the binding of ion(s) to site(s) in the channel. Such effects of the permeant ion on the conformational state of the channel are not so surprising in light of studies on other calcium binding proteins. Once again turning to the example of troponin C, the binding of a calcium ion to its binding site affects the conformation of the binding sites as well as other parts of the protein (10). Extending the comparison, it is reasonable to speculate that the conformation of the divalent binding site in the calcium channel in the absence of calcium would be different from that found in a channel occupied by calcium. Thus the apparent large pore size of the calcium channel in the absence of calcium (15) may in part reflect another conformational state of the binding site.

It was expected that if a few channels were incorporated together into a large planar lipid bilayer, they would diffuse apart and behave independently. In fact, tests for nonrandom and nonindependent behavior of calcium channels did not reveal any obvious discrepancies. However, a small fraction of the events did exhibit behavior that was consistent with channel–channel interactions (see Figs. 6 and 7). These observations stress the importance of viewing an ion channel, not only as an individual molecule, but a part of a group of molecules that may interact or interconvert (5). Therefore, another possible mechanism of calcium channel modulation may be by affecting the interactions between calcium channels.

ACKNOWLEDGMENTS

I thank Drs. Jennings F. Worley, III and Jane Talvenheimo for comments on the manuscript. This work was supported by grants from the American Heart Association, National (#84–879), Maryland Greater and Miami Chapters. The author is an Established Investigator of the American Heart Association.

REFERENCES

1. Affolter, H. and Coronado, R. (1985). *Biophys. J.*, **48**, 341–347.
2. Almers, W. and McCleskey, E. W. (1984). *J. Physiol.* (London), **353**, 585–608.
3. Blaustein, M. P. and Nelson, M. T. (1982). In Carafoli, E., Ed., *Membrane Transport of Calcium*, Academic, New York, pp. 217–233.

4. Byerly, L., Bryant Chase, P., and Stimers, J. (1985). *J. Gen. Physiol.*, **85**, 491–518.
5. Conti, F., Hille, B., and Nonner, W. (1984). *J. Physiol.* (London), **353**, 199–230.
6. Dalgarno, D., Klevit, R., Levine, B., and Williams R. (1984). *Trends Parmacol. Sci.*, 266–271.
7. Drapeau, P. and Nachshen, D. (1984). *J. Physiol.* (London), **348**, 493–510.
8. Ehrlich, B. E., Finkelstein, A., Forte, M., and Kung, C. (1984). *Science*, **225**, 427–428.
9. Hagiwara, S. and Byerly, L. (1981). *Ann. Rev. Neurosci.*, **4**, 69–125.
10. Herzberg, O. and James, M. (1985). *Nature*, **313**, 653–659.
11. Hess, P. and Tsien, R. W. (1984). *Nature*, **309**, 453–456.
12. Horn, R. (1984). In Stein, W. D., Ed., *Ion Channels: Molecular and Physiological Aspects*, Academic, New York, pp. 53–97.
13. Krueger, B. K., Worley, J. F., III, and French, R. J. (1983). *Nature*, **303**, 172–175.
14. Latorre, R., Vergara, C., and Hidalgo, C. (1982). *Proc. Natl. Acad. Sci.* (USA), **79**, 805–809.
15. McCleskey, E. W. and Almers, W. (1984). *Biophys. J.*, **47**, 66a.
16. Moczydlowski, E., Garbar, S., and Miller, C. (1984). *J. Gen. Physiol.*, **84**, 665–686.
17. Moore, K. E. and Johnston, C. A. (1982). In Mueller, E. E. and Macleod, R. M., Eds., *Neuroendocrine Perspectives*, Vol. 1, Elsevier Biomedical Press, Amsterdam, pp. 23–68.
18. Nachshen, D. A. and Blaustein, M. P. (1979). *Molec. Pharmacol.*, **16**, 579–586.
19. Neher, E. and Steinbach, J. H. (1978). *J. Physiol.* (London), **277**, 153–176.
20. Neher, E. (1983). *J. Physiol.* (London), **339**, 663–678.
21. Nelson, M. T. (1983). *Soc. Neurosci. Abstr.*, **9**, Part 1, 508.
22. Nelson, M. T. (1984). In Bronner, F. and Peterlik, M., Eds., *Epithelial Calcium and Phosphate Transport: Molecular and Cellular Aspects*, Alan R. Liss, pp. 59–64.
23. Nelson, M. T., French, R. J., and Krueger, B. K. (1984). *Nature*, **308**, 77–80.
24. Nelson, M. T. (1986). *J. Gen. Physiol.* **87**, 201–222.
25. Nelson, N., Anholt, R., Lindstrom, J., and Montal, M. (1980). *Proc. Natl. Acad. Sci.* (USA), **77**, 3057–3061.
26. Nilius, B., Hess, P., Lansman, J. B., and Tsien, R. W. (1985). *Nature*, **316**, 443–446.
27. Nowycky, M., Fox, A., and Tsien, R. W. *Nature*, **316**, 440–443.
28. Orozco, C., Suarez-Isla, B., Froehlich, J., and Heller, P. (1985). *Biophys. J.*, **47**, 57a.
29. Sariban-Sohraby, S., Latorre, R., Burg, M., Olans, L., and Benos, D. (1984). *Nature*, **308**, 80–82.
30. Siami, Y. and Kung, C. (1982). *Science*, **218**, 153–156.
31. Smith, J. S., Coronado, R., and Meissner, G. (1985). *Nature*, **316**, 446–449.
32. Tsien, R. W. (1983). *Ann. Rev. Physiol.*, **45**, 341–358.
33. Wilson, D. L., Morimoto, K., Tsuda, Y., and Brown, A. (1983). *J. Membrane Biol.*, **72**, 309-324.

CHAPTER 14

CHARACTERIZATION OF PARTIAL REACTIONS IN THE CATALYTIC AND TRANSPORT CYCLE OF SARCOPLASMIC RETICULUM ATPase

GIUSEPPE INESI
University of Maryland School of Medicine
Department of Biological Chemistry
Baltimore, Maryland

Vesicular fragments of sarcoplasmic reticulum (SR) membrane were introduced as an experimental system by Hasselbach and Makinose (18,19) and independently by Ebashi and Lipmann (11). It is shown in Fig. 1 that differential centrifugation of a skeletal muscle homogenate separates a microsomal fraction containing mostly a 110,000 M_γ protein from other cellular components. This fraction is identified with vesicular fragments of SR membrane. Its main protein component is densely assembled, and each 110,000-molecular-weight chain includes a polar segment protruding from the cytoplasmic plane of the membrane and a hydrophobic segment intruding into the membrane. While the polar segments of the protein molecules can be visualized by electron microscopy as distinct entities in negatively stained preparations, the hydrophobic segments appear aggregated within the membrane bilayer in freeze-fracture preparations (Fig. 2). The ATPase protein can be purified (29), and, if suitable precautions are taken to avoid denaturation (Fig. 3), it can be demonstrated that each 110,000-molecular-weight polypeptide chain includes one catalytic site and two high-affinity column binding sites. (1,42). The specific functional feature of the SR vesicles is to accumulate calcium from the medium into

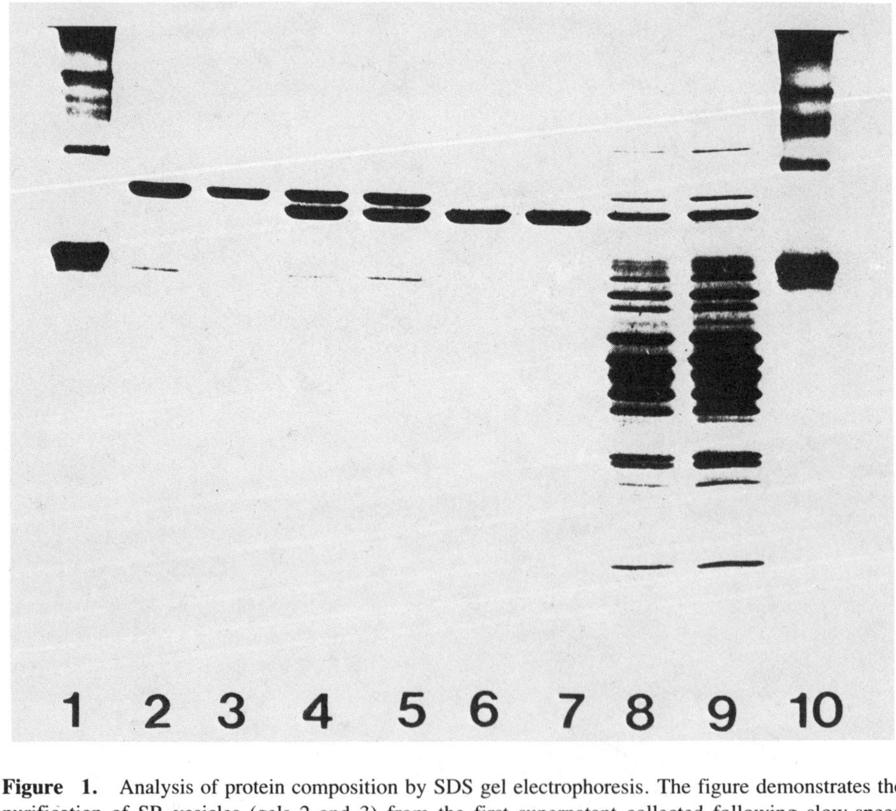

Figure 1. Analysis of protein composition by SDS gel electrophoresis. The figure demonstrates the purification of SR vesicles (gels 2 and 3) from the first supernatant collected following slow-speed centrifugation of a muscle homogenate (gels 8 and 9). (From ref. 45.)

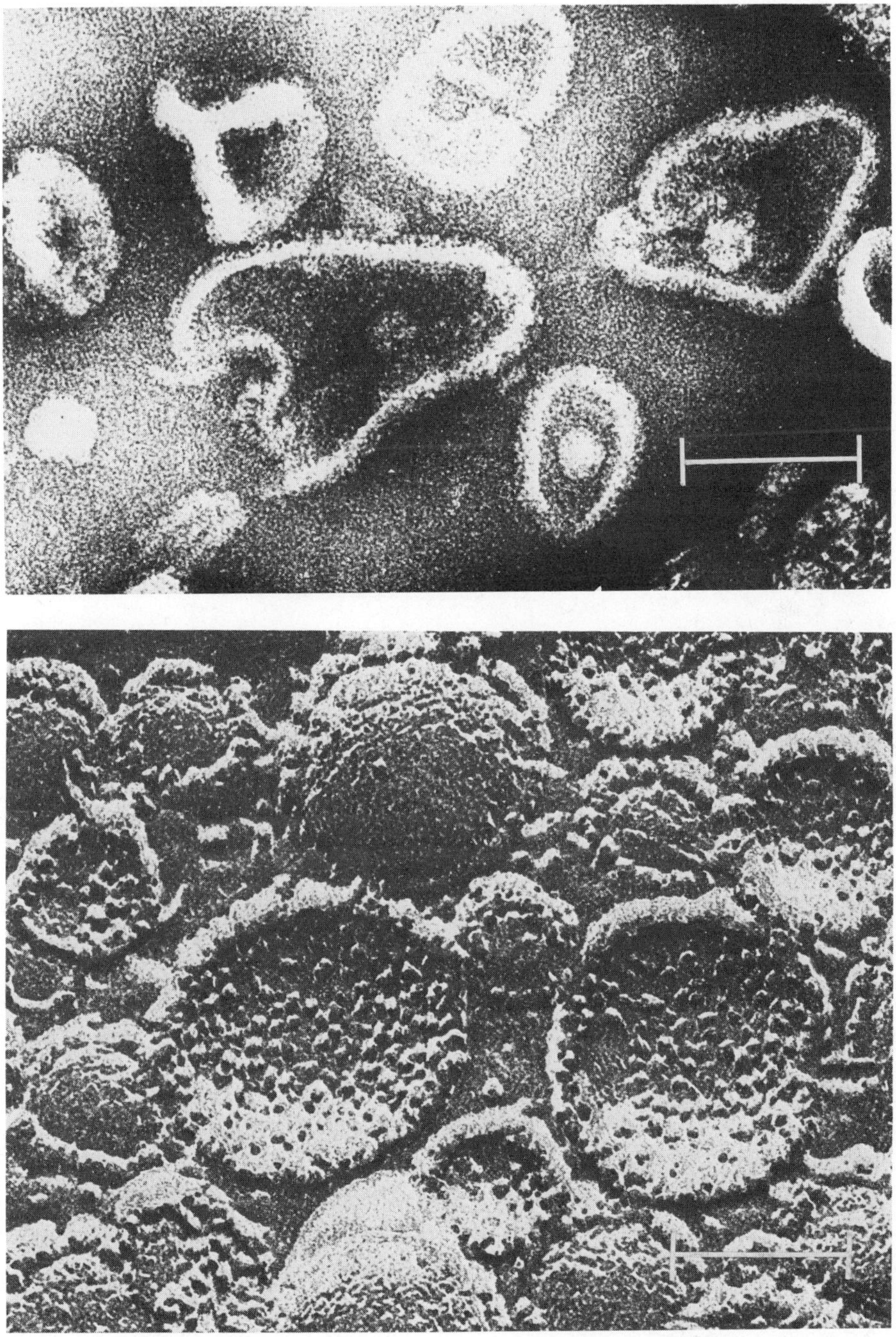

Figure 2. Electron-microscopic visualization of negatively stained (top) and freeze-fractured (bottom) SR vesicles. The polar portions of the amphiphilic ATPase molecules protrude from the outer surface of the membrane (top), while the nonpolar portions of the ATPase protein intrude into the hydrophobic interior of the membrane bilayer and are visualized on the concave freeze-fractured faces (bottom). Magnification markings correspond to 0.1 μM. (Courtesy of Dr. Donald Scales.)

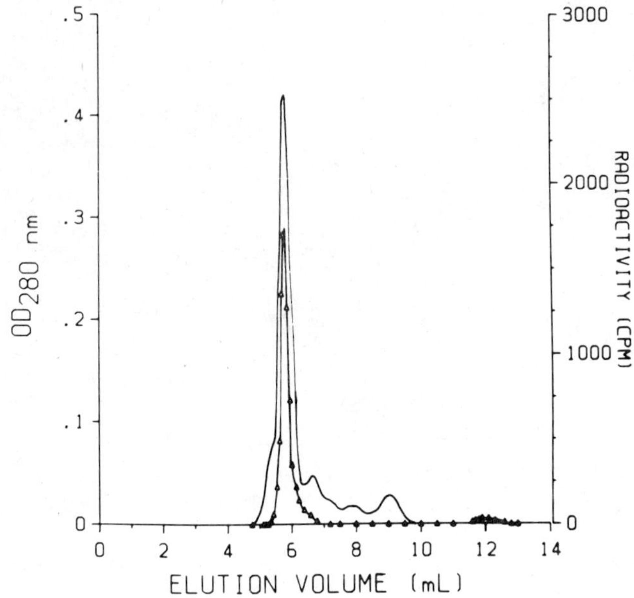

Figure 3. A molecular-sieve HPLC peak demonstrating the purification of SR ATPase and its phosphorylation by incorporation of radioactivity-labeled ATP terminal phosphate. (From ref. 1.)

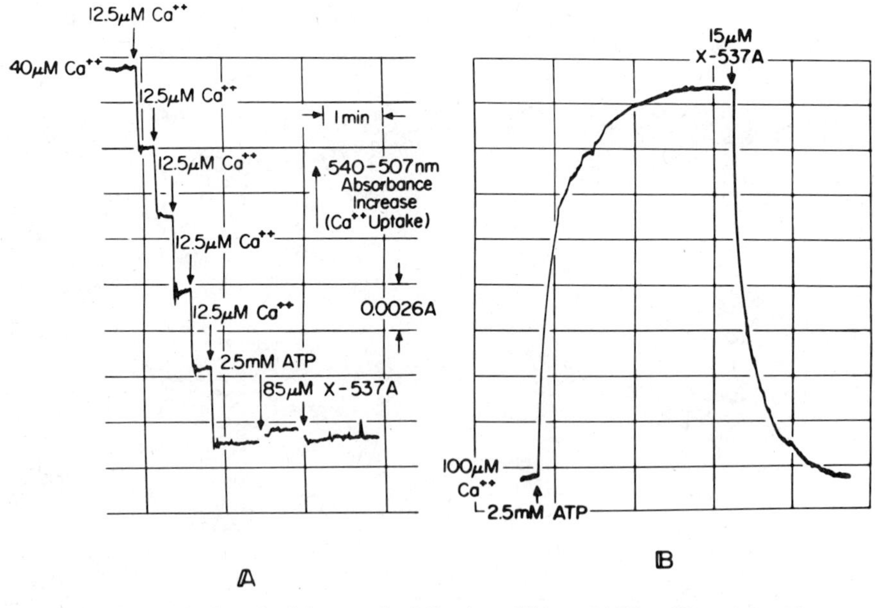

Figure 4. Demonstration of calcium uptake following addition of ATP to SR vesicles. The resulting transmembrane gradient of calcium ion collapses readily when the passive permeability of the membrane is increased by the addition of a suitable ionophore. (From ref. 40.)

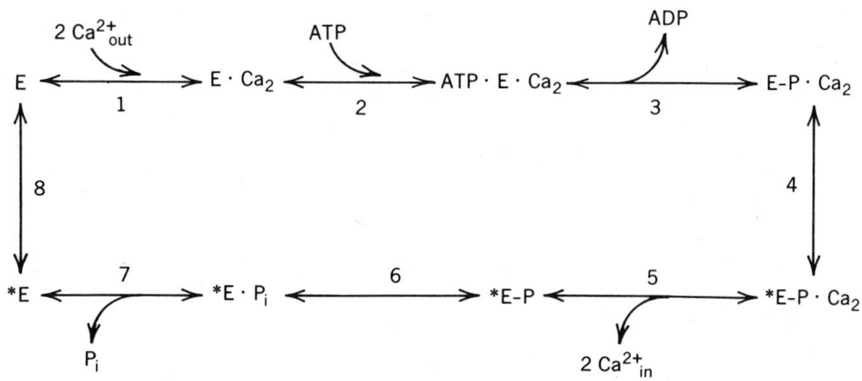

Scheme 1

the lumen on addition of ATP. Thereby, a transmembrane gradient of calcium ion is formed, which collapses rapidly if the passive permeability of the membrane to calcium is increased by the addition of a suitable ionophore (Fig. 4).

Adenosine triphosphate utilization by the SR ATPase is coupled to calcium transport by a mechanism (8,22) including several partial reactions (Scheme 1): calcium binding to high-affinity sites producing enzyme activation, formation of phosphorylated enzyme intermediate by incorporation of the ATP terminal phosphate, change in orientation and reduction in the affinity of the enzyme for calcium, vectorial release of calcium, hydrolytic cleavage of the phosphoenzyme, and return of the enzyme to a new cycle. The cycle is fully reversible, as demonstrated by ATP $\leftrightarrow$ P_i exchange, ATP synthesis, and coupled ejection of calcium from the lumen of the vesicles into the medium. An important feature of the SR vesicles as an experimental system for studies of enzyme catalysis and coupling mechanism is that several of the partial reactions can be measured and values for equilibrium and/or kinetic constants obtained thereby.

1. CHARACTERIZATION OF PARTIAL REACTIONS

Calcium binding to the SR ATPase may be considered as the first partial reaction of the cycle inasmuch as binding of this divalent cation is a requirement for enzyme activation in the absence of ATP, and calcium binding can be measured by equilibration of SR vesicles in the presence of various concentrations of calcium ion and radioactive tracer. Two or three classes of calcium binding sites of various affinities are then found (Fig. 5). The class of the highest affinity is of interest, since its binding isotherm matches very well the calcium concentration dependence of ATPase activity (Fig. 6). At neutral pH, and in the presence of magnesium and potassium concentrations used in studies of enzymatic activity, the apparent dissociation constant for calcium binding to these specific sites is approximately 1 μM.

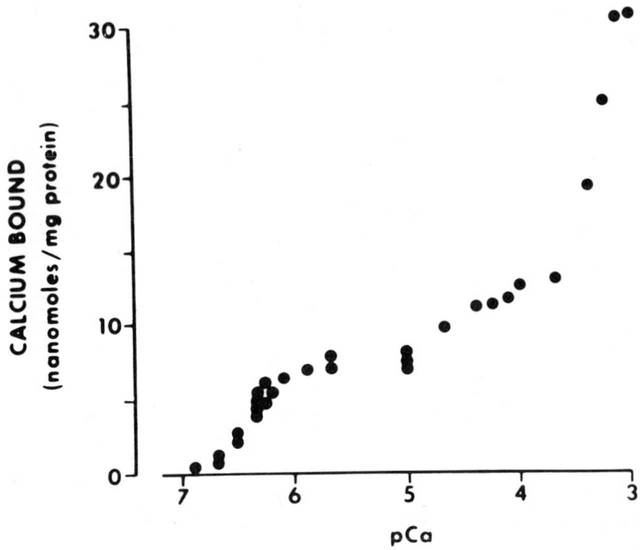

Figure 5. Calcium binding to SR vesicles. Sites of various affinities were detected by equilibration of radioactive calcium with SR vesicles in chromatography columns. (From ref. 23.)

Adenosinetriphosphate utilization by the calcium activated enzyme includes formation of a phosphorylated enzyme intermediate by transfer of the ATP terminal phosphate to an aspartyl residue at the catalytic site (2,6,30,48). Kinetic resolution of enzyme phosphorylation with ATP can be obtained by rapid quench methods (14,44). Initial velocities of enzyme phosphorylation following addition of various concentrations of ATP to SR vesicles preincubated with calcium are shown in Fig. 7. Extrapolation of the initial velocity of phosphorylation to infinite ATP concentration yields a maximal velocity that, divided by the concentration of catalytic sites, provides an estimate of the phosphorylation rate constant. The value for this constant is approximately 200 sec^{-1}. The reverse rate constant for this reaction has been

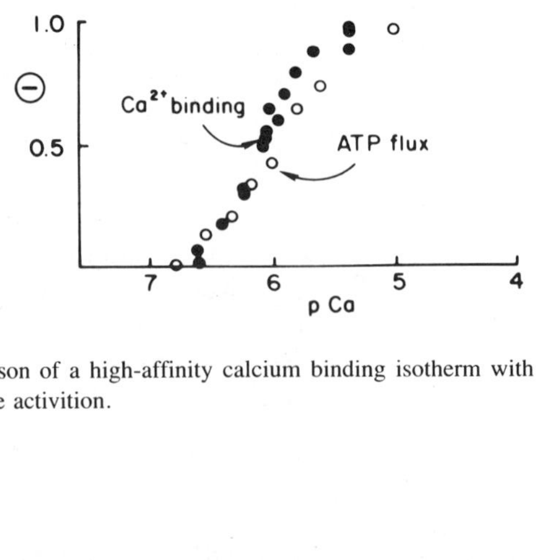

Figure 6. Comparison of a high-affinity calcium binding isotherm with the calcium concentration dependence of ATPase activition.

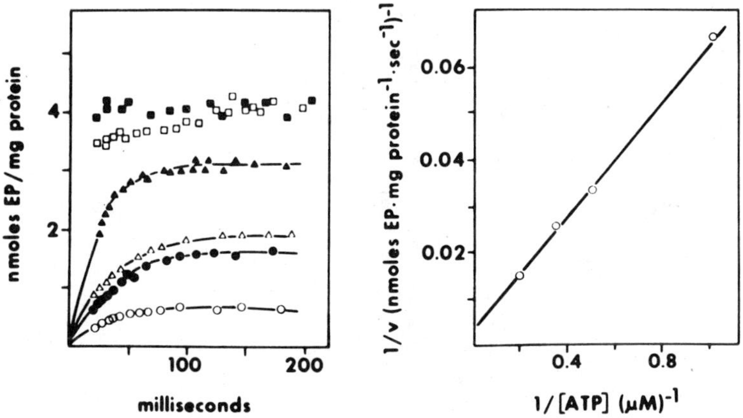

Figure 7. The initial phase of enzyme phosphorylation following addition of ATP. Extrapolation of the initial velocities (right panel) to infinite ATP concentration allows an estimate of the maximal velocity and (based on the catalytic site stoichiometry) a first-order constant. (From ref. 44.)

estimated to be approximately 400 sec^{-1} (39). The dissociation constant of the catalytic site for ATP is approximately 10^{-5} *M* (39).

The measurable events occurring in the pre-steady state, following addition of ATP to enzyme preincubated with calcium are illustrated in Fig. 8. It is clear that as a consequence of phosphoenzyme formation, a burst of calcium uptake occurs (28). This burst, which is demonstrated by the inability of lanthanum or EGTA (added with the quenching medium) to displace calcium bound to the specific sites on the enzyme,

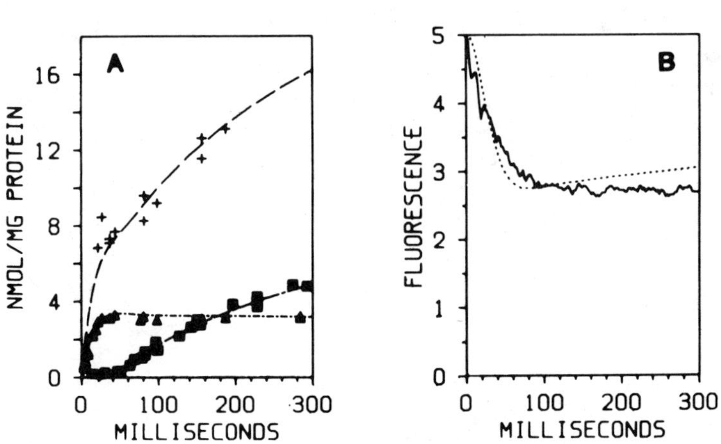

Figure 8. Pre-steady-state kinetics following addition of ATP to SR vesicles preincubated with calcium. In the left panel, the initial ATPase phosphorylation (▲) is rapidly followed by an initial burst of calcium internalization (+) and then by P_i production (■). The right panel shows a decrease of intrinsic fluorescence consequent to enzyme phosphorylation. (From ref. 12.)

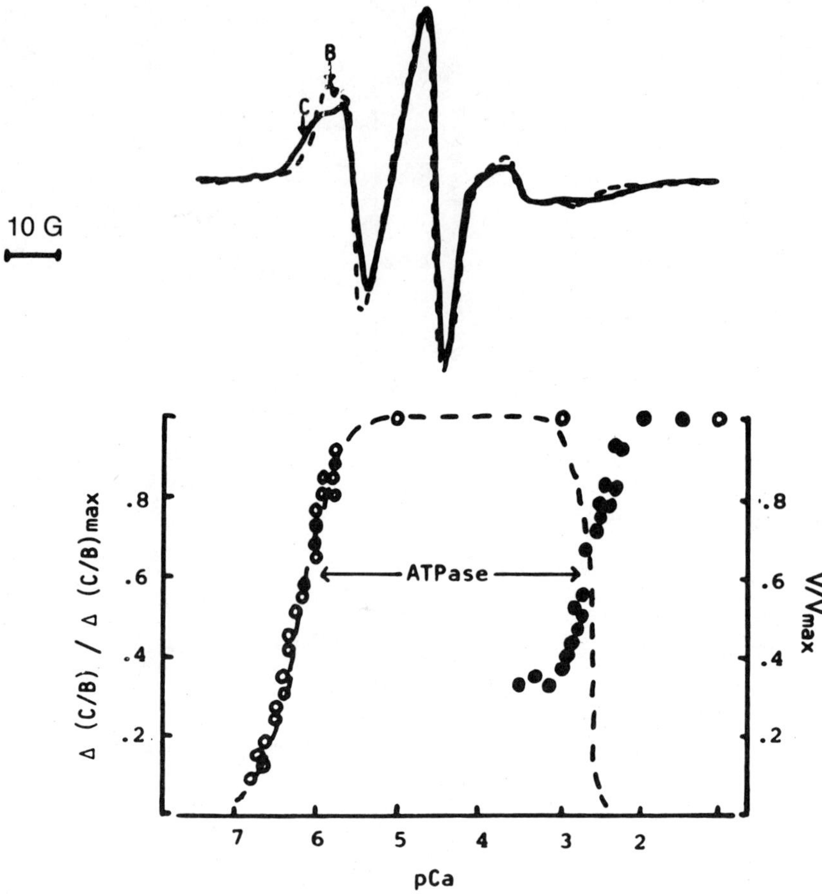

Figure 9. Changes in calcium binding affinity following enzyme phosphorylation. The occupancy of specific transport sites in the nonphosphorylated (○) and phosphorylated (●) enzyme was monitored by measurements of a spin-label parameter. (From ref. 5.)

is interpreted as an internalization or ''occlusion'' (9) of the divalent cation within the structure of the enzyme. The burst involves 2 moles of calcium per mole of phosphoenzyme and occurs very rapidly following enzyme phosphorylation. Then, after a kinetic delay (approximately 20 sec^{-1}) related to translation and dissociation of calcium ion to the opposite side of the membrane (24), the phosphoenzyme undergoes hydrolytic cleavage and proceeds to subsequent cycles. Steady-state production of P_i and calcium transport begins thereafter, retaining Ca^{2+}/P_i ratio equal to 2, in optimal conditions.

It can be shown independently that in addition to internalization, the specific calcium sites undergo a three-order-of-magnitude reduction in affinity for the divalent cation. Direct measurement of low affinity binding with radioactive tracer is

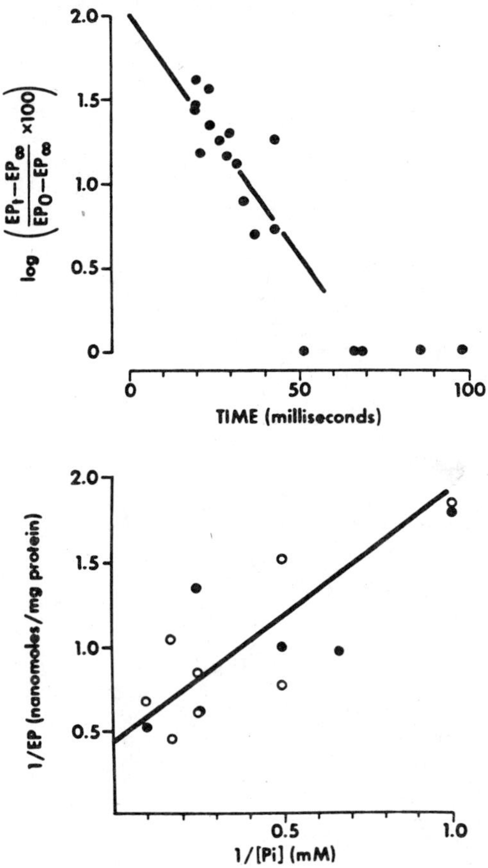

Figure 10. Kinetics of phosphoenzyme formation following addition of P_i to SR ATPase in the absence of calcium (top). The lower panel shows the P_i concentration dependence of the levels of phosphoenzyme under identical conditions. These experiments yield both equilibrium and kinetic constants for phosphorylation of SR ATPase with P_i. (From ref. 24.)

very difficult because of the occurrence of nonspecific binding in the presence of high (m*M*) Ca^{2+} concentrations. On the other hand, the phosphorylation-induced reduction in affinity can be demonstrated by monitoring a spectroscopic signal related to calcium occupancy of the specific sites in the nonphosphorylated and the phosphorylated enzymes (Fig. 9).

Finally, the hydrolytic cleavage of phosphoenzyme and its reverse reaction [i.e., enzyme phosphorylation of P_i (32)] can be studied by both equilibrium and kinetic experimentation (Fig. 10). At neutral pH, and in presence of magnesium and potassium concentrations used for ATPase activity, the equilibrium constant for this reaction is approximately 1 and the forward and reverse rate constants are approximately 60 sec^{-1}.

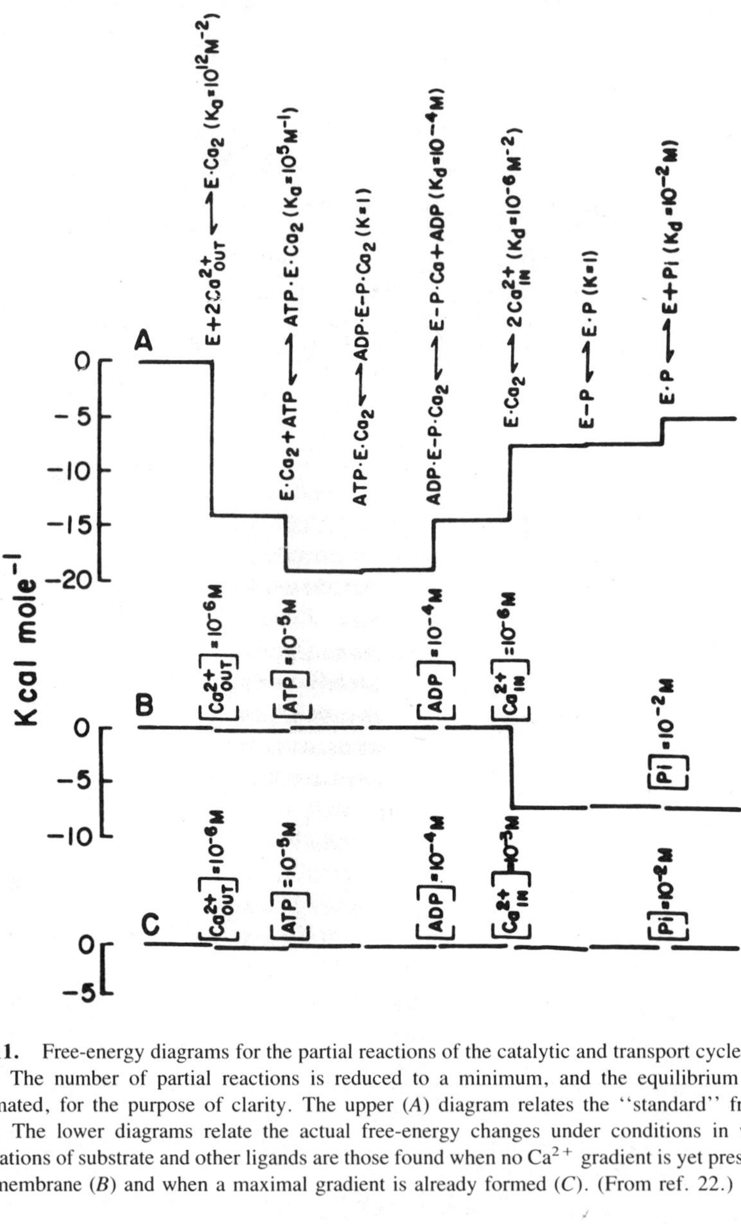

Figure 11. Free-energy diagrams for the partial reactions of the catalytic and transport cycle of the SR ATPase. The number of partial reactions is reduced to a minimum, and the equilibrium constants approximated, for the purpose of clarity. The upper (*A*) diagram relates the "standard" free-energy changes. The lower diagrams relate the actual free-energy changes under conditions in which the concentrations of substrate and other ligands are those found when no Ca^{2+} gradient is yet present across the SR membrane (*B*) and when a maximal gradient is already formed (*C*). (From ref. 22.)

With a set of apparent equilibrium constants for the partial reactions, it is then possible to construct a free-energy diagram for the catalytic and transport cycle of the SR ATPase (Fig. 11). The requirement that is met here is that, in standard conditions, the overall free-energy drop corresponds to the standard free energy of ATP hydrolysis. Of course, since the cycle includes binding and dissociation reactions, the free-energy changes for these reactions must be corrected for the ligand concentrations present under any particular experimental condition. Generally the free-energy change for each reaction is much reduced by this correction. In particular, it is intellectually satisfying to see that if the calculation is made for a case soon after addition of ATP (when the concentration of Ca^{2+} inside the vesicles is still equal to that of the outside medium), the overall free-energy drop is present allowing for a rapid forward flux of substrate. On the contrary, when the concentration of calcium inside the vesicles reaches its maximal level, no overall free-energy drop is observed; consequently, no additional net forward flow is expected.

Simple schemes are a useful qualitative guide for visualization of the sequence of principal reactions and can even be utilized to represent thermodynamic relationships in the catalytic-transport cycle. It should be strongly emphasized, however, that there is no instance in which detailed experimental kinetics could be fitted satisfactorily with analysis based on a reaction scheme such as a Scheme 1. On the contrary, it is by introducing further details into each of the partial reactions, and setting up imaginative analysis for each set of demanding experiments, that new mechanistic features can be discovered and clarified. An example of this type of approach can be appreciated by considering why the level of phosphoenzyme is not dependent on pH (within the 6–8 range) when phosphorylation is obtained with ATP, while it is pH dependent when phosphorylation is obtained with P_i (even with P_i is expressed as a single ionization species). It is shown in Fig. 12 that a very good fit of these data is obtained (26) with a scheme including participation of two enzyme residues

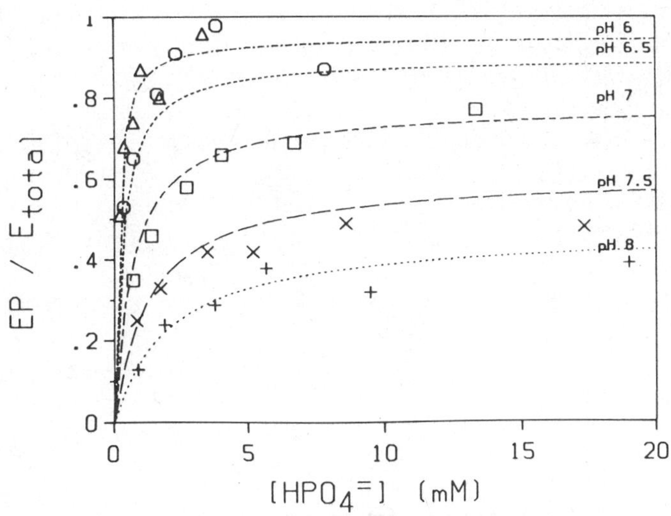

Figure 12. Phosphoenzyme levels following equilibration of SR vesicles with various concentrations of P_i in the absence of calcium and at varying pH. The experimental points are expressed here as functions of a single phosphate ionization species, and the connecting lines are derived from the analysis described in ref. 26.

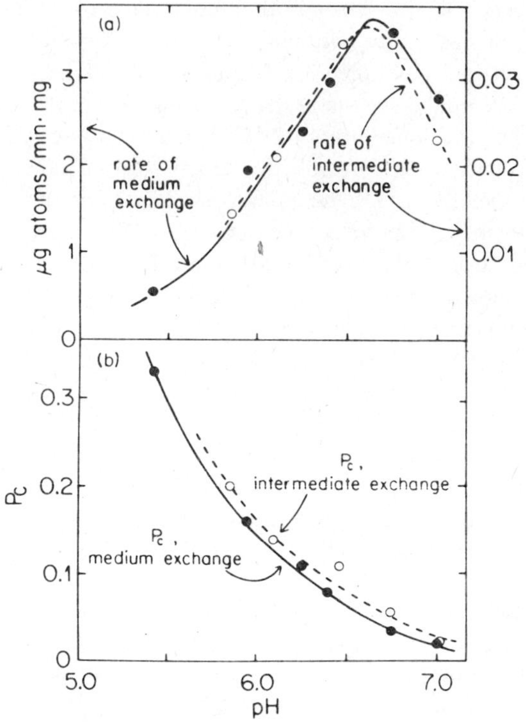

Figure 13. The pH dependence of medium and of intermediate $P_i \rightleftharpoons$ HOH exchange catalyzed by SR ATPase. (From ref. 33.)

dissociating protons with pK_a near neutrality. In their dissociated form these residues would assist hydrolytic cleveage of the phosphoenzyme by withdrawing protons from hydrolytic water, thereby favoring attack by the water oxygen on the phosphorus atom. In their protonated form, these residues would favor phosphoenzyme formation by supplying protons for formation of water during the phosphorylation reaction with P_i. At neutral pH, at which half of these residues are protonated and half are dissociated, maximal rates of phosphorylation and hydrolytic cleveage are predicted, as observed experimentally with measurements of oxygen exchange between water and phosphate (Fig. 13). On the other hand, these two residues are not involved in enzyme phosphorylation with ATP, since this is a phosphoryl transfer reaction with no direct participation of water. In agreement with this mechanism, it is possible to inactivate selectively these residues with DCC (41), with consequent inhibition of enzyme phosphorylation with P_i, and no inhibition of enzyme phosphorylation with ATP (Fig. 14).

It must be understood that no universal scheme can be used in the analysis of different experiments under different conditions, and reaction schemes must be adapted for any given analysis of partial reactions. Simulation of experimental data from the overall cycle requires a complete list of detailed reactions, with assignment

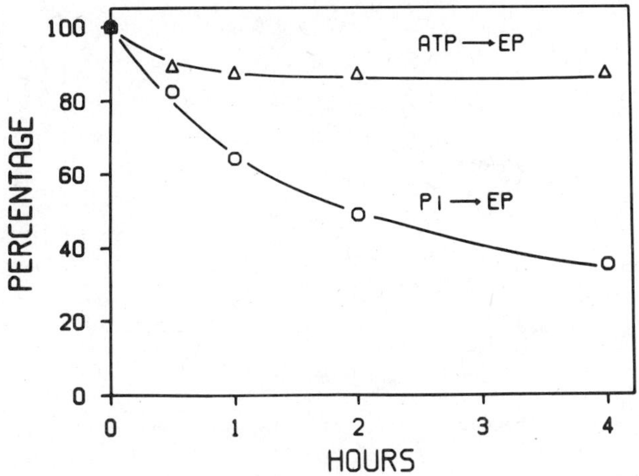

Figure 14. Selective inhibition of phosphoenzyme formation with P_i, in SR ATPase incubated for increasing periods of time with dicyclohexylcarbodiimide (DCC). Note that the enzyme still retains its ability to be phosphorylated by ATP. (From ref. 41.)

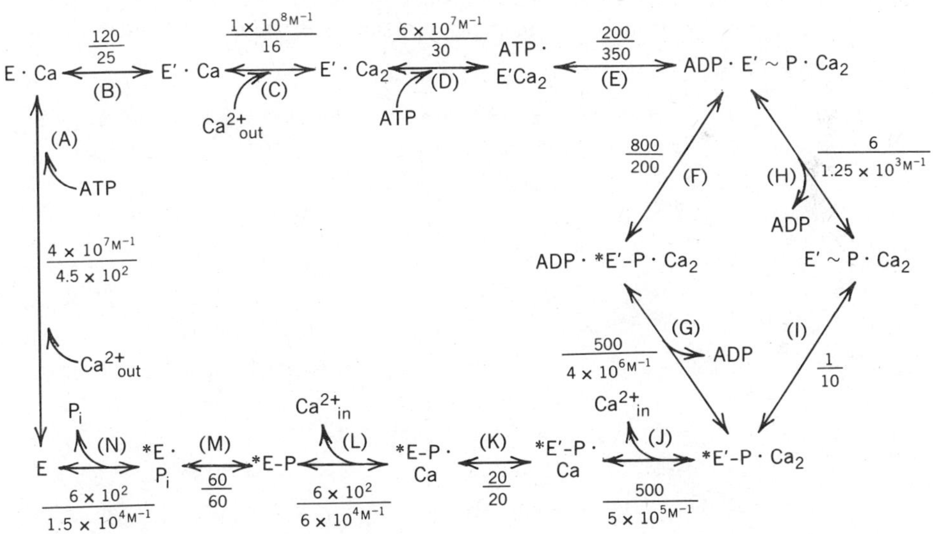

Scheme 2. Rate constants are expressed in sec^{-1}. The upper constant is forward, and the lower is reverse, in the clockwise flow of the cycle. The constants for reactions A, B, C, D, E, M, and N are derived from experimental measurements, directly or indirectly. The other constants are best fits and are likely to need adjustment.

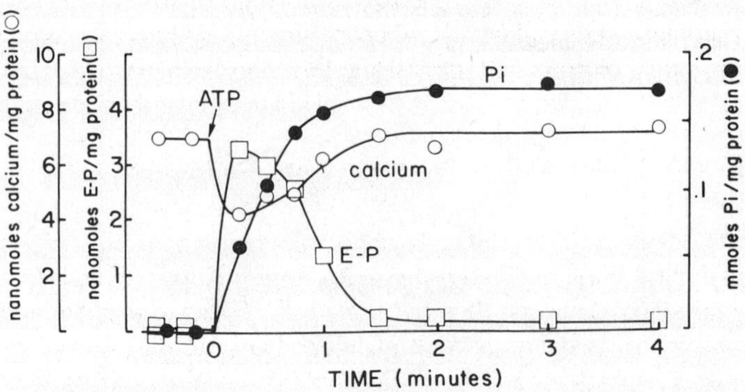

Figure 15. Release of calcium bound to transport site, following phosphorylation of SR ATPase with ATP. As the phosphoenzyme is cleaved, calcium is bound again. (From ref. 47.)

of bidirectional rate constants compatible with the expected overall free-energy change. Scheme 2 is an example of the reaction schemes that we have recently used (38) in our laboratory for this type of analysis. Compared to Scheme 1, Scheme 2 considers a sequential mechanism for calcium binding and also considers a branched pathway for calcium occlusion and ADP release following enzyme phosphorylation with ATP. Most importantly, Scheme 2 makes some commitments with regard to the bidirectional rate constant of the partial reactions, thereby permitting testing of the postulated mechanism. Although schemes such as these may not offer a unique solution, and some of the constant need adjustment, they determine unambiguously whether a postulated mechanism is possible and state precisely the implications of any proposed mechanism.

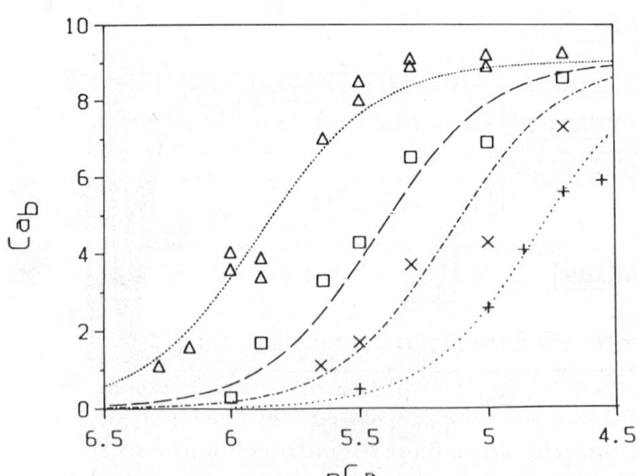

Figure 16. Calcium binding to the transport sites of SR ATPase at equilibrium in the absence of ATP. The high-affinity noted in the control experiment (△) is reduced by the addition of increasing concentrations of vanadate: 5 μM (□), 20 μM (×) and 100 μM (+). (From ref. 26.)

2. MUTUAL EXCLUSION OF CALCIUM AND PHOSPHORYLATION SITE OCCUPANCY

Having considered the entire catalytic cycle and its kinetic regulation, we now focus on the coupling mechanism of catalysis and active transport. It was shown that the orientation of the specific calcium sites is changed (Fig. 8), and their affinity decreased as a consequence of enzyme phosphorylation (Fig. 9). It can be also demonstrated that calcium does in fact dissociate from the enzyme following phosphorylation and is bound again when the phosphoenzyme is cleaved (Fig. 15). Such a phosphorylation-induced calcium dissociation is the basic feature of the coupling mechanism since it can be readily understood that when dissociation and binding occur cyclically and vectorially in vesicles bound by a sealed membrane, a transmembrane gradient of calcium ion is formed.

A most important observation with regard to this coupling is that enzyme phosphorylation with ATP is not strictly necessary to produce the reduction in affinity of the calcium sites. In fact, such a reduction can be produced simply by vanadate (Fig. 16), which forms a strong complex with the enzyme, probably as an analog of a pentacoordinated transition state of phosphate in the phosphorylation reaction. Even more interesting is the observation (7) that if enzyme phosphorylation with P_i (rather than ATP) is rendered favorable by inclusion of organic solvents in the medium, calcium is released as the enzyme is phosphorylated (Fig. 17). It is thereby demonstrated that the fundamental coupling mechanism of catalysis and active transport is a mutual exclusion of occupancy of calcium and phosphorylation sites. The mutual exclusion is not produced by physical overlap of the sites since identical calcium binding isotherms are obtained (Fig. 18) in both the absence and the presence of AMPPNP, and ATP analog that occupies the catalytic site (including the terminal

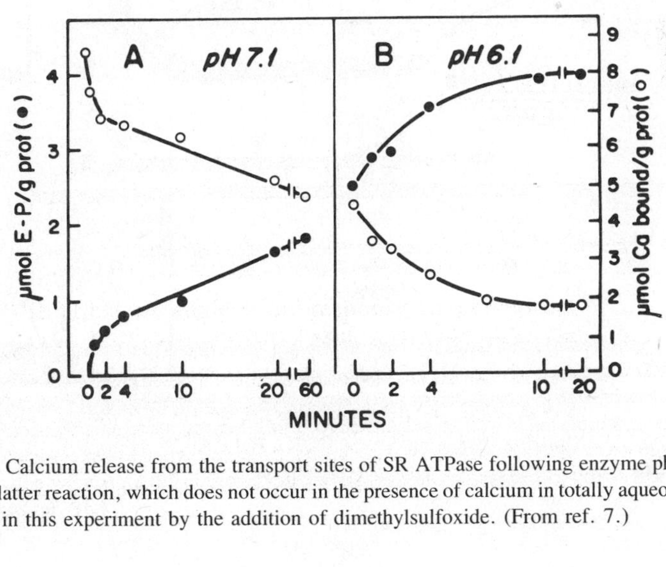

Figure 17. Calcium release from the transport sites of SR ATPase following enzyme phosphorylation with P_i. The latter reaction, which does not occur in the presence of calcium in totally aqueous media (32), was favored in this experiment by the addition of dimethylsulfoxide. (From ref. 7.)

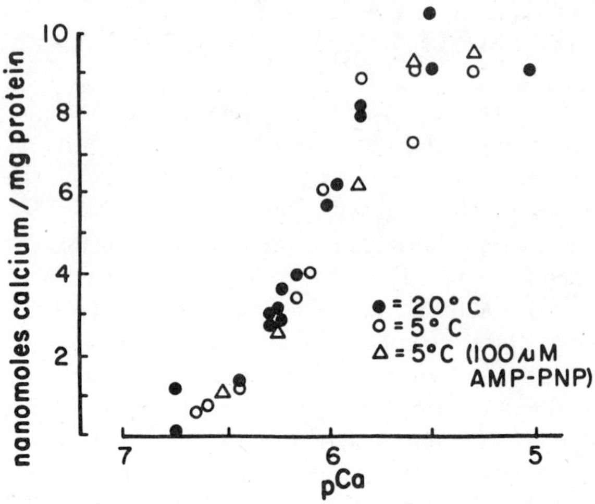

Figure 18. Calcium binding to the transport sites of SR ATPase in both the absence and the presence of the ATP analog AMPPNP. This analog binds to the catalytic site but it is not utilized by SR ATPase at significant rates. (From ref. 47.)

phosphate portion), without actually producing covalent phosphorylation of the site. Measurements of energy transfer between suitable labels also suggests a rather large distance between the catalytic and the calcium sites (19). It is also apparent that while the catalytic site is readily accessible on the outer surface of the vesicles, calcium binding occurs in a crevice within the ATPase protein that is protected from the influence of phospholipid surface charge (42). The reciprocal influence of phosphorylation and calcium sites must then be a long-range effect, produced by ligand-induced changes in protein conformation.

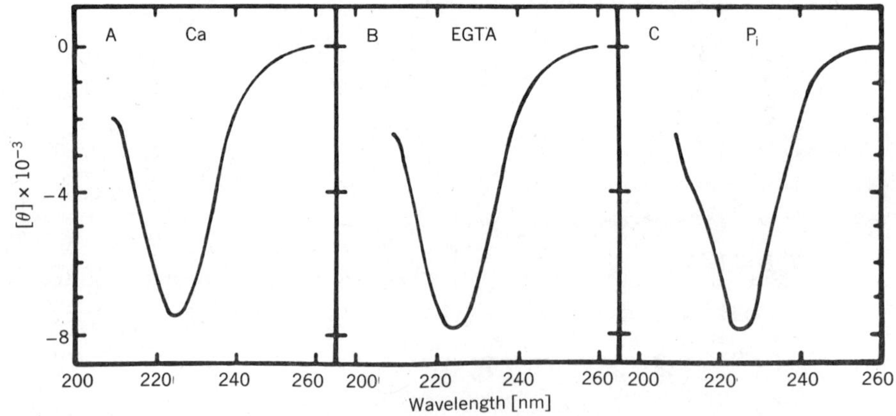

Figure 19. Circular dichroism spectra showing no change in the ellipticity of SR ATPase in the presence of calcium concentrations saturating the transport sites, in the absence of calcium (EGTA), and following phosphorylation of the catalytic site. (From ref. 36.)

3. LIGAND-INDUCED CONFORMATIONAL CHANGES IN SR ATPASE

We then consider two main perturbations of specific relevance: (1) calcium occupancy of its specific sites and (2) phosphorylation of the catalytic site. A very interesting finding in this regard is that neither of these perturbations produces a significant change in protein ellipticity, as revealed by circular dichroism (Fig. 19). This is in definite contrast with the change produced by calcium on regulatory proteins such as calmodulin (Fig. 20).

In spite of the constant ellipticity, several experimental findings indicate that the SR ATPase undergoes ligand-induced conformational changes. With respect to calcium, the cooperativity of the binding isotherms (Fig. 21) is consistent with a conformational change in the mechanism of cooperative binding and enzyme activation (23). The change in conformation is also revealed by measurements of intrinsic fluorescence (Fig. 22) (10,12,15) and spin-label spectroscopic parameters (Figs. 23 and 24). Furthermore, calcium binding is known to produce changes in the kinetics of –SH reactivity (35).

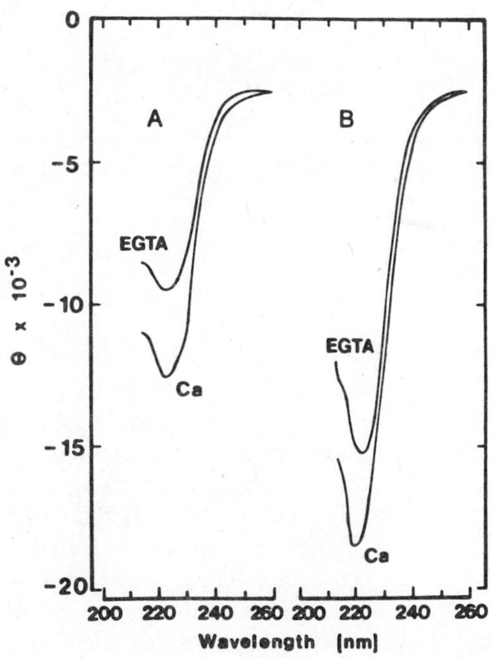

Figure 20. Calcium-induced change in the ellipticity of calmodulin (*A*). The same measurement was obtained (*B*) when SR ATPase was present in the second cavity of a double cell, demonstrating that the presence of the ATPase suspension does not interfere optically with detection of the ellipticity changes. (From ref. 36.)

$$E + \mathrm{Ca}^{2+} \rightleftharpoons E \cdot \mathrm{Ca}$$

$$E \cdot \mathrm{Ca} = \frac{K[E_{\mathrm{tot}}][\mathrm{Ca}^{2+}]}{1 + K[\mathrm{Ca}^{2+}]}$$

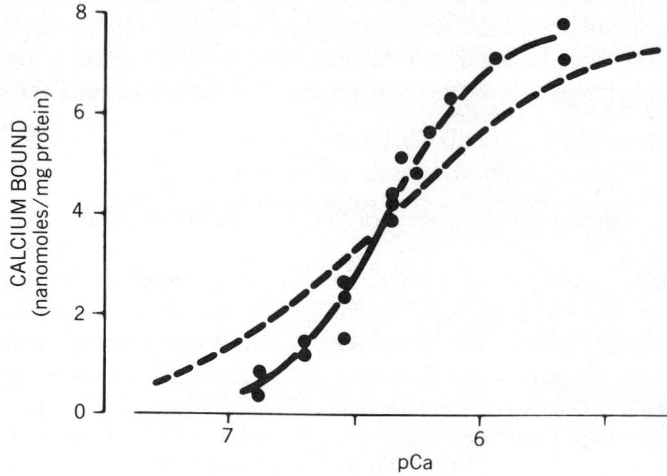

$$E + \mathrm{Ca}^{2+} \overset{K_1}{\rightleftharpoons} E \cdot \mathrm{Ca} + \mathrm{Ca}^{2+} \overset{K_2}{\rightleftharpoons} E \cdot \mathrm{Ca}_2$$

$$\frac{\text{Calcium bound}}{\text{Enzyme unit}} = \frac{K_1[\mathrm{Ca}^{2+}] + 2K_1K_2[\mathrm{Ca}^{2+}]^2}{1 + K_1[\mathrm{Ca}^{2+}] + K_1K_2[\mathrm{Ca}^{2+}]^2}$$

Figure 21. Calcium binding isotherm for the transport sites of SR ATPase. The experimental points were fitted satisfactorily with an equation based on cooperative binding of two calcium ions (bottom). Unsatisfactorily fitting was otained with an equation based on a single, independent site binding mechanism (top). (From ref. 23.)

With respect to phosphorylation of the catalytic site, it was shown (4) by time-resolved X-ray diffraction studies that a significant change in the SR membrane electron density profile follows a single cycle of ATPase phosphorylation. Another indication of a conformational change can be obtained much more easily by simple visual inspection, if the ATPase of SR vesicles in suspension is phosphorylated in the presence of low concentrations (insufficient to produce solubilization) of detergent. Under these conditions, ATPase phosphorylation is immediately followed by clearing of the turbidity (45). It can be monitored by measurements of optical density that the clearing effect is simultaneous with formation of phosphoenzyme and reverses when the phosphoenzyme is hydrolyzed (Fig. 25). The clearing effect, then, is due to a phosphorylation-dependent and reversible conformational change that favors interaction of the enzyme with detergent.

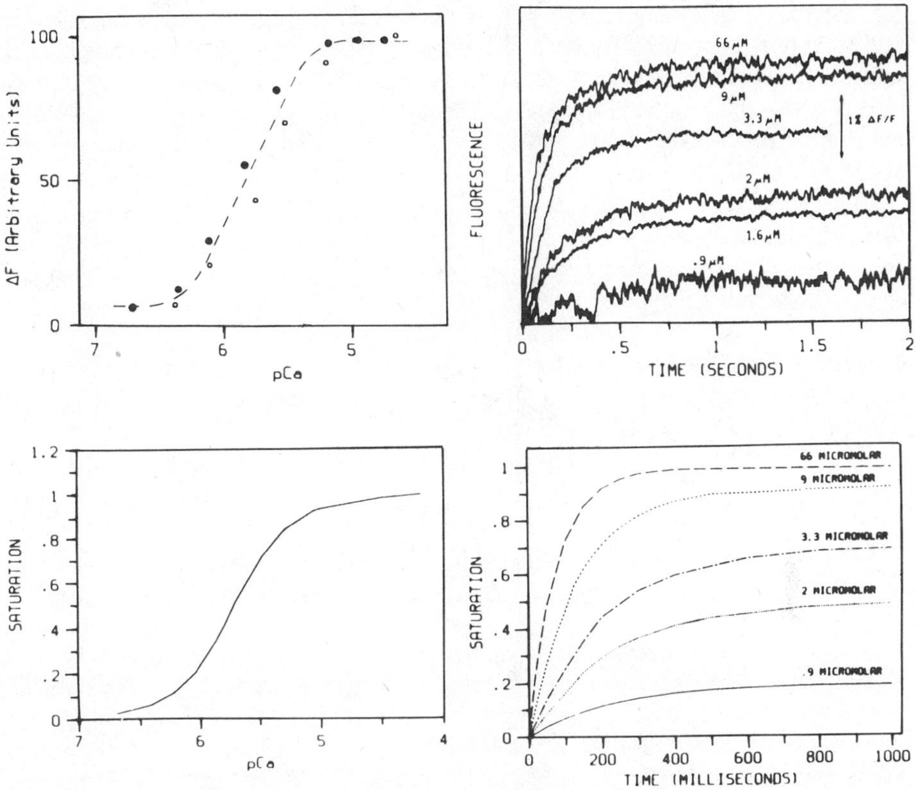

Figure 22. Equilibrium (left) and kinetic (right) measurements of calcium binding to SR ATPase, monitored by changes of intrinsic fluorescence. The experimental data (top) are simulated (bottom) assuming a cooperative binding mechanism. (From ref. 12.)

Another conformational effect of ATPase phosphorylation can be demonstrated (44) by monitoring the fluorescence of TNP nucleotide analogs (Fig. 26). These analogs manifest a very low fluorescence intensity in aqueous solution. Their fluorescence is slightly increased when they bind to nonphosphorylated ATPase and greatly increased when they bind to phosphorylated ATPase (Fig. 27). The effect is observed when the enzyme is phosphorylated with ATP in the presence of calcium or with P_i in the absence of calcium (Fig. 28). A lower effect is observed when the catalytic site is occupied by vanadate (Fig. 28). These studies, as well as measurements of fluorescence lifetime, demonstrate a phosphorylation-dependent conformational change in the environment of the fluorophore, producing a local reduction in polarity and motional freedom.

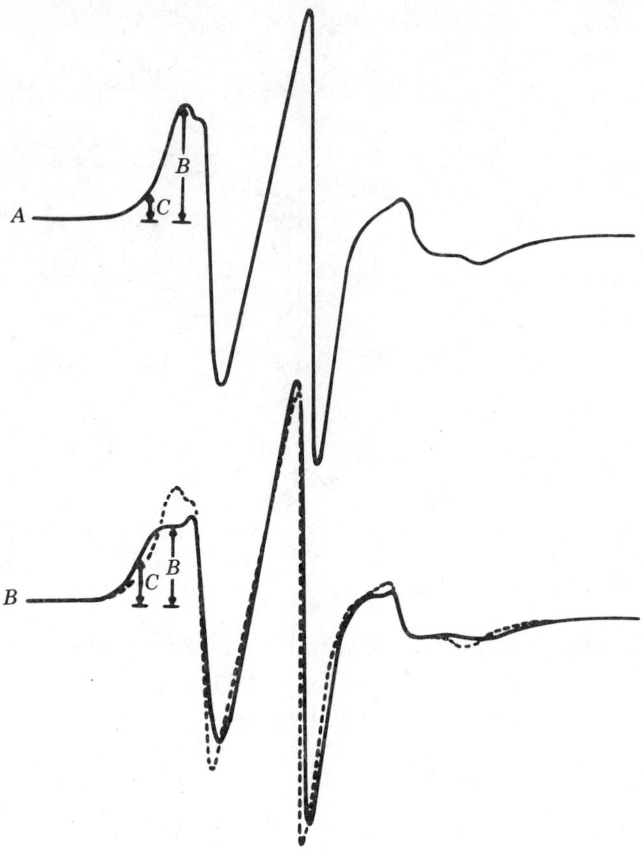

Figure 23. Electron paramagnetic resonance (EPR) spectra of spin-labeled SR ATPase, in the absence (top) and in the presence (bottom) of calcium. (From ref. 5.)

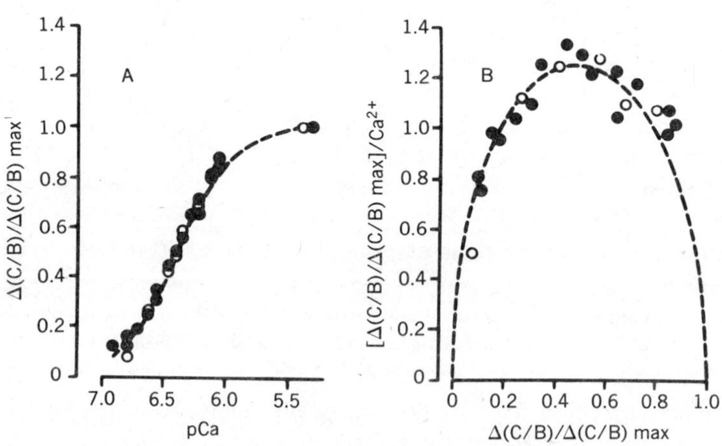

Figure 24. Titration of the SR ATPase calcium binding sites, as revealed by an EPR spectral parameter. These plots are identical to those obtained by direct measurements of calcium binding with radioactive tracer. (From ref. 5.)

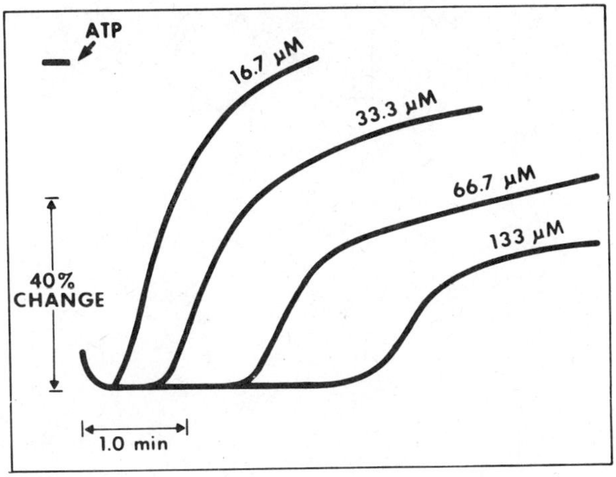

Figure 25. Reduction in the turbidity of suspensions of SR vesicles following addition of ATP, in the presence of suboptimal detergent concentrations. The turbidity returns to the original level when the ATP is exhausted. (From ref. 46.)

4. CONCLUSIONS

The basic feature of the coupling mechanism in the catalytic and transport cycle of SR ATPase is a reciprocal influence of the phosphorylation and calcium sites whereby phosphorylation of the catalytic site destabilizes occupancy of the calcium sites, and vice versa (22,27). This reciprocal effect is produced by means of ligand-induced conformational changes. Because of the lack of detection of ellipticity changes, it must be inferred that the ligand-induced effects must involve reorientation of whole peptide segments, rather than large changes in secondary structure. It is interesting to consider that in the case of the phosphorylation reaction, the conformational change may be triggered by inversion of the oxygen atoms of the phosphoryl moiety (Fig. 28), through their interaction with basic residues at the catalytic site. The effect would be produced partially by the pentacoordinated transition state (analogous to that produced by vanadate; see Fig. 29), and would be maximal on completion of the phosphorylation reaction (Fig. 28). Reversal of steric inversion and the related conformational affect is then produced as a consequence of hydrolytic phosphoenzyme cleavage.

ACKNOWLEDGMENT

The author is grateful to James Bishop, Mary Kirtley, Mark Kurzmack, and Robert Nakamoto for helpful discussions and suggestions. Several experimental findings reproduced in this chapter were obtained with the financial support of the National Institutes of Health (HL27867) and the Muscular Dystrophy Association of America.

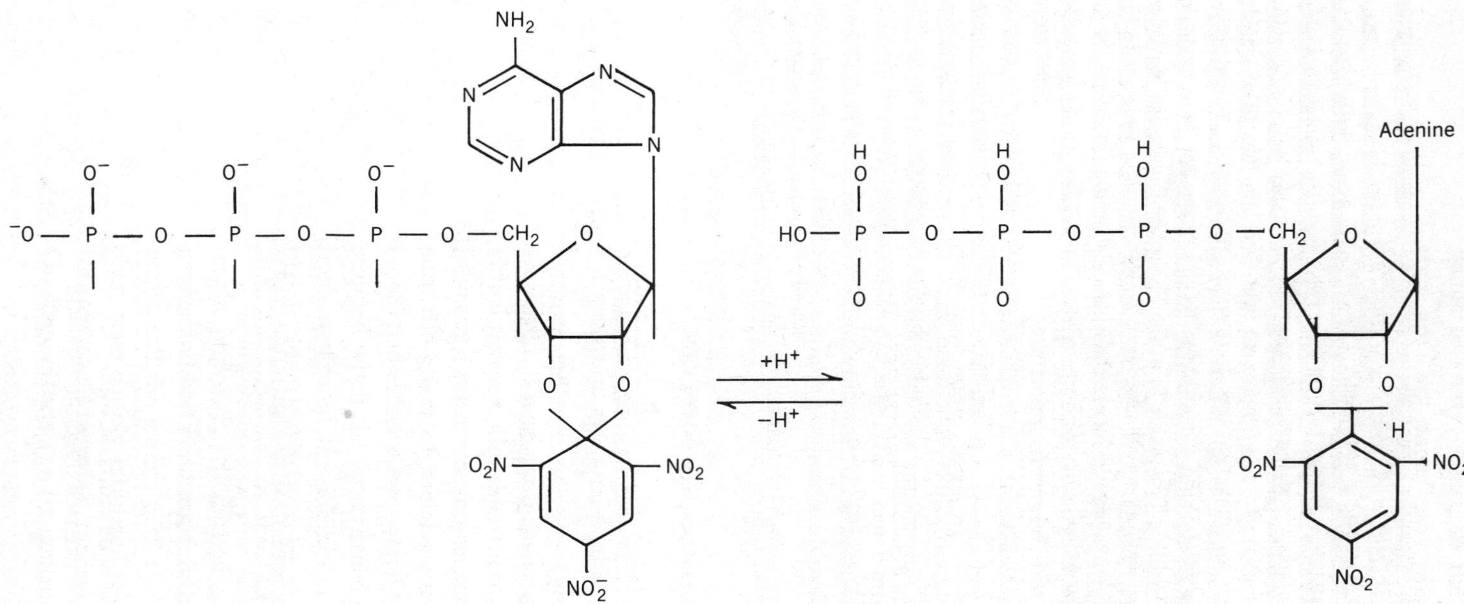

Structure proposed for TNP-ATP at acidic and basic pH.

Figure 26. Structure of the ATP analog 2′–3′–0–(2,4,6–trinitrophenyl) adenosine 5′–triphosphate (TNP-ATP).

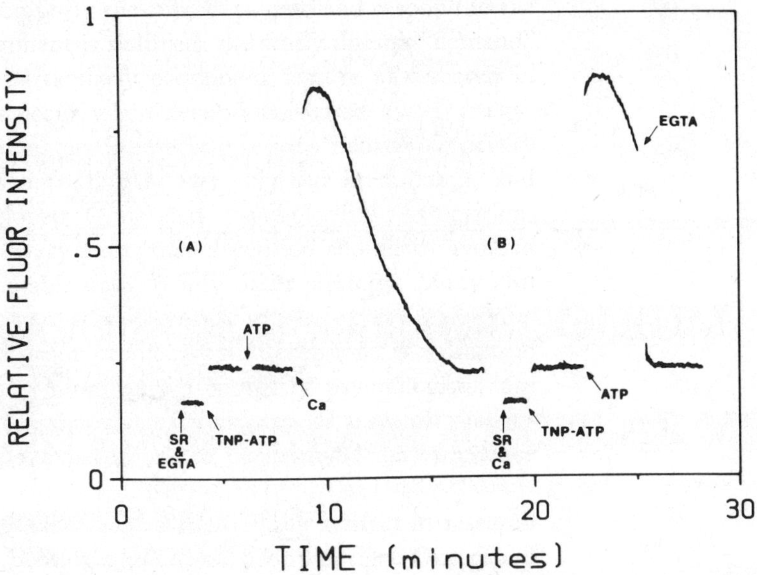

Figure 27. The fluorescence of TNP–ATP bound to SR ATPase increases severalfold when the enzyme is phosphorylated with ATP in the presence of calcium. (From ref. 45.)

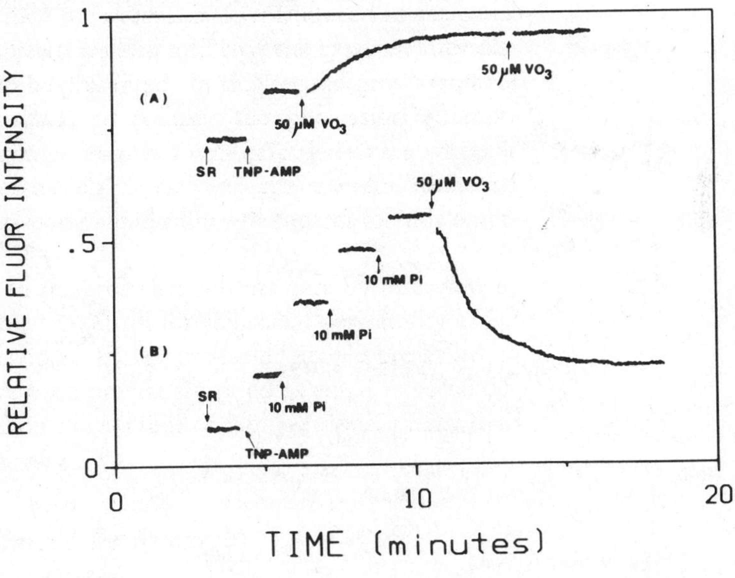

Figure 28. The fluorescence of TNP–nucleotide is also greatly increased when bound to SR ATPase phosphorylated with P_i in the absence of calcium. A slower and lesser effect is obtained when vanadate, instead of P_i is added. (From ref. 37.)

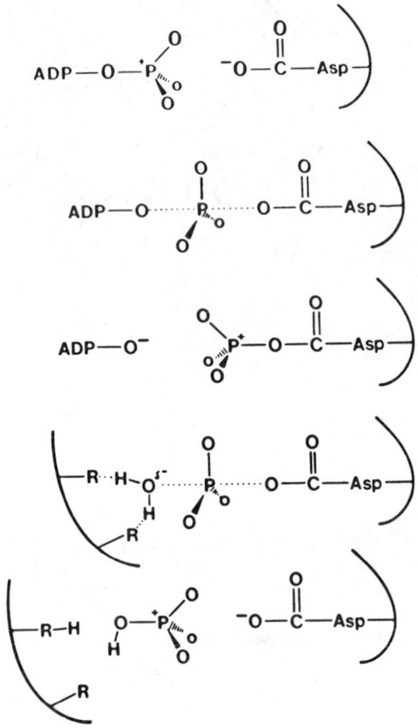

Figure 29. Steric representation of the phosphoryl transfer and hydrolytic reactions shown in Fig. 13. Double inversion of the oxygen atoms in the phosphoryl moiety is illustrated here.

REFERENCES

1. Barrabin, H., Scofano, H., and Inesi, G. (1984). *Biochemistry*, **23**, 1542–1548.
2. Bastide, F., Meissner, G., Fleischer, S., and Post, R. L. (1973). *J. Biol. Chem.*, **248**, 8385–8391.
3. Berman, M. (1982). *Biochim. Biophys. Acta*, **694**, 95–121.
4. Blasie, J., Herbette, L., Pascoline, D., Vita, V., Pierce, D., and Scarpa, A. (1985). *Biophys. J.*, **48**, 9–18.
5. Coan, C., Verjovski-Almeida, S., and Inesi, G. (1979). *J. Biol. Chem.*, **254**, 2968–2974.
6. Degani, C. and Boyer, P. (1973). *J. Biol. Chem.*, **248**, 8222–8226.
7. deMeis, L. and Inesi, G. (1985). *Biochemistry*, **24**, 922–925.
8. deMeis, L. and Vianna, A. (1979). *Ann. Rev. Biochem.*, **48**, 275–292.
9. Dupont, Y. (1980). *Eur. J. Biochem.*, **109**, 231–238.
10. Dupont, Y. (1976). *Biochem. Biophys. Res. Commun.*, **71**, 544–550.
11. Ebashi, S. and Lipmann, F. (1962). *J. Cell Biol.*, **14**, 389–400.
12. Fernandez—Belda, F., Kurzmack, M., and Inesi, G. (1984). *J. Biol. Chem.*, **259**, 9687–9698.
13. Froehlich, J. and Heller, P. (1985). *Biochemistry*, **24**, 126–136.
14. Froehlich, J. and Taylor, E. (1975). *J. Biol. Chem.*, **250**, 2013–2021.

15. Guillain, F., Champeil, P., Lacapere, J., and Gingold, M. (1981). *J. Biol. Chem.*, **256**, 6140–6147.
16. Hasselbach, W. (1974). In Boyer, P., Ed., *The Enzymes,* Vol. 10, Academic, New York, pp. 431–467.
17. Hasselbach, W., Beil, F., and Rauch, B. (1979). *Topics Curr. Chem*, **78**, 1–56.
18. Hasselbach, W. and Makinose, M. (1961). *Biochem. Zeitschrift*, **333**, 518–528.
19. Hasselbach, W. and Makinose, J. (1963). *Biochem. Zeitschrift*, **339**, 94–111.
20. Highsmith, S. and Murphy, A. (1984). *Biol. Chem.*, **259**, 14651–14656.
21. Ikemoto, N. (1982). *Ann. Rev. Physiol.*, **44**, 297–317.
22. Inesi, G. (1985). *Ann. Rev. Physiol.*, **47**, 573–601.
23. Inesi, G., Kurzmack, M., Coan, C., and Lewis, D. (1980). *J. Biol. Chem.*, **255**, 3025–3031.
24. Inesi, G., Kurzmack, M., Kosk-Kosicka, D., Lewis, D., Scofano, H., and Guimaraes-Motta, H. (1982). *Z. Naturforschung, Part C*, **37**, 685–691.
25. Inesi, G., Kurzmack, M., and Verjovski-Almeida, S. (1978). *Ann. N. Y. Acad. Sci.*, **307**, 224–227.
26. Inesi, G., Lewis, D., and Murphy, A. (1984). *J. Biol. Chem.*, **259**, 996–1003.
27. Jencks, W. (1980). *Adv. Enzymol.*, **51**, 75–106.
28. Kurzmack, M., Verjovski-Almeida, S., and Inesi, G. (1977). *Biochem. Biophys. Res. Commun.*, **78**, 772–776.
29. MacLennan, D. (1970). *J. Biol. Chem.*, **245**, 4508–4518.
30. Makinose, M. (1969). *Eur. J. Biochem.*, **10**, 74–82.
31. Martonosi, A. (1972). In Hokin, L., Ed., *Metabolic Pathways*, Vol. 6, Academic, New York, pp. 317–349.
32. Masuda, H. and deMeis, L. (1973). *Biochemistry*, **12**, 4581–4585.
33. McIntosh, D. and Boyer, P. (1983). *Biochemistry*, **22**, 2867–2875.
34. Moller, J., Andersen, J., and LeMaire, M. (1982). *Molec. Cell. Biochem.*, **42**, 83–107.
35. Murphy, A. (1978). *J. Biol. Chem.*, **253**, 385–389.
36. Nakamoto, R. and Inesi, G. (1986). *FEBS Lett.* **194**, 258–262.
37. Nakamoto, R. and Inesi, G. (1984). *J. Biol. Chem.*, **259**, 2961–2970.
38. Nakamura, Y., Kurzmack, M., and Inesi, G. (1986). *J. Biol. Chem.* **261**, 3090–3097.
39. Pickart, C. and Jencks, W. (1984). *J. Biol. Chem.*, **259**, 1629–1643.
40. Scarpa, A., Baldassare, J., and Inesi, G. (1972). *J. Gen. Physiol.*, **60**, 735–749.
41. Scofano, H., Barrabin, H., Lewis, D., and Inesi, G. (1985). *Biochemistry*, **24**, 1025–1029.
42. Scofano, H., Barrabin, H., Inesi, G., and Cohen, J. (in press). *Biochim. Biophys. Acta.*
43. Tanford, C. (1983). *Ann. Rev. Biochem.*, **52**, 379–409.
44. Verjovski—Almeida, S., Kurzmack, M., and Inesi, G. (1970). *Biochemistry*, **17**, 5006–5013.
45. Watanabe, T. and Inesi, G. (1982). *J. Biol. Chem.*, **257**, 11510–11516.
46. Watanabe, T. and Inesi, G. (1982). *Biochemistry*, **21**, 3254–3259.
47. Watanabe, T., Lewis, D., Nakamoto, R., Kurzmack, M., Fronticelli, C., and Inesi, G. (1981). *Biochemistry*, **20**, 6617–6625.
48. Yamamoto, T. and Tonomura, Y. (1968). *J. Biochem.* (Tokyo), **64**, 137–145.

CHAPTER 15

STRUCTURE OF Ca^{2+}–ATPase IN SARCOPLASMIC RETICULUM

ANTHONY MARTONOSI[1]
LASZLO DUX[2]
KENNETH A. TAYLOR[3]
H. P. TING-BEALL[3]
SANDOR VARGA[1]
PETER CSERMELY[1]
NANDOR MÜLLNER[1]
SANDOR PAPP[1]
ISTVAN JONA[1]

[1]*Department of Biochemistry*
State University of New York
Upstate Medical Center
Syracuse, New York
[2]*Institute of Biochemistry*
University Medical School
Szeged, Hungary
[3]*Department of Anatomy*
Duke University Medical Center
Durham, North Carolina

Supported by research grants from the NIH (AM 26545 to A. Martonosi, GM 30598 to K. A. Taylor, and GM 27804 to H. P. Ting-Beall), the National Science Foundation (to A. Martonosi) and by the Muscular Dystrophy Association (to K. A. Taylor and A. Martonosi). K. A. Taylor is an Established Investigator of the American Heart Association. Please send all correspondence to Anthony Martonosi at SUNY Upstate Medical Center, Department of Biochemistry, Syracuse, New York 13210.

The ATP-dependent transport of calcium by the sarcoplasmic reticulum is initiated by the binding of calcium and ATP to the Ca^{2+} transport ATPase in the E_1 conformation (43). This is followed by the phosphorylation of an aspartyl group in the active site and the conversion of the enzyme from the E_1 into the E_2 form, accompanied by the transfer of calcium across the membrane. The cycle is completed by the release of Ca^{2+} on the luminal side and the hydrolysis of the phosphoenzyme intermediate, yielding inorganic phosphate (Scheme 1). The alternation between the two distinct conformations of Ca^{2+}–ATPase during the transport cycle is supported by changes in the affinity of the enzyme for Ca^{2+}, Mg^{2+}, ATP, and ADP (36), the differences in the reactivity of protein functional groups (30), the responses of fluorescence (17), and electron-spin resonance probes (31) at predicted stages in the Ca^{2+} transport process. There are no net changes in the polypeptide backbone conformation on transition between the E_1 and E_2 states (15); therefore, the change in conformation presumably involves a hinge type or relative sliding motion of domains.

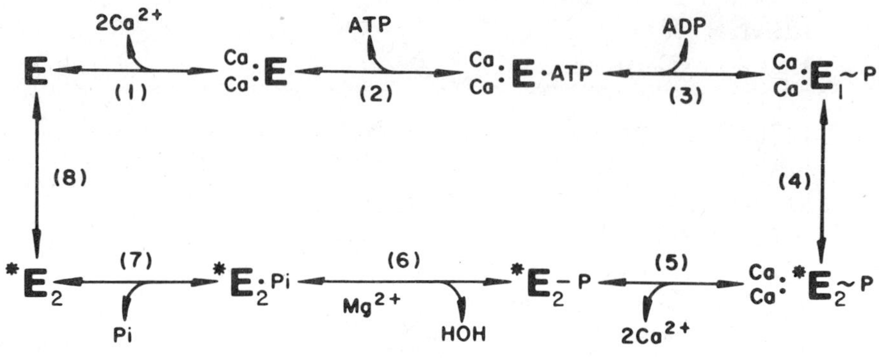

Scheme 1.

Vanadate (V) ions stabilize the Ca^{2+}–ATPase in the E_2 conformation (50, 51). Therefore, the Ca^{2+}–ATPase crystals induced by vanadate in a Ca^{2+}-free medium (Figs. 1 and 2) are expected to reflect the geometry of ATPase–ATPase interactions characteristic of the E_2 conformation (19–23). The E_2-type crystals induced by vanadate ions consist of dimer chains of ATPase molecules forming an oblique surface lattice with a space group of P_2 (55) that is consistent with ATPase dimers as structural units. The formation of E_2 crystals is promoted by inside positive membrane potential (22).

A different crystal form of Ca^{2+}–ATPase develops in sarcoplasmic reticulum vesicles exposed to 10^{-5}–10^{-4} M Ca^{2+} (Fig. 3) or 10^{-6}–10^{-5} M lanthanides (Figs. 4 and 5) at a slightly alkaline pH (25). Calcium and lanthanides are presumed to bind to Ca^{2+}–ATPase in the E_1 conformation. Therefore, the crystals formed under these conditions presumably arise from interaction between Ca^{2+}–ATPase molecules in the E_1 form. The space group of the Ca^{2+}- or lanthanide-induced crystals is P_1, and the unit cell dimensions are consistent with ATPase monomers as structural units (25). The formation of E_1 crystals is promoted by inside negative membrane potential.

Since the E_1 and the E_2 conformations represent the structures of the Ca^{2+} pump at the beginning and the end of the Ca^{2+} transport cycle, high-resolution analysis of these crystals by reconstruction techniques, together with the static and time-resolved structure studies of the same system by x-ray and neutron diffraction techniques (6,7), may provide the much needed structural information about the molecular basis of Ca^{2+} translocation.

In this chapter, some recent observations on the structure of the Ca^{2+}–ATPase are presented on the basis of three-dimensional reconstruction of the E_1 and E_2-type crystals of Ca^{2+}–ATPase.

1. E_2-TYPE CRYSTALS OF Ca^{2+}–ATPase

Two-dimensional crystalline arrays of Ca^{2+}–ATPase molecules develop on treatment of sarcoplasmic reticulum vesicles or reconstituted membranes containing the purified Ca^{2+} transport ATPase with 5 m*M* vanadate, at 2°C, in a medium of 0.1 *M* KCl, 10 m*M* imidazole, pH 7.4, and 0.5 m*M* EGTA. Unphysiological pH (pH 5 or 9), elevated temperature (37°C), and high ionic strength (0.6*M* KCl) interfere with crystallization. Vesicles with extensive crystalline arrays assume elongated tubular shapes (Fig. 1) with average diameters of 50–70 nm, suggesting that the crystalline arrays are best accommodated by a cylindrical surface contour. The rows of ATPase particles are oriented diagonally to the long axis of the tubules at an angle of 57°, and superimposition of the front and rear images of the flattened cylinders creates a diamondlike pattern of varied regularity. At the end regions of the cylinders and in spherical vesicles, the lattice is either absent or irregular, presumably because the hemispherical surface contours represent unfavorable geometry for crystallization. In regular arrays the rows of negatively stained particles are arranged in pairs, forming dimer chains, that are separated from neighboring chains by wide bands of negative stain. Occasionally, the helical lattice unwinds at the end of the tubules, revealing chains of ATPase dimers separated from neighboring chains by variable distances (Fig. 2). Interaction between ATPase molecules within the chain appears to be stronger than between them, suggesting that the crystalline lattice develops in the following three stages: (1) formation of ATPase dimers, (2) linear polymerization of ATPase dimers into tetramers and longer dimer chains by a process reminiscent of fibrous condensation (47), and (3) lateral association of dimer chains into a two dimensional right-handed helical lattice that surrounds the cylinders (Fig. 6). Different types of bonds appear to be involved in the association of monomers into dimers (bond A), of dimers into dimer chains of diverse lengths (bonds B and C), and of dimer chains into extended two-dimensional lattice (bond D). The existence of monomers, dimers, and tetramers of Ca^{2+}–ATPase was observed in native membranes and in detergent solubilized sarcoplasmic reticulum by

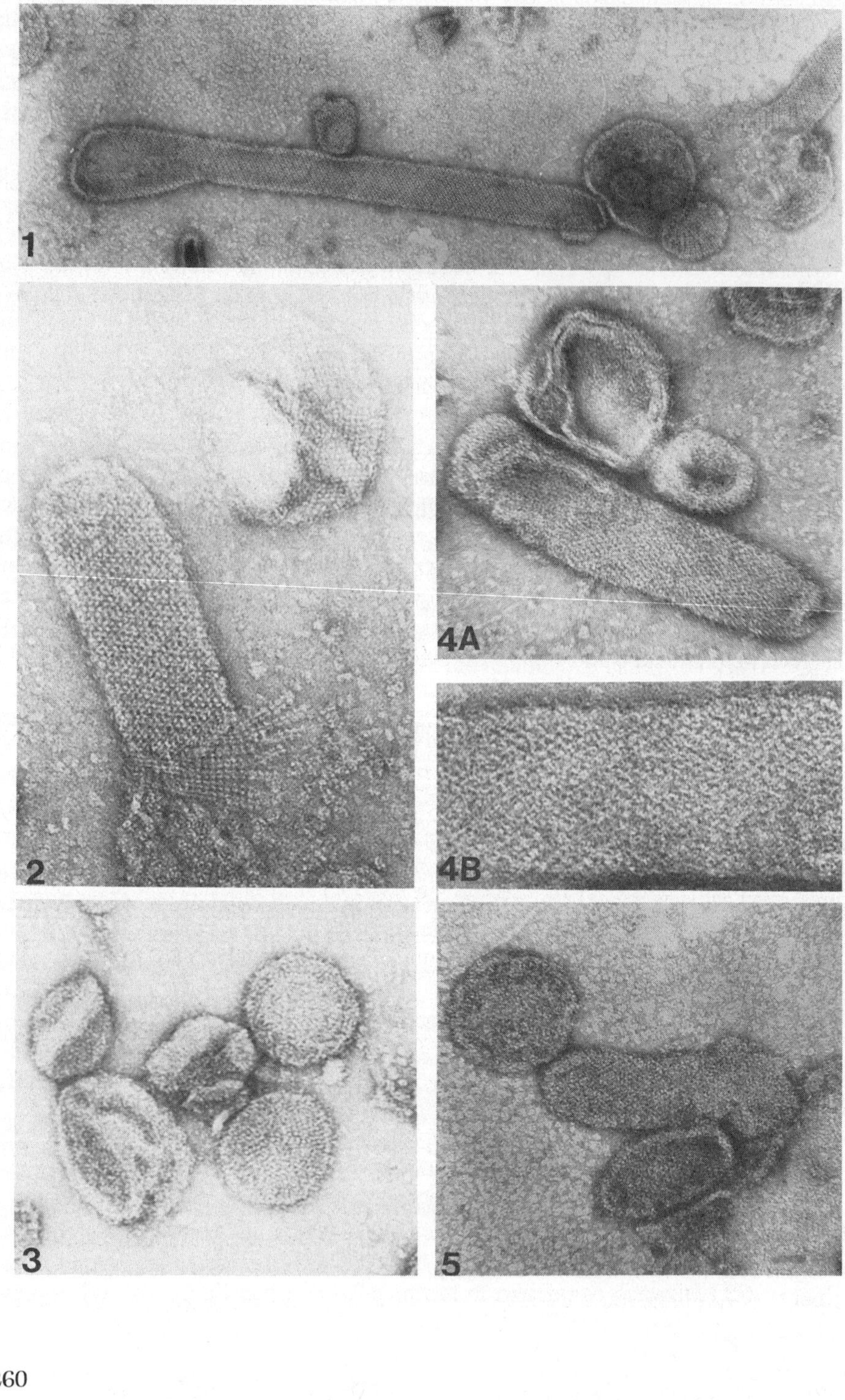
1
2
3
4A
4B
5

ultracentrifugation, exclusion chromatography, target inactivation, fluorescence energy transfer, and electron microscopy (26,40,43,45). Long dimer chains and extended crystalline arrays are observed only after treatment of sarcoplasmic reticulum with Na_3VO_4 that stabilizes the E_2 conformation.

Digital Fourier transforms of images of flattened tubules (55) yielded lattice constants of $a = 66$ Å, $b = 114$ Å, and $\gamma = 78°$ (Fig. 7 and Table 1). The average structure calculated by Fourier synthesis is consistent with ATPase dimers as structural units of the crystals. Tilt-view reconstruction of the three-dimensional structure of Ca^{2+}–ATPase according to Henderson and Unwin (28) provides a clear picture of the interaction of the ATPase molecules into dimers, dimer chains, and their lateral association into ordered arrays by defining new features of the ATPase that were not apparent in two-dimensional projections (56). The Ca^{2+}–ATPase molecules are linked together to form dimers by a stain-excluding bridge above the cytoplasmic surface of the bilayer (Fig. 8). The bridge presumably represents interacting domains on the adjacent surfaces of two Ca^{2+}–ATPase molecules. Immediately beneath the bridge there is a stain-filled region of about 18-Å width and 43-Å height that extends throughout the opposing cytoplasmic faces of the ATPase molecules. This arrangement implies that the only area of interaction between the cytoplasmic portion of the two Ca^{2+}–ATPase molecules within the dimer is at the region of the bridge. Additional interactions may occur in the lipid bilayer, but these are not visualized by negative staining techniques. The gap under the bridge is likely to be accessible in the native membrane to ATP, calcium, and other solutes of relatively low molecular weight that interact with Ca^{2+}–ATPase.

Figure 1. Vanadate-induced Ca^{2+}–ATPase crystals in sarcoplasmic reticulum. Sarcoplasmic reticulum vesicles were incubated in 0.1 *M* KCl, 10 m*M* imidazole, pH 7.4, 5 m*M* $MgCl_2$, 0.5 m*M* EGTA, and 5 m*M* Na_3VO_4 for 48 hr at 2°C. The diamond pattern arises from superimposition of images from the front and rear surfaces of the collapsed cylinder. Magnification: ×71,400.

Figure 2. Separation of Ca^{2+}–ATPase ''chains'' in a disrupted vanadate-treated sarcoplasmic reticulum vesicle. Sarcoplasmic reticulum vesicles were treated with Na_3VO_4 for 48 hr, as described in Fig. 1. At the end of the crystalline tubule Ca^{2+}–ATPase ''chains'' are seen unwinding from crystalline arrays, which consist of rows of Ca^{2+}–ATPase dimers. The width of the flattened crystalline tubule is about 1000 Å. Magnification: × 143,000.

Figure 3. Ca^{2+}–ATPase crystals induced by Ca^{2+}. Sarcoplasmic reticulum vesicles were incubated with Chelex chelating resin (100 mg wet resin/mg protein) at 2°C overnight. The Chelex particles were removed by centrifugation at low speed and the crystallizaton was induced in a medium of 0.1 *M* KCl, 10 m*M* imidazole, pH 8.0, 5 m*M* $MgCl_2$, and 0.1 m*M* $CaCl_2$ by incubation for 16 hr at 2°C. Protein concentration: 1 mg /ml. Negative staining with 1% uranyl acetate. Magnification × 140,000.

Figure 4. Ca^{2+}–ATPase crystals induced by $GdCl_3$. (*A*) Sarcoplasmic reticulum vesicles were incubated in 0.1 *M* KCl, 10 m*M* imidazole, pH 7.2, 5.0 m*M* $MgCl_2$, and 10^{-6} *M* $GdCl_3$ at 2°C for 48 hr, and samples were taken for negative staining with 1% uranyl acetate (magnification × 140,000); (*B*) a portion of the vesicles seen in part *A* at higher magnification (× 280,000).

Figure 5. Ca^{2+}–ATPase crystals induced by $LaCl_3$. Sarcoplasmic reticulum vesicles were incubated in 0.1 *M* KCl, 10 m*M* imidazole, pH 7.2, 5.0 m*M* $MgCl_2$ in the presence of 10^{-6} *M* $LaCl_3$ at 2°C for 5 days. Samples were stained with 1% uranyl acetate. Magnification: × 140,000.

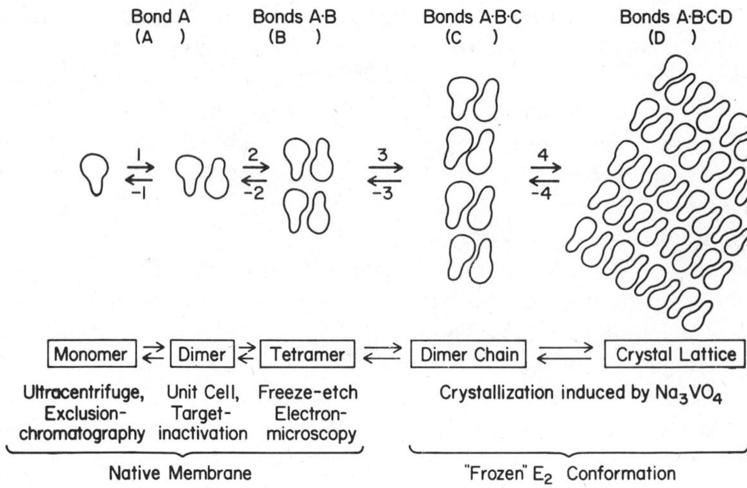

Figure 6. Hypothetical scheme of ATPase–ATPase interactions. For details, see text.

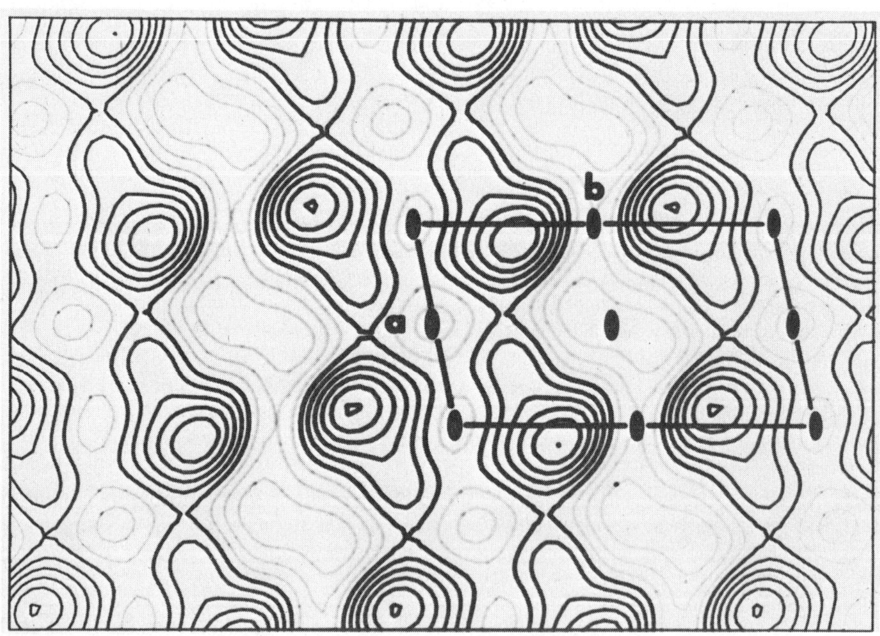

Figure 7. Low-dose electron density map of vanadate-induced Ca^{2+}–ATPase crystals from rabbit sarcoplasmic reticulum, negatively stained with uranyl acetate. The unit cell and its symmetry elements and cell axes are identified. The dimers form long chains that are wound around the tubules in a right-handed helix separated from neighboring dimer chains by wider bands of negative stain. (From ref. 55.)

TABLE 1. CRYSTALLOGRAPHIC PROPERTIES OF Ca^{2+}–ATPase CRYSTALS

Putative Conformation	Inducer	Unit Cell Dimensions a (Å)	b (Å)	γ (°)	Space Group	Structural Unit
E_1	La	71.9	51.1	104	P_1	Monomer
	Gd	61.7	54.4	111	P_1	Monomer
	Pr	66.8	50.4	115	P_1	Monomer
	Ca	—	—	—	(P_1)	Monomer
E_2	EGTA + V	65.9	114.4	77.9	P_2	Dimer
	EGTA + P_i	—	—	—	(P_2)	Dimer

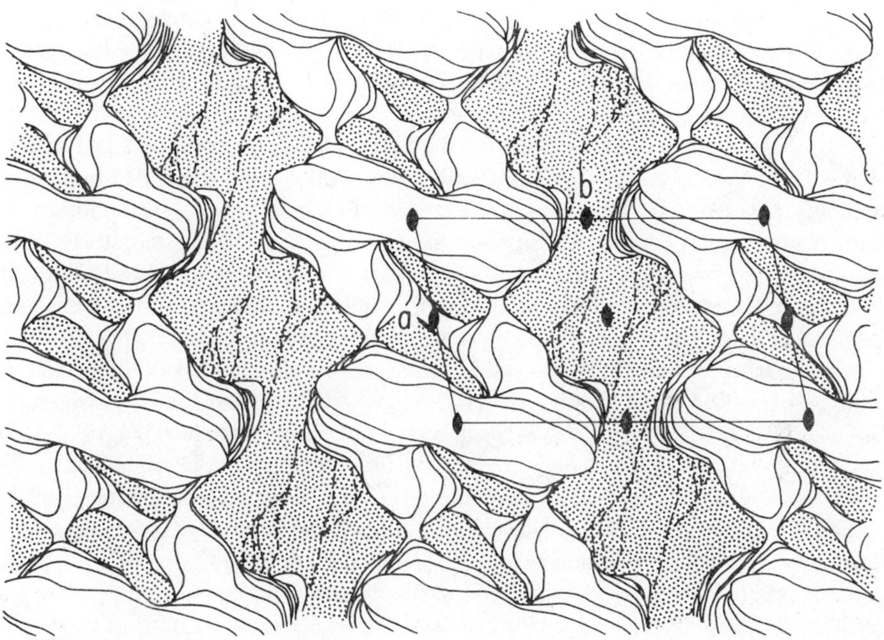

Figure 8. Surface contours of Ca^{2+}–ATPase viewed normal to the membrane plane. Within the dimer chain, lines of increased density connect the main domains of the two ATPase molecules that constitute a dimer. These connecting bridges run parallel with the axis of the tubules. Lines of stain excluding density connect opposing lobes of the ATPase molecules in the two neighboring dimers within the dimer chain. The lobe projecting from the main mass of Ca^{2+}–ATPase contour is located about 15 Å above the cytoplasmic surface of the bilayer. Lines a and b are the two unit cell axes and the pointed elipses are at the location of two-fold rotation axes. Stippling has been added to clarify the cytoplasmic surface of the molecule and to give some indication of the surface of the lipid bilayer. Dashed lines within the stippled area are positive density features within the lipid bilayer. (Fiau ref 56)

The Ca^{2+}–ATPase profiles are asymmetric, with two clearly defined domains (Figs. 7 and 8). The larger domain contains the polypeptide region involved in bridge formation and extends through the lipid bilayer, anchoring the ATPase molecule into the membrane. The smaller domain, or lobe, projects in a direction nearly parallel with the axis of the ribbons and links the ATPase dimers to form dimer chains. The lobe is separated from the bilayer surface by about 16 Å. These features of the cytoplasmic domain of the Ca^{2+}–ATPase molecule are very similar to what has been observed in a Fourier–Bessel reconstruction of the Ca^{2+}–ATPase tubules from scallop sarcoplasmic reticulum (10).

The length of the crystalline tubules varies over a wide range but usually exceeds 3000 Å (Fig. 1). The shaping of a membrane surface of this length would require ATPase dimer chains of at least 5400-Å length, the length needed to advance an actual distance of 3000 Å. This explains why sarcoplasmic reticulum vesicles that contain only short dimer chains do not form tubules but retain a near-spherical profile. The number of dimer chains is presumably defined by the number of nucleation sites and the growth kinetics of the dimer chains; the kinetics, in turn, will be influenced by the concentration of the Ca^{2+}–ATPase molecules that are in appropriate conformation for crystallization. The structural features of the Ca^{2+}–ATPase revealed by three-dimensional reconstruction may help to explain the tubular shape and helical structure of the Ca^{2+}–ATPase crystals. The range of diameters of the crystalline tubules is about 600–900 Å, and the crystalline arrays generally cover nearly the whole surface of the tubules, with the exception of the hemispherical end portions. These observations imply that the crystal lattice is best accommodated on cylindrical surfaces of a fairly narrow range of radii. With the increase in diameter of the tubules, the number of dimer chains per tubule must also increase, because the pitch of the right-handed helix of the dimer chains varies only within a narrow range ($57.6 \pm 2.1°$). Based on the dimensions of the unit cell, tubules of 600–900-Å diameter will be covered by 9–13 dimer chains of Ca^{2+}–ATPase, respectively. Changes in tubule circumference are expected to occur in quanta of about 213 Å (34-Å radius) in order to bring ATPase molecules in neighboring dimer chains in proper orientation. Obviously, some features of the ATPase molecules and the ATPase dimers limit the number of possible helical structures. The bridge that links two Ca^{2+}–ATPase molecules into a dimer is always oriented parallel to the axis of the tubules, perhaps because changes in the radius of curvature of the membrane place little or no strain on the structure of the dimer in this orientation. If the bridges were oriented circumferentially, the structure of the dimer would be forced to adapt to the curvature of the membrane surface. This might result in changes in the orientation of the hydrophobic segments of the ATPase molecules within a dimer and could alter intermolecular distances by as much as 20% near the luminal side of the bilayer. The close relationship between the curvature of the membrane surface and the formation of the crystal lattice is best illustrated by the observation that during crystallization of Ca^{2+}–ATPase the originally spherical vesicles are progressively transformed into elongated tubules. These observations imply that the most favorable energy state for the formation of Ca^{2+}–ATPase crystals is provided by an orientation of the crystal lattice in which the bridge that links the ATPase dimers is aligned parallel with the axis of the tubules.

The studies that have been discussed so far used the negative staining technique that visualizes only the cytoplasmic portion of the Ca^{2+}–ATPase molecule (55,56). The cytoplasmic structures seen in these maps have a volume of about 71,000 $Å^3$, corresponding to a molecular weight of about 57,000, that represents approximately two-thirds of the molecular weight of Ca^{2+}–ATPase. In negatively stained membrane crystals, the intramembranous region of the Ca^{2+}–ATPase is not visualized. The next major step is determination of the three-dimensional structure of Ca^{2+}–ATPase from electron micrographs of frozen hydrated unstained membranes

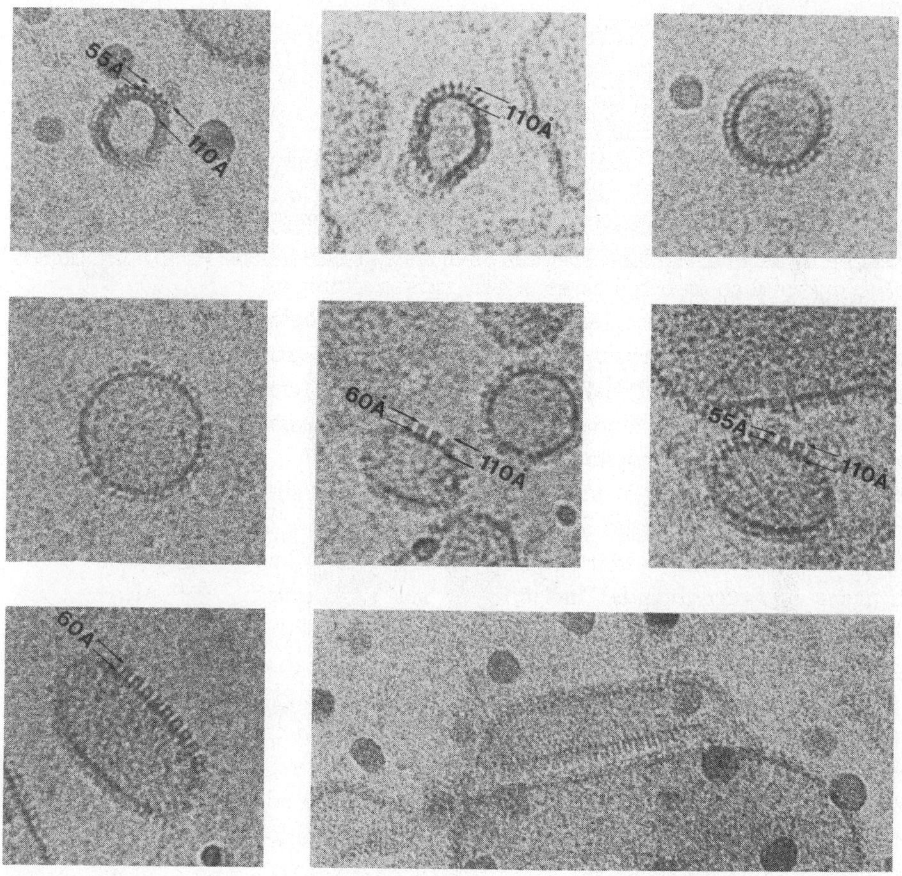

Figure 9. Monovanadate-induced E_2 crystals preserved in amorphous ice. The membranes were frozen and viewed in amorphous ice without staining as described in refs. 54 and 57. The tubules are not flattened by drying and illustrate the morphology of native helical cylinders. Protein is represented by the darker features of the image. Views down the *b* axis (upper panel) illustrate the lobe feature of the structure, and views down the *a* axis (lower panels) show a bridge that connects Ca^{2+}–ATPase monomers to form the dimer. Magnification: ×229,320.

(54) that reveal the intramembranous, as well as the cytoplasmic regions of the protein (Fig. 9). These studies are in progress (29), and the results will be reported elsewhere in the near future.

2. CONDITIONS AFFECTING FORMATION OF E_2 TYPE Ca^{2+}–ATPase CRYSTALS IN PRESENCE OF Na_3VO_4: RELATIONSHIP BETWEEN CRYSTAL FORM AND ENZYME CONFORMATION

2.1 Effect of Ca^{2+}

In a concentration sufficient to saturate the high-affinity Ca^{2+} binding site of Ca^{2+}–ATPase (1-10 μ*M*), Ca^{2+} prevents the formation of E_2 type Ca^{2+}–ATPase crystals and "cracks" the crystals that were formed previously in the presence of 0.5 m*M* EGTA and 5 m*M* Na_3VO_4 (Table 2) (19,21). In low concentration, Ca^{2+} also protects the Ca^{2+} transport ATPase against inhibition by Na_3VO_4. The formation of two-dimensional E_2 crystals requires that most of the enzyme molecules assume the E_2 conformation; the inhibition of crystallization by Ca^{2+} may be attributed to the stabilization of the E_1 enzyme form by calcium.

2.2 Effect of Nucleoside Phosphates

The formation of E_2 type Ca^{2+}–ATPase crystals is inhibited by ATP (5 m*M*), but not by ADP, AMP, 5-adenylylimidodiphosphate (AMPPNP) or adenylylmethylene-diphosphonate (AMPPCP) (21). We did not observe any effect on the crystal

TABLE 2. EFFECT OF Ca^{2+} ON FORMATION AND STABILITY OF Ca^{2+}–ATPase CRYSTALS[a]

[EGTA] (m*M*)	Added [Ca] (m*M*)	Free [Ca^{2+}] (μ*M*)	Crystallization Index *A*	Crystallization Index *B*
0.5	0.00	<0.01	64.32 ± 6.11	54.78 ± 2.57
0.5	0.045	<0.01	66.66 ± 3.56	59.04 ± 7.74
0.5	0.10	0.015	51.78 ± 1.57	48.60 ± 9.47
0.5	0.20	0.035	50.11 ± 9.64	46.39 ± 2.77
0.5	0.30	0.130	50.26 ± 8.71	56.98 ± 6.91
0.5	0.45	1.80	0.00	4.90 ± 1.03

[a]Microsomes were incubated in 0.1 *M* KCl, 10 m*M* imidazole, pH 7.4, 5 m*M* $MgCl_2$, 0.5 m*M* EGTA, and 5 m*M* Na_3VO_4 for 2 hr at 2°C in the presence of added Ca^{2+} (*A*) or in a Ca-free medium (*B*). Samples of the *A* series were negatively stained with 1% uranyl acetate and viewed. To samples of series *B* Ca^{2+} was added at indicated concentration for an additional 4 hr, followed by negative staining. Free [Ca^{2+}] was measured by Ca^{2+}-specific microelectrode.

structure when 5 m*M* ATP, ADP, AMPPNP, or AMPPCP were added to microsomes previously crystallized with vanadate in the absence of nucleotides. These observations suggest that binding of ATP to the enzyme interferes with crystallization, presumably by changing the structure of the protein; once crystals are formed, ATP cannot disrupt the crystal lattice, either because the ATP binding site becomes inaccessible, or because the E_2 conformation of the enzyme is stabilized by the crystal structure against disruption.

2.3 Effect of Inorganic Phosphate

Vanadate is an analogue of inorganic phosphate. It is plausible to assume that vanadate inhibits Ca^{2+}–ATPase and promotes its crystallization by forming an inactive E_2–vanadate complex at the inorganic phosphate binding site. In fact, inorganic phosphate (1 m*M* at pH 6) mimicks the effect of vanadate and promotes the crystallization of Ca^{2+}–ATPase in a Ca^{2+}-free medium (21). The crystal form produced by inorganic phosphate was usually less regular but otherwise similar to the vanadate-induced Ca^{2+}–ATPase crystals. Crystals were not produced by 1 m*M* inorganic phosphate at pH 7, where the affinity of P_i for the enzyme is low.

TABLE 3. EFFECT OF ION SUBSTITUTION ON CRYSTALLIZATION OF Ca^{2+}–ATPase IN SARCOPLASMIC RETICULUM VESICLES.[a]

	Vanadate Concentration (m*M*)		Crystallization Index 15 sec After Ion Substitution			
Sets	Solution 1	Solution 2	KG → KG	KG → CCl	CCl → CCl	CCl → KG
1	0 →	5.0	10.90	0.00	0.00	42.30
2	0.1 →	5.0	0.00	5.06	0.00	82.85
3	0.5 →	5.0	0.00	—	1.13	56.19
4	1.0 →	5.0	8.74	—	4.05	79.30
5	5.0 →	5.0	68.89	3.57	45.63	55.69

[a]Sarcoplasmic reticulum vesicles (10 mg protein/ml) were preequilibrated in solution 1 containing either K–glutamate (KG) or choline chloride (CCl) as the principal salt and Na_3VO_4 at the indicated concentrations for 16 hr at 2°C. The suspensions were then diluted tenfold into solution 2 containing 5 m*M* Na_3VO_4 and either K–glutamate (KG) or choline chloride (CCl) in various combinations. Samples were taken for negative staining 15 sec after dilution into solution 2. The expected direction of the transmembrane potential is as follows: KG KG and CCl CCl, no potential; KG CCl, inside negative potential; CCl KG inside positive potential. Rapid crystallization of Ca^{2+}–ATPase was observed on dilution from CCl into KG medium (sets 1–4), which is attributable to the generation of inside positive potential. Transfer of precrystallized vesicles (set 5) from KG into CCl medium disrupted the preformed ATPase crystals within 15 sec, while other ion substitutions within set 5 had little or no effect. Significant crystallization was retained on transfer of precrystallized vesicles (5 m*M* Na_3VO_4) from KG into KG or from CCl into KG medium without vanadate (not shown).

2.4 Effect of Membrane Potential on Formation of Ca^{2+}–ATPase Crystals in the Presence of Na_3VO_4

The rate of vanadate-induced crystallization is increased more than tenfold by inside positive membrane potential generated by substitution of potassium glutamate for choline chloride (Table 3) or potassium methanesulfonate + valinomycin for sodium methanesulfonate (Table 4) in the incubation medium (22). Negative potential caused the transient disruption of preformed Ca^{2+}–ATPase crystals, followed by slow reappearance of the lattice after the potential was dissipated (22). Using oxonol VI and 3,3′-diethyl-2,2′-thiadicarbocyanine [di-*S*-C_2(5)] to monitor inside positive and inside negative membrane potentials, respectively (4), we found that conditions that promote the crystallization of Ca^{2+}–ATPase molecules slightly diminished the magnitude and duration of membrane potential generated by ion substitution (5). The effect of vanadate-induced crystallization on membrane potential implies some increase in the premeability of sarcoplasmic reticulum for ions in the crystalline state. It is not known whether this is due to the stabilization of the E_2 conformation of the enzyme or the development of crystalline arrays, per se. The physiological significance of ATPase–ATPase interactions in the regulation of the permeability of sarcoplasmic reticulum remains to be further explored (39).

2.5 Effect of proteolysis on Formation of Ca^{2+}–ATPase Crystals

Tryptic cleavage of the Ca^{2+} transport ATPase at the T_1 site into two major fragments of 57,000 and 52,000 daltons did not interfere with the vanadate induced formation of

TABLE 4. EFFECT OF MEMBRANE POTENTIAL GENERATED BY K^+ GRADIENT AND VALINOMYCIN ON THE CRYSTALLIZATION OF Ca^{2+}–ATPase[a]

Set Number	Conditions	Crystallization Index +Valinomycin	Crystallization Index −Valinomycin	Expected Potential (+Valinomycin)
1	Na–$MeSO_4$(O) → K–$MeSO_4$ (5)	41.65	4.05	+
2	K–$MeSO_4$(O) → Na–$MeSO_4$ (5)	3.96	1.22	−
3	K–$MeSO_4$(O) → K–$MeSO_4$ (5)	10.04	—	0
4	Na–$MeSO_4$(O) → Na–$MeSO_4$ (5)	4.62	—	0

[a]Sarcoplasmic reticulum vesicles (10 mg protein /ml) equilibrated in media containing 10 m*M* imidazole, pH 7.4, 5 m*M* $MgCl_2$, 0.5 m*M* EGTA, and either 0.15 *M* Na–methanesulfonate (sets 1 and 4) or 0.15 *M* K–methanesulfonate (sets 2 and 3) were diluted into media containing 5 m*M* Na_3VO_4 and 0.15 *M* Na–methanesulfonate (sets 2 and 4) or 0.15 *M* K–methanesulfonate (sets 1 and 3) with or without 1 μ*M* valinomycin. The concentration of Na_3VO_4 is indicated by (0) and (5), corresponding to zero and 5 m*M* Na_3VO_4, respectively. Negative staining was initiated 15 sec after dilution. Significant increase in the rate of crystallization was observed in set 1 in the presence of valinomycin due to the generation of inside positive potential. Nigericine (1 μ*M*) in the dilution medium abolished this effect of valinomycin.

membrane crystals, indicating that the two major proteolytic fragments interact with preservation of their conformation required for crystallization (20). This is in accord with earlier observations that limited cleavage of the enzyme by trypsin did not interfere with ATPase activity and Ca^{2+} transport. The ability of Ca^{2+}–ATPase to crystallize was lost after further cleavage of the A fragment at the T_2 site into the A_1 and A_2 subfragments (20), that is known to be accompanied by loss of Ca^{2+} uptake. Vanadate (0.1–5 m*M*) completely inhibited the secondary cleavage of Ca^{2+}–ATPase by trypsin at the T_2 site (20), suggesting that the susceptibility of the T_2 site is influenced by the conformation of the enzyme, the formation of the crystal lattice, or a direct effect of vanadate bound to the T_2 region of the enzyme.

2.6 Effect of Chemical Modification on Morphology of E_2–Type Crystalline ATPase Aggregates

Chemical modification of Ca^{2+}–ATPase by guanidino-group reagents (2,3-butanedione and phenylglyoxal) prevents the formation of E_2 crystals of Ca^{2+}–ATPase by vanadate (61). Amino group reagents (fluorescein 5′-isothiocyanate, pyridoxal phosphate, and fluorescamine) inhibit ATPase activity, but not the crystallization of Ca^{2+}–ATPase (13,61). The crystallization of Ca^{2+}–ATPase is also inhibited by treatment of the sarcoplasmic reticulum vesicles with phospholipase C or A_2, detergents, or SH-group reagents (61). These observations imply that the formation of E_2 crystals is exquisitely sensitive to changes affecting either the lipid phase or the protein.

3. Ca^{2+}–ATPase CRYSTALS IN SARCOPLASMIC RETICULUM OF FAST AND SLOW SKELETAL AND CARDIAC MUSCLES

The proportion of vesicles containing E_2-type Ca^{2+}–ATPase crystals in microsome preparations isolated from rat muscles of different fiber types (semimembranosus, levator ani, extensor digitorum longus, diaphragm, soleus, cardiac) correlates well with Ca^{2+}–ATPase content and Ca^{2+}-modulated ATPase activity (24). This implies that the concentration of Ca^{2+}–ATPase in sarcoplasmic reticulum membranes of fast and slow skeletal or cardiac muscles differs only slightly, and the low Ca^{2+} transport activity of crude microsome preparations isolated from red skeletal and cardiac muscles may be due largely to the presence of contaminating non-sarcoplasmic reticulum membrane elements. Since the crystalline arrays of Ca^{2+}–ATPase extend over much of the surface of the vesicles, the ATPase content of sarcoplasmic reticulum membranes in adult muscles of diverse fiber type composition is close to physical saturation. Therefore, the sarcoplasmic reticulum content of any given muscle in adult animals is regulated by changing the total amount of sarcoplasmic reticulum of relatively constant ATPase content rather than by changing the concentration of Ca^{2+}–ATPase in the membrane. As the Ca^{2+} transport ATPase is cotranslationally synthesized on polysomes bound to the surface of the sarcoplasmic reticulum (36,37), the synthesis and insertion of Ca^{2+}–ATPase may be inhibited when the membrane is fully saturated by the enzyme. The average distance between ATPase molecules in the membrane is only about 100 Å; therefore, it is possible that the hydrophilic head portions of the ATPase molecules interfere with the attachment of polysomes to the membrane, causing the inhibition of ATPase synthesis when the ATPase concentration rises above a certain level.

4. MEMBRANE CRYSTALS OF Ca^{2+}–ATPase IN SARCOPLASMIC RETICULUM OF DEVELOPING MUSCLE

The vanadate-induced crystallization of Ca^{2+}–ATPase was analyzed in sarcoplasmic reticulum vesicles isolated between 10 and 28 days of development from pectoralis muscles of chicken (18). After exposure to Na_3VO_4 in a Ca^{2+}-free medium, Ca^{2+}–ATPase crystals begin to appear on portions of the surface of a few vesicles isolated at 18 days of development. Thereafter, the number of vesicles containing Ca^{2+} ATPase crystals rapidly increases and after 1 week of postnatal development (28 days) reaches the adult level of about 30% of the vesicle population (18). The Ca^{2+}–ATPase content of the sarcoplasmic reticulum membrane progressively increases from the phospholipid-rich endoplasmic reticulum in 10-day-old embryos, to the fully developed sarcoplasmic reticulum of the adult animals (8,35–37), in which the Ca^{2+}–ATPase content approaches 40% of the total membrane mass. The vanadate-induced crystallization of the Ca^{2+}–ATPase can be described as a form of fibrous condensation (47). In processes of this kind, the formation of protein polymers occurs only if the concentration of protein exceeds a certain minimal, so-called critical, concentration. Above this concentration, the protein molecules associate into well-defined fibrous polymers. It seems that in developing sarcoplasmic reticulum, the concentration of Ca^{2+} transport ATPase first reaches its critical concentration at 18 days of development, resulting in the formation of small crystalline patches of Ca^{2+}–ATPase arrays. These crystalline patches expand rapidly with continued insertion of ATPase molecules, until the membrane becomes physically saturated with Ca^{2+}–ATPase. On the basis of earlier data (8,58), the critical Ca^{2+}–ATPase concentration for the formation of Ca^{2+}–ATPase crystals is 1000–2000 ATPase molecules/μm^2 membrane surface area. Saturation of the membrane by Ca^{2+}–ATPase requires about 20,000–30,000 ATPase molecules/μm^2 surface area (26,41,42). Even in sarcoplasmic reticulum isolated from fully developed chicken muscle, about 70% of the vesicles do not develop identifiable Ca^{2+}–ATPase crystals and may originate from structures other than sarcoplasmic reticulum. The identification of the origin of these membranes is an important task for the future.

5. MEMBRANE CRYSTALS OF Ca^{2+}–ATPase IN SARCOPLASMIC RETICULUM OF NORMAL AND DYSTROPHIC HUMAN, MOUSE, AND CHICKEN MUSCLE

The dimensions of Ca^{2+}–ATPase crystals are similar in sarcoplasmic reticulum membranes isolated from normal and genetically dystrophic human, mouse, and chicken muscles (23). The slightly lower Ca^{2+} transport activity of microsomes isolated from dystrophic muscles is presumably attributable to greater contamination of dystrophic microsome preparations by membranes of nonsarcoplasmic reticulum origin (38).

6. MECHANISM OF EFFECT OF MONOVANADATE AND OLIGOVANADATES ON CRYSTALLIZATION OF Ca^{2+}–ATPase: BINDING OF VANADIUM (V) OLIGOANIONS TO SARCOPLASMIC RETICULUM

The rate of crystallization of Ca^{2+}–ATPase is markedly enhanced by increasing the vanadate concentration of the medium of 5 m*M* (19,21). This is surprising since essentially complete inhibition of ATPase activity was observed at less than 0.5 m*M* vanadate concentration. These observations suggested that inhibition of ATPase activity through the formation of a kinetically stable E_2–V complex may be a required but not sufficient condition for rapid crystallization to occur. At least two additional mechanisms may be involved in the massive increase of the rate of crystallization of Ca^{2+}–ATPase at millimolar Na_3VO_4 concentrations:

1. Vanadate may bind to low-affinity sites on Ca^{2+}–ATPase, stabilizing the conformation that is most compatible with the formation of crystalline arrays.
2. The Ca^{2+}–ATPase may interact with one or more of the large number of oligomeric vanadate species that accumulate in aqueous solutions containing millimolar concentrations of vanadium (V) at neutral pH (2,9,11). The tetrahedrally coordinated vanadium dimer and tetramer interconvert rapidly with monomeric vanadate (9), while the octahedrally coordinated decavanadate is kinetically inert. There is good evidence that phosphotransferases, such as phosphofructokinase (12), adenylate kinase (48), and hexokinase (9), are inhibited only by the large, octahedrally coordinated decameric vanadium polyanion. In these systems the monomeric vanadate anion or its lower tetrahedral polyanionic species is ineffective. In the case of phosphorylase, all oligomeric forms are inhibitory, while monomeric vanadate activates the enzyme slightly (53).

We correlated the binding of monovanadate and decavanadate to sarcoplasmic reticulum vesicles with their effects on the crystallization of the Ca^{2+}–ATPase and the rate of Ca^{2+}-activated ATP hydrolysis. The results suggest that both monovanadate and decavanadate inhibit the Ca^{2+}-stimulated ATP hydrolysis but that decavanadate is particularly effective in promoting the crystallization of Ca^{2+}–ATPase (Fig. 10) (60). Therefore, in addition to the formation of the stable E_2–vanadate intermediate, coordination of the large octahedral decavanadate at or

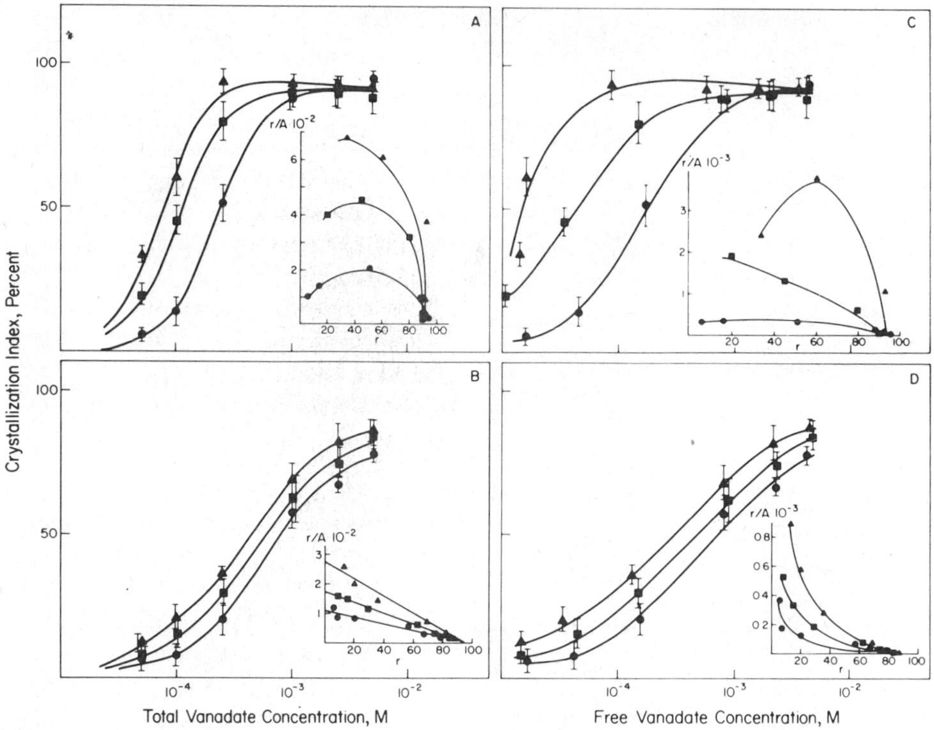

Figure 10. Relationship between the concentration of monovanadate and decavanadate and rate of crystallization of Ca^{2+}–ATPase in sarcoplasmic reticulum. Sarcoplasmic reticulum vesicles were incubated at 4°C in 0.1 *M* KCl, 10 m*M* imidazole, pH 7.4, 5 m*M* $MgCl_2$, 0.5 m*M* EGTA, and Na_3VO_4 at the concentrations indicated. Samples were taken after 4 hr (●——●), 24 hr (■——■), and 4 days (▲——▲) for electron microscopy and negatively stained with 1% uranyl acetate, pH 4.3. The crystallization index denotes the number of vesicles with crystalline regions on their surface expressed as a percentage of the total number of vesicles. On average, 630 vesicles were counted for each data point and the standard error denotes the range of variation of the counts made by four independent observers. The crystallizaton indices obtained with decavanadate (*A*) or monovanadate (*B*) are plotted against the total molar vanadate concentration. The data of *A* and *B* were replotted in *C* and *D*, respectively, against the free vanadate concentration measured in a separate set of experiments, under conditions similar to those used in the electron-microscope studies. The concentration of decavanadate (*A*, *C*) and monovanadate (*B*, *D*) are all expressed as molar monovanadate to facilitate comparison. Inserts are the Scatchard plots of the same data sets; *A* is the total or free vanadate concentration, and *r* is the crystallization index.

near the active site may be required to stabilize the enzyme conformation that is most suitable for the crystallization of Ca^{2+}–ATPase. The affinity of vanadate binding and the molar amount of vanadium bound at equilibrium is much greater with decavanadate than with monovanadate. The binding data can be rationalized in terms of one binding site (site 1) per ATPase molecule for monovanadate and two sites per ATPase for decavanadate (13,60). The occupancy of site 1 by mono- or decavanadate may be responsible for the stabilization of the E_2 conformation that leads to the crystallization of Ca^{2+}–ATPase. Site 1 is presumably related to the binding site for inorganic phosphate. Competition of one half of the decavanadate binding with fluorescein isothiocyanate (FITC) implies that the second decavanadate binding site (site 2) may overlap with the ATP binding site of the Ca^{2+}–ATPase (13). Since FITC-labeled ATPase readily crystallizes in the presence of mono- or decavanadate in a Ca^{2+}-free medium, occupancy of site 2 is not required for the crystallization of Ca^{2+}–ATPase.

The affinity of mono-, di-, tetra-, hexa-, and decavanadates for Ca^{2+}–ATPase can be conveniently measured by ^{51}V-NMR spectroscopy (14). The binding affinity progressively increases with the length of the oligovanadate chain. The greater affinity of the highly charged oligovanadates as compared with monovanadate to the enzyme is best explained if electrostatic forces play an important role in the interaction. This, in turn, suggests that the binding sites for oligovanadates in Ca^{2+}–ATPase, as in adenylate kinase (48), may be rich in basic amino acids. These observations emphasize the complex composition of vanadate solutions generally used for biochemical studies and stress the need for consideration of the binding of di-, tetra,- hexa-, and decavanadate species at multiple and overlapping sites in the interpretation of the effects of vanadate on the Ca^{2+} transport ATPase of sarcoplasmic reticulum.

7. CRYSTALLIZATION OF Ca^{2+}–ATPase IN E_1 CONFORMATION BY Ca^{2+} AND LANTHANIDE IONS

Two-dimensional crystalline arrays of Ca^{2+}–ATPase molecules develop in sarcoplasmic reticulum vesicles exposed to calcium (Fig. 3) or lanthanide ions (Figs. 4 and 5) at a moderately alkaline pH (25). The Ca^{2+}- or lanthanide-induced crystals are presumed to represent the E_1 conformation of the Ca^{2+}–ATPase, and their crystal form is clearly different from the earlier described E_2 crystals induced by Na_3VO_4 in the presence of EGTA (55). Analysis of the crystalline arrays by negative staining or freeze–fracture electron microscopy reveals obliquely oriented rows of particles corresponding to individual Ca^{2+}–ATPase molecules (25). Computer analysis of the negatively stained lanthanide-induced crystalline Ca^{2+}–ATPase arrays shows that the molecules are arranged in a P_1 lattice (Fig. 11), in contrast to the P_2 lattice obtained for the E_2-type crystals induced by vanadate. The cell constants for the gadolinium-induced Ca^{2+}–ATPase crystals are a = 61.7 Å, b = 54.4 Å, and γ =

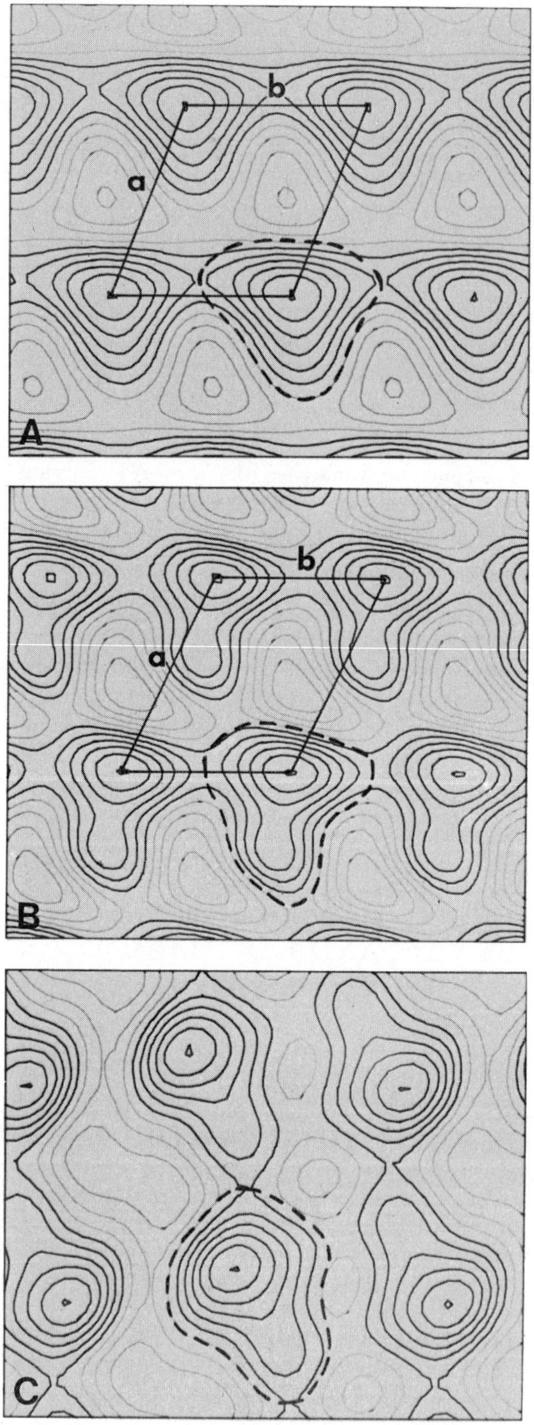

Figure 11. Isodensity contour maps of E_1 and E_2 crystals in projection at 30-Å resolution. (*A*) Gadolinium-induced E_1 crystals. The unit cell is drawn and *a* and *b* axes are labeled. One molecule of Ca^{2+}–ATPase is outlined with the dashed line. Data used was an average of six crystals. (*B*) Praseodymium-induced E_1 crystals. Both E_1 crystal forms analyzed have approximately pear-shaped Ca^{2+} AtPase profiles in projection that are similar to that observed in vanadate-induced crystals. Data used are an average of 16 crystals. (*C*) Projection derived from vanadate-induced crystals (55) with one Ca^{2+}–ATPase molecule outlined with dashed line. In the vanadate crystals Ca^{2+}–ATPase molecules are arranged pairwise in antiparallel chains. Map scale: 0.382 mm/Å.

111° (Table 1). Similar cell constants were obtained for the crystals induced by lanthanum, praseodymium, and Ca^{2+}. The unit cell dimensions of E_1 crystals are consistent with a single Ca^{2+}–ATPase monomer per unit cell, in contrast to the ATPase dimers that are the structural units of the E_2-type crystals.

These structural differences can also be observed in freeze-fracture electron microscopy. The E_1-type crystals shown in Fig. 12*A* display rows of individual particles of about 60-Å diameter on the concave *P* face, oriented obliquely (25,59). The size of the particles appears smaller than the 75–85 Å reported in earlier studies (49,58). These differences can be explained by the smaller amount of metal used for replication in the recent experiments. The ridges of particles seen on the *P* face of the E_2 crystals (Fig. 12*B* and ref. 49) are not evident in the E_1 crystals. The ridges represent intramembranous contacts between ATPase molecules located in adjacent dimer chains in the area beneath the stain filled grooves seen in negatively stained

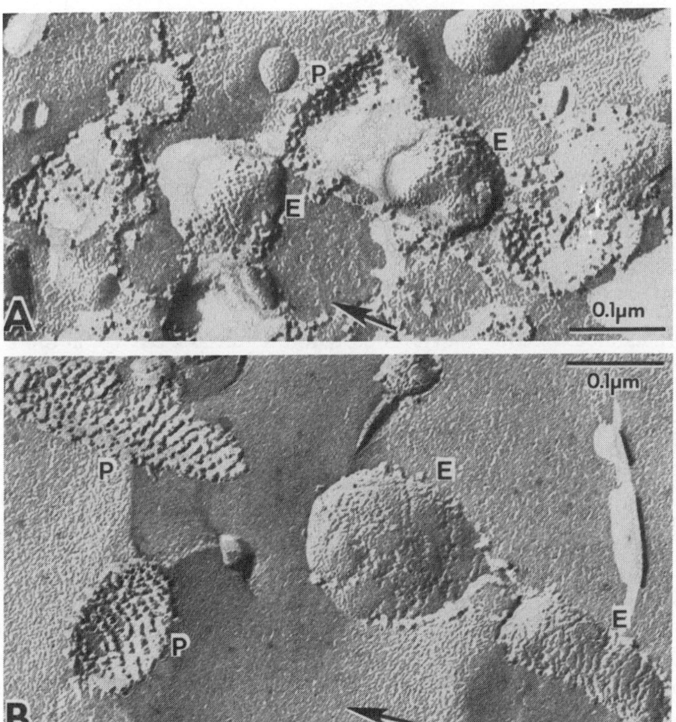

Figure 12. Freeze-fracture electron micrographs of Ca^{2+}–ATPase crystals. Sarcoplasmic reticulum vesicles were incubated in 5 m*M* decavanadate (*A*) or 10^{-5} *M* $PrCl_3$ (*B*) as described in the text and fixed in 2% glutaraldehyde overnight. The fixed vesicles were ultrarapidly frozen in liquid propane without any cryoprotectant, fractured in Cryofract 250 (Reichert-Jung, Paris, France) at −183°C and 2.5×10^{-9} Torr, and replicated immediately with Pt/C at a 20° angle followed by C at a 90° angle. The thickness of Pt and C was monitored to be 11 and 110 Å, respectively. Cytoplasmic and luminal fracture faces are labeled as *P* and *E*, and the arrows indicate the directions of shadowing. Magnification: ×127,500.

surface views (29,59). Therefore, these observations indicate differences between the E_2 and E_1 states with respect to the interaction of polypeptides in the lipid bilayer.

8. CONDITIONS AFFECTING FORMATION OF E_1-TYPE CRYSTALS

The optimum Ca^{2+} concentration required for crystallization is in the range 10^{-4}–10^{-5} M and the optimum lanthanide concentration in the range 6–8 μM. Slightly alkaline pH (pH 8) favored crystallization. Optimum crystallization in the E_1 form required 0.1 M KCl and 5 mM $MgCl_2$. No crystals were obtained with 6 μM praseodymium at 0 or at 0.6 M KCl concentration in the presence of 5 mM $MgCl_2$, and only few crystals were formed with 0 or 20 mM $MgCl_2$ in the presence of 0.1 M KCl. Adenosinetriphosphate (5 mM) inhibited the Ca^{2+}-induced crystallization and disrupted E_1 crystals previously formed in the presence of 100 μM $CaCl_2$.

9. EFFECT OF Ca^{2+} AND LANTHANIDES ON TRYPTIC DIGESTION PROFILES OF Ca^{2+}–ATPase

Trypsin preferentially cleaves Ca^{2+}–ATPase of sarcoplasmic reticulum at the T_1 cleavage sites into two major fragments designated as fragment A and fragment B. Further cleavage of fragment A at the T_2 cleavage site yields fragments A_1 and A_2. Previously we demonstrated that exposure of sarcoplasmic reticulum vesicles to vanadate in a Ca^{2+}-free medium inhibits the cleavage of Ca^{2+}–ATPase by trypsin at the T_2 cleavage site, without interference with the initial cleavage at the T_1 site (20). The inhibition of tryptic cleavage by vanadate at the T_2 site is assumed to reflect the stabilization of the E_2 conformation of Ca^{2+}–ATPase. This effect of vanadate is extremely sensitive to trace Ca^{2+} contamination. Even with Chelex-treated microsomes and buffer solutions, about 10^{-5} M EGTA is required in addition to 5 mM Na_3VO_4 to prevent the cleavage of fragment A into the A_1 and A_2 subfragments. Addition of Ca^{2+} (Fig. 13) or praseodymium (Fig. 14) released the inhibition caused by vanadate and EGTA, as shown by the appearance of the A_1 and A_2 fragments at cation concentrations in slight excess over EGTA (25). We assume that Ca^{2+} and praseodymium shift the equilibrium from the E_2 in favor of the E_1 conformation and thereby facilitate the attack of trypsin at the T_2 site. Other lanthanides, with slight variations in effectiveness, gave similar results (25). These effects may be due to either binding of lanthanides to the high-affinity binding site of the Ca^{2+}–ATPase or to a site distinct from that of Ca^{2+} (32). The Ca^{2+} released from EGTA by lanthanides may contribute to the observations, and further experiments are needed to resolve this ambiguity.

Prior cleavage of Ca^{2+}–ATPase by trypsin into the two major fragments, A and B, did not interfere with the formation of E_1 crystals by praseodymium, but further cleavage of the ATPase at the T_2 site inhibited the crystallization (25).

10. EFFECT OF Ca^{2+} AND LANTHANIDES ON CONFORMATION OF Ca^{2+}–ATPase: FLUORESCENCE OF TRYPTOPHAN AND FLUORESCEIN 5′-ISOTHIOCYANATE

The effects of Ca^{2+}, gadolinium, lanthanum, and praseodymium on the fluorescence of tryptophan and covalently bound fluorescein were analyzed in native and

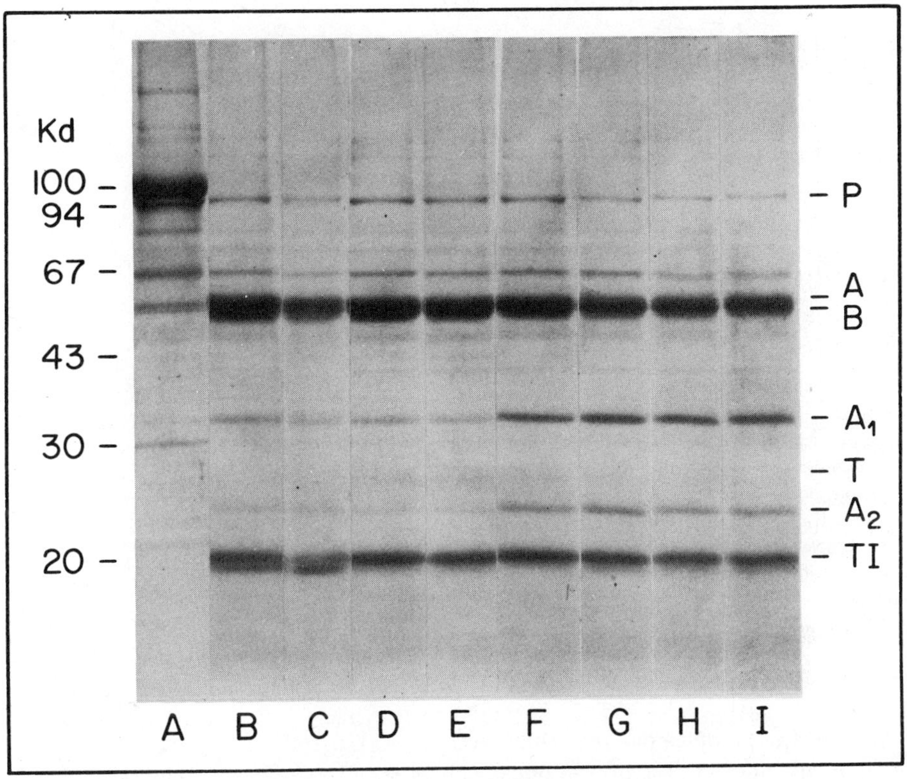

Figure 13. Effect of [Ca^{2+}] on the tryptic cleavage of Ca^{2+}–ATPase in the presence of Na_3VO_4. Sarcoplasmic reticulum vesicles (2 mg protein/ml) were pretreated with Chelex; 5 m*M* $MgCl_2$, 0.1 *M* Na_3VO_4, and 5×10^{-5} *M* EGTA were added together with $CaCl_2$ in final concentrations (*M*) of 0 (*B*); 10^{-6} (*C*); 5×10^{-6} (*D*); 10^{-5} (*E*); 5×10^{-5} (*F*); 10^{-4} (*G*); 5×10^{-4} (*H*); and 10^{-3} (*I*) and incubated for 30 min at 25°C. The tryptic digestion and the gel electrophoresis of the samples were carried out as follows. The digestion was initiated by trypsin (0.1 mg/ml) at 25°C. The cleavage was stopped 5 min later by soybean trypsin inhibitor in twofold excess over trypsin (0.2 mg/ml). The samples were solubilized and the proteins separated by sodium dodecyl sulfate polyacrylamide-gradient gel electrophoresis (SDS–PAGE) (6–18% gradient) as described earlier (20). In a concentration equal to or exceeding that of the EGTA (50 μ*M*, lanes *F*–*I*) Ca^{2+} can overcome the inhibitory effect of VO_4 on trypsin cleavage at the second (T_2) cleavage site, and the tryptic digestion of Ca^{2+}–ATPase proceeds with the formation of A_1 and A_2 subfragments in lanes *F*–*I*. Lane *A*, control sample, without trypsin and trypsin inhibitor; P, phosphorylase; T, trypsin; TI, trypsin inhibitor.

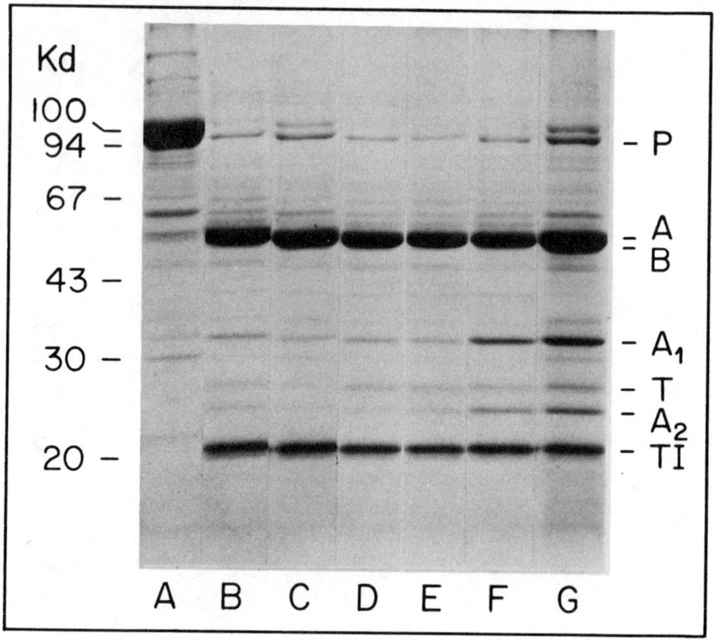

Figure 14. Effect of Pr^{3+} on tryptic cleavage of the Ca^{2+}–ATPase in the presence of Na_3VO_4. The experimental procedure was as described in the legend to Fig. 13 except that Ca^{2+} was replaced by $PrCl_3$ in concentrations of 10^{-6} *M* (*C*), 10^{-5} *M* (*D*), 5×10^{-5} *M* (*E*), 10^{-4} *M* (*F*), and 10^{-3} *M* (*G*) in the incubating media. Lane *A*: control sample without trypsin inhibitor, EGTA, or Pr. Lane *B*: control sample, without Pr^{3+} but with trypsin inhibitor (TI) and EGTA. In the presence of 5×10^{-5} *M* EGTA, 10^{-4} *M* Pr^{3+} was able to convert the enzyme from the VO_4-induced E_2 into the E_1 conformation, as reflected by the appearance of A_1 and A_2 subfragments in lanes *F* and *G*. The relatively high concentration of Pr (10^{-4} *M*) required to alter the digestion pattern, as compared with the 6–8 μ*M* needed for crystallization, is attributable in part to the presence of 50 μ*M* EGTA, that chelates much of the added Pr^{3+}.

FITC-labeled sarcoplasmic reticulum vesicles (33). Both Ca^{2+} and lanthanides increased the fluorescence of tryptophan (Fig. 15) and decreased the fluorescence of FITC (Fig. 16); the dependence of these effects on cation concentration is consistent with the involvement of the high-affinity Ca^{2+} binding sites of Ca^{2+}–ATPase in the cation-induced fluorescence changes and in the crystallization of the Ca^{2+}–ATPase by Ca^{2+} and lanthanides. The effect of Ca^{2+}, lanthanides, and vanadate on the fluorescence of tryptophan and FITC are readily interpretable in terms of the two major conformations of Ca^{2+}–ATPase (E_1 and E_2) that are assumed to alternate during Ca^{2+} transport. Therefore, these observations provide support to the suggestion that the vanadate induced crystals of Ca^{2+}–ATPase represent the E_2, while the Ca^{2+}- and lanthanide-induced crystals the E_1 conformation of the enzyme (33).

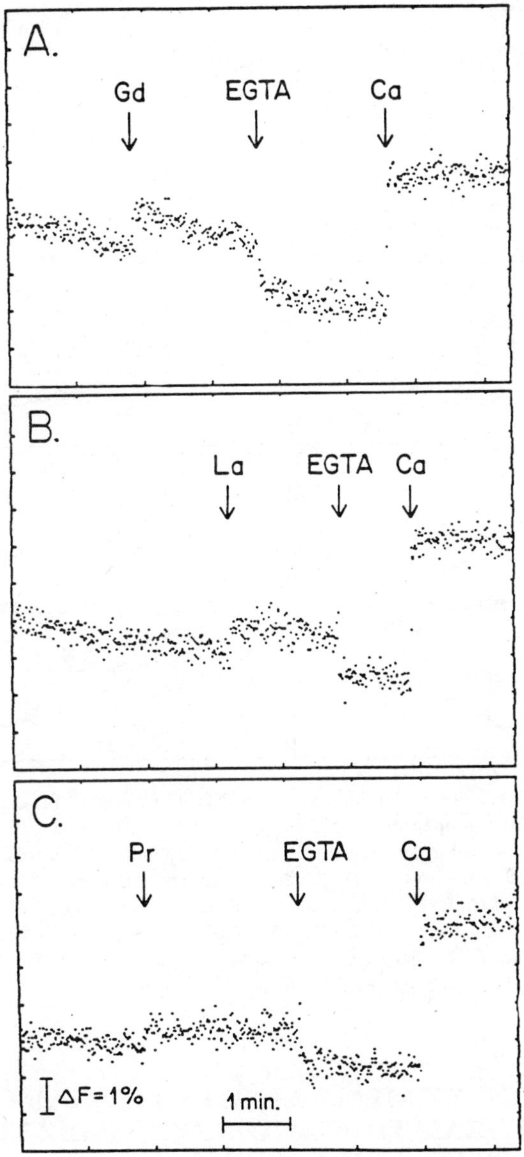

Figure 15. Effects of lanthanides on fluorescence of tryptophan in sarcoplasmic reticulum vesicles after treatment with Chelex chelating resin. Sarcoplasmic reticulum vesicles (0.1 mg protein/ml) were treated with Chelex for 10 hr. To the Chelex-treated vesicles suspended in 0.1 *M* KCl, 30 m*M* Tris–MOPS, pH 7.0 and 5 m*M* $MgCl_2$, lanthanides (5 μ*M*), EGTA (10μ*M*), and $CaCl_2$ (100 μ*M*) were added as indicated in the figure. The fluorescence of tryptophan was measured at 330 nm by use of 295-nm light for excitation. The changes in fluorescence intensity (Δ fluorescence) are expressed as percent of starting value. Temperature, 10°C. (*A*) Top panel, $GdCl_3$; (*B*) middle panel, $LaCl_3$; (*C*) lower panel, $PrCl_3$.

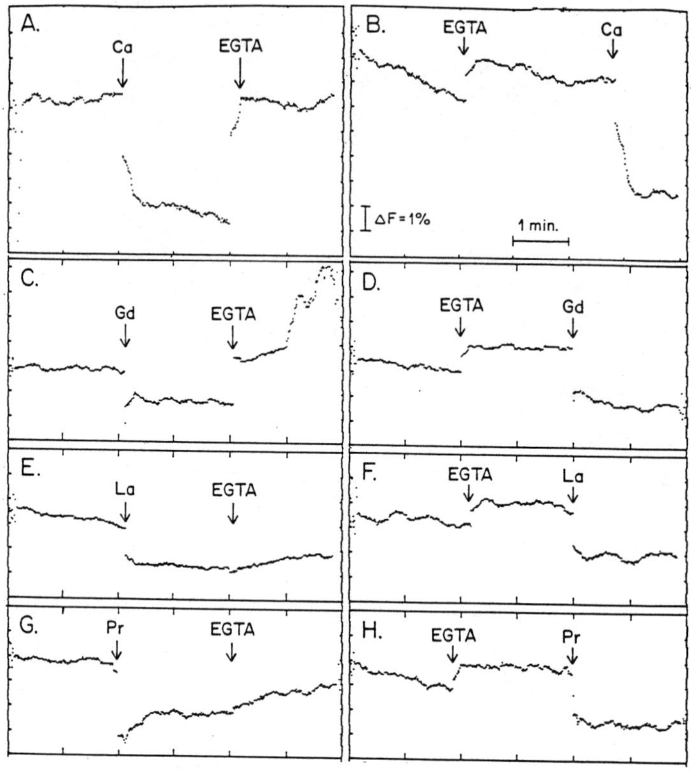

Figure 16. Effect of lanthanides on the fluorescence of FITC–microsomes. FITC-labeled microsomes (10 nmol FITC/mg protein) were treated with Chelex-100 for 16 hr at 2°C to reduce Ca^{2+} contamination. The microsomes were suspended in Chelex-treated, 0.1 *M* KCl, 30 m*M* MOPS solution to which 5 m*M* $MgCl_2$ was added just before the experiment. Protein concentration: 0.1 mg/ml. The fluorescence was measured by use of 500-nm light for excitation and 517.5 nm for emission (slits 2 and 4 nm, respectively). The following additions were made: (*A*) 10 μ*M* Ca, 15 μ*M* EGTA; (*B*) 5 μ*M* EGTA, 15 μ*M*Ca; (*C*) 5 μ*M* Gd, 10 μ*M* EGTA; (*D*) 5 μ*M* EGTA, 10 μ*M* Gd; (*E*) 5 μ*M* La, 10 μ*M* EGTA; (*F*) 5 μ*M* EGTA, 10 μ*M* La; (*G*) 5 μ*M* Pr, 10 μ*M* EGTA; (*H*) 5 μ*M* EGTA, 10 μ*M* Pr. Temperature: 10°C.

11. EFFECT OF MEMBRANE POTENTIAL ON CONFORMATION AND CRYSTALLIZATION OF Ca^{2+}–ATPase IN E_1 FORM

Inside positive membrane potential generated by dilution of sarcoplasmic reticulum vesicles from choline chloride into a K–glutamate medium promoted the formation of E_2 crystals of Ca^{2+}–ATPase in the presence of EGTA and vanadate (5,22). Substitution of K–glutamate for choline chloride produced negative potential and disrupted the preformed E_2 crystals of Ca^{2+}–ATPase (22). These observations suggested that inside positive potential stabilizes the E_2, while negative potential stabilizes the E_1 conformation of the ATPase. If these assumptions are correct, it is expected that negative potential would promote the formation of the E_1 crystals of Ca^{2+}–ATPase in the presence of 6 μM praseodymium. This is indeed the case. Significant crystallization of Ca^{2+}–ATPase was obtained within 30 sec after dilution of the vesicles equilibrated with K–glutamate into a choline chloride medium containing 6 μM praseodymium chloride. Similar degree of crystallization in vesicles diluted from choline chloride into choline chloride or from K–glutamate into

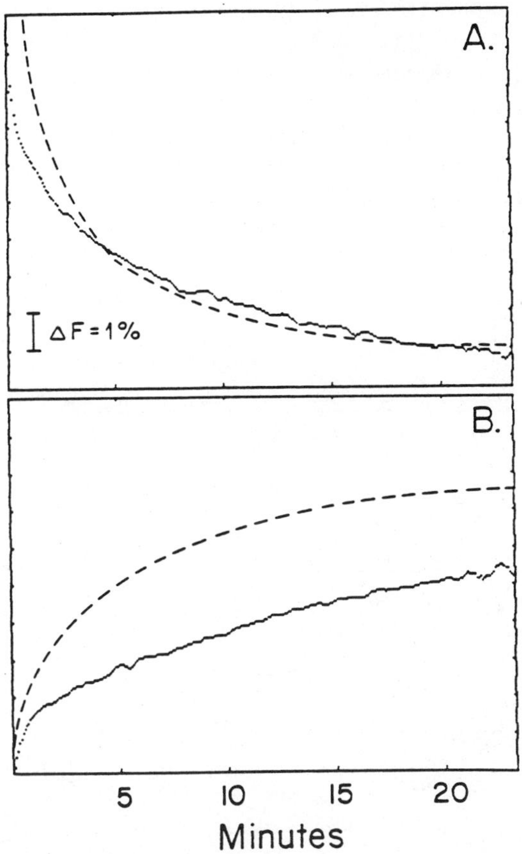

Figure 17. Effect of membrane potential on the fluorescence of FITC covalently bound to the Ca^{2+}–ATPase. Sarcoplasmic reticulum vesicles were labeled with FITC at a ratio of 10 nmol/mg protein and dialyzed overnight against two changes (500 volumes each) of either K–glutamate (KG) or choline chloride (CC) medium. Protein concentration: 20 mg/ml. (*A*) Inside positive potential was generated by 190-fold dilution of CC microsomes into KG medium (top panel). Dilution of KG microsomes into KG medium served as control. (*B*) Inside negative potential was generated by 190-fold dilution of KG microsomes into CC medium (bottom panel). Dilution of CC microsomes into CC medium served as control. The changes in FITC fluorescence were measured by using excitation and emission wavelengths of 500 and 517.5 nm, with slits of 2 and 4 nm, respectively. Dotted lines: dilution medium contained no EGTA. Broken lines: dilution medium contained 0.5 mM EGTA.

K–glutamate medium required several hours of incubation. Since ion substitution of this type generates inside negative potential, we propose that negative potential promotes the formation of E_1 crystals, presumably by stabilizing the E_1 conformation of Ca^{2+}–ATPase (25).

Inside-positive membrane potential generated by transfer of vesicles from choline chloride into K–glutamate medium increases the fluorescence of tryptophan and of FITC covalently bound to Ca^{2+}–ATPase (Figs. 17 and 18). Transfer of vesicles from K–glutamate into choline chloride medium generates negative potential and decreases both tryptophan and FITC fluorescence (Figs. 17 and 18) (33). The fluorescence transients caused by either type of ion substitution declined with the dissipation of membrane potential over a period of several minutes. To an estimated potential of 50–60 mV, FITC-labeled Ca^{2+}–ATPase responds with a fluorescence change of 7–9%. The corresponding changes in tryptophan fluorescence are much smaller, usually of the order of 1%. Therefore, the analysis of tryptophan signals is difficult and generally requires averaging.

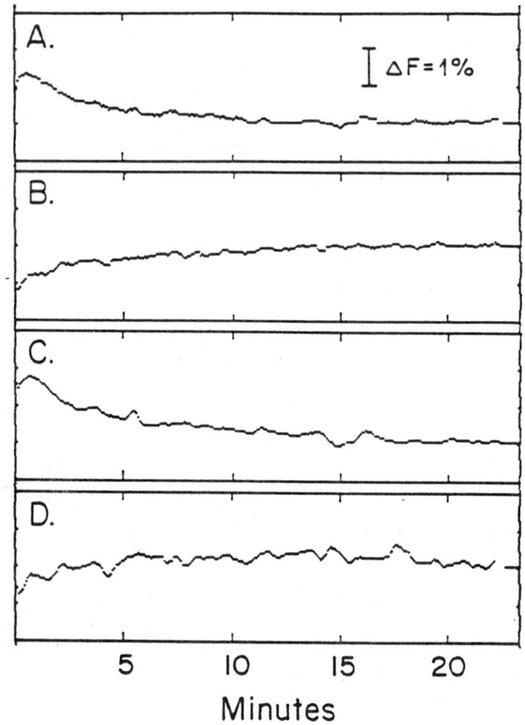

Figure 18. Effect of membrane potential on tryptophan fluorescence. Sarcoplasmic reticulum vesicles equilibrated in choline chloride medium were diluted into K–glutamate medium to generate inside positive potential (*A* and *C*). Inside negative potential was generated by dilution of K–glutamate vesicles into choline chloride medium (*B* and *D*). The dilution media of *A* and *B* contained $\simeq 10^{-5}$ *M* free Ca^{2+}; to the dilution media of *C* and *D*, 0.5 m*M* EGTA was added to lower the free Ca^{2+} concentration below 10^{-8} *M*. The curves in *A* and *B* represent averages of four separate experiments.

It is noteworthy that the fluorescence responses of FITC and tryptophan to membrane potential are of the same sign, while the two probes respond with changes of opposite sign to the binding of Ca^{2+} and lanthanides to the sarcoplasmic reticulum (1,33,51). The disparate responses of tryptophan and the covalently bound FITC to potential and to the binding of cations imply that one or both of these responses are significantly influenced by local changes in the environment of the probes. Therefore, without additional independent criteria it is difficult to provide a unifying interpretation of all responses of both probes to calcium and to potential in terms of an equilibrium between two hypothetical conformational states, E_1 and E_2, of Ca^{2+}–ATPase. Either there are more than two conformations sensed by one or both probes, or the response of one or both probes contains significant local contribution that acts independently of the overall conformation of the protein. Local effects on tryptophan fluorescence would be expected if the binding site of calcium and lanthanides were in the vicinity of the tryptophan that is sampled.

The physiological significance of these observations is that through an effect on the conformation of Ca^{2+}–ATPase, membrane potential may influence the kinetics of ATP-dependent Ca^{2+} transport and the permeability properties of the sarcoplasmic reticulum membrane (39,40). For example, Ca^{2+} release from sarcoplasmic reticulum generates inside-negative membrane potential; this, in turn, promotes the subsequent reaccumulation of Ca^{2+} by an electrogenic process that restores electroneutrality across the membrane (4,39).

12. REVERSIBLE INTERCONVERSION OF Ca^{2+}–ATPase BETWEEN E_1 AND E_2 CRYSTAL FORMS

By proper selection of experimental conditions, reversible interconversion between the E_1 and E_2 crystal forms can be observed (25). The E_1-type crystals were formed by incubation of Chelex-treated sarcoplasmic reticulum vesicles with 6 μM $PrCl_3$ for 16 hr at 2°C. Addition of 0.5 m*M* EGTA to these vesicles caused the rapid disappearance of the E_1 crystals, and typical E_2-type crystals formed 15 hr after the addition of 5 m*M* Na_3VO_4. The vanadate-induced E_2 crystals were disrupted by 1 m*M* $PrCl_3$, and a somewhat disorganized E_1 lattice reappeared. Aggregation of the vesicles in the presence of $PrCl_3$ made the observation of these crystals difficult. In a concentration sufficient to chelate praseodymium (5 m*M*), EGTA dissolved the microscopic vesicle aggregates, destroyed the E_1 lattice, and together with the 5 m*M* vanadate already present, caused the reformation of the E_2 crystals. Glutaraldehyde fixation of the E_1 crystals induced by praseodymium stabilized the E_1-type arrays that persisted in fixed preparations, even after addition of 0.5 m*M* EGTA and 5 m*M* Na_3VO_4. These experiments clearly demonstrate that reversible interconversion between the E_1 and E_2 crystal forms of Ca^{2+}–ATPase is possible and provide a striking illustration of the visible structural differences between the E_1 and E_2 crystals on the same microsome preparation.

13. COMPARISON OF Ca^{2+}–ATPase CRYSTALS WITH CRYSTALS OF Na, K–ATPase

In view of the structural similarities between Ca^{2+}–ATPase and the (Na^+ + K^+) ATPase, a comparison of their crystallographic properties seems worthwhile.

The Ca^{2+}–ATPase forms monomer-type (P_1) crystals in the E_1 and dimer-type (P_2) crystals in the E_2 conformation. The two crystal forms appear to be specifically related to the two major conformations of the Ca^{2+}–ATPase and do not coexist in significant amounts in native sarcoplasmic reticulum membranes.

Crystallization of the (Na^+ + K^+)–ATPase in the presence of vanadate was first observed by Skriver and her colleagues (27,34,52) and was subsequently described by several other laboratories (16,44,63). Crystals of (Na^+ + K^+)-ATPase are also induced by either Ca^{2+} or lanthanides (3,62). Surprisingly, the (Na^+ + K^+)–ATPase isolated from kidney medulla forms primarily P_1-type crystals under all conditions tested so far. These include media containing Na^+ + K^+, with or without vanadate, inorganic phosphate, Mg^{2+}, Ca^{2+}, gadolinium, lanthanum, or praseodymium. The P_1-type (Na^+ + K^+)–ATPase crystals coexist with smaller amounts of P_2 crystalline arrays, but the relative amounts of the two types of crystals do not show clear correlation with the postulated conformation of the (Na^+ + K^+)–ATPase. Although fine-structural analysis may reveal subtle differences between the monomer-type (Na^+ + K^+)–ATPase crystals obtained under various conditions, thus far there is no indication that such differences exist.

14. PERSPECTIVES

The crystallization of Ca^{2+}–ATPase in sarcoplasmic reticulum membranes and reconstituted ATPase vesicles in the presence of vanadate and lanthanide ions offers a powerful new approach for the analysis of the three-dimensional structure of the Ca^{2+}–ATPase. If the different crystal forms obtained with vanadate and lanthanide ions reflect the different conformational states of the enzyme in successive stages of the transport cycle, we may soon be able to identify the structural changes that represent the physical basis of Ca^{2+} translocation. Further resolution of the structure to reveal polypeptide chain folding will require the development of three-dimensional Ca^{2+}–ATPase crystals that are suitable for x-ray diffraction studies.

ACKNOWLEDGMENTS

We thank Drs. T. J. Beeler, M.-H. Ho, K. Kustin, G. C. Levy, and Shigeaki Miyamoto for their contributions to the general problem discussed in this chapter.

REFERENCES

1. Andersen, J. P., Møller, J. V. and Jorgensen, P. L. (1982). *J. Biol. Chem.*, **27**, 8300–8307.
2. Baes, C. F. and Mesmer, R. E. (1976). *The Hydrolysis of Cations,* Wiley, New York.
3. Beall, H. C., Hastings, D. F., Dux, L., Ting-Beall, H. P., and Martonosi, A. N. (1986). *Biophys. J.* **49**, 35a.

9. Boyd, D. W. and Kustin, K. (1984). *Adv. Inorg. Biochem.*, **6**, 311–365.
10. Castellani, L., Hardwicke, P., and Vibert, P. (1985). *J. Mol. Biol.* **185**, 579–594.
11. Chasteen, N. D. (1983). *Struct. Bonding*, **53**, 105–138.
12. Choate, G. and Mansour, T. E. (1979). *J. Biol. Chem.*, **254**, 11457–11462.
13. Csermely, P., Varga, S., and Martonosi, A. (1985). *Eur. J. Biochem.*, **150**, 455–460.
14. Csermely, P., Martonosi, A., Levy, G. C., and Ejchart, A. J. (1985). *Biochem. J.*, **230**, 807–815.
15. Csermely, P., Katopis, C., Papp, S., Wallace, B. A., and Martonosi, A. (1986). *Biophys. J.* **49**, 562a.
16. Demin, V. V., Barnakov, A. N., Lunev, A. V., Dzhandzhugazyan, K. N., Kuzin, A. P., Modyanov, N. N., Hovmoller, S., and Farrants, G. (1984). *Biol. Membranes*, **1**, 831–838.
17. Dupont, Y., Bennett, N., and Lacapere, J. J. (1982). *Ann. N. Y. Acad. Sci.*, **402**, 569–572.
18. Dux, L. (1985). *FEBS Lett.*, **183**, 177–181.
19. Dux, L. and Martonosi, A. (1983). *J. Biol. Chem.*, **258**, 2599–2603.
20. Dux, L. and Martonosi, A. (1983). *J. Biol. Chem.*, **258**, 10111–10115.
21. Dux, L. and Martonosi, A. (1983). *J. Biol. Chem.*, **258**, 11896–11902.
22. Dux, L. and Martonosi, A. (1983). *J. Biol. Chem.*, **258**, 11903–11907.
23. Dux, L. and Martonosi, A. (1983). *Muscle Nerve*, **6**, 566–573.
24. Dux, L. and Martonosi, A. (1984). *Eur. J. Biochem.*, **141**, 43–49.
25. Dux, L., Taylor, K. A., Ting-Beall, H. P., and Martonosi, A. (1985). *J. Biol. Chem.*, **260**, 11730–11743.
26. Franzini-Armstrong, C., Ferguson, D., Castellani, L., and Hardwicke, P. M. D. (1985). *Proceedings of the 50th Jubilee Congress of the Hungarian Physiological Society*, p. 137.
27. Hebert, H., Jørgensen, P. L., Skriver, E., and Maunsbach, A. B. (1982). *Biochim. Biophys. Acta*, **689**, 571–574.
28. Henderson, R. and Unwin, P. N. T. (1975). *Nature*, **257**, 28–32.
29. Ho, M.-H., Taylor, K. A., and Martonosi, A. N.. (1986). *Biophys. J.* **49**, 570a.
30. Ikemoto, N. (1982). *Ann. Rev. Physiol.*, **44**, 297–317.
31. Inesi, G., Watanabe, T., Coan, C., and Murphy, A. (1982). *Ann. N. Y. Acad. Sci.*, **402**, 515–534.
32. Itoh, N. and Kawakita, M. (1984). *J. Biochem.* (Tokyo), **95**, 661–669.
33. Jona, I. and Martonosi, A. (1986). *Biochem. J.* **234**, 363–371.
34. Jorgensen, P. L., Skriver, E., Hebert, H., and Maunsbach, A. B. (1982). *Ann. N. Y. Acad. Sci.*, **402**, 207–255.
35. Martonosi, A. (1975). *Biochim. Biophys. Acta*, **415**, 311–333.
36. Martonosi, A. (1982). In Cheung, W. Y., Ed., *Calcium and Cell Function* Vol. 3, Academic, New York, pp. 38–102.
37. Martonosi, A. (1982). *Ann. Rev. Physiol.*, **44**, 337–355.

38. Martonosi, A. (1982). In Schotland, D. L., Ed., *Disorders of the Motor Unit*, Wiley, New York, pp. 565–583.
39. Martonosi, A. (1984). *Physiol. Rev.*, **64**, 1240–1320.
40. Martonosi, A., Nakamura, H., Jilka, R. L., and Vanderkooi, J. M. (1977). In Semenza, G. and Carafoli, E., Eds., *Biochemistry of Membrane Transport*, Springer-Verlag, Berlin, pp. 405–415.
41. Martonosi, A., Roufa, D., Ha, D. B., and Boland, R. (1980). *Fed. Proc.*, **39**, 2415–2421.
42. Martonosi, A. N., Dux, L., Terjung, R. L., and Roufa, D. (1982). *Ann. N. Y. Acad. Sci.*, **402**, 485–514.
43. Martonosi, A. and Beeler, T. J. (1983). In Peachey, L. D. and Adrian, R. H., Eds., *Handbook of Physiology. Skeletal Muscle*, American Physiological Society, Bethesda, MD, pp. 417–485.
44. Mohraz, M. and Smith, P. R. (1984). *J. Cell Biol.*, **98**, 1836–1841.
45. Møller, J. V., Andersen, J. P., and Le Maire, M. (1982). *Molec. Cell. Biochem.*,**42**, 83–107.
46. O'Neil, S. G., Rhoads, D. B., and Racker, E. (1979). *Biochem. Biophys. Res. Commun.*, **89**, 845–850.
47. Oosawa, F. and Asakura, S. (1975). *Thermodynamics of Polymerization of Protein*, Academic, New York.
48. Pai, E. F., Sachsenheimer, W., Schirmer, R. H., and Schulz, G. E. (1977). *J. Molec. Biol.*, **114** , 37–45.
49. Peracchia, C., Dux, L., and Martonosi, A. N. (1984). *J. Muscle Res. Cell Motil.*, **5**, 431–442.
50. Pick, U. (1982). *J. Biol. Chem.*, **257**, 6111–6119.
51. Pick, U. and Karlish, S. J. D. (1982). *J. Biol. Chem.*, **257**, 6120–6126.
52. Skriver, E., Maunsbach, A. B., and Jorgensen, P. L. (1981), *FEBS Lett.*, **131**, 219–222.
53. Soman, G., Chang, Y. C., and Graves, D. J. (1983). *Biochemistry*, **22**, 4994–5000.
54. Taylor, K. A. and Glaeser, R. M. (1976). *J. Ultrastruct. Res.*, **55**, 448–456.
55. Taylor, K. A., Dux, L., and Martonosi, A. (1984). *J. Molec. Biol.*, **174**, 193–204.
56. Taylor, K. A., Dux, L., and Martonosi, A. (1986). *J. Molec. Biol.* **187**, 417–427.
57. Taylor, K. A., Dux, L., Varga, S., Ting-Beall, H. P., and Martonosi, A. (in press). *Methods Enzymol.*
58. Tillack, T. W., Boland, R., and Martonosi, A. (1974). *J. Biol. Chem.*, **249**, 624–633.
59. Ting-Beall, H. P., Dux, L., and Martonosi, A. (1986). *Biophys. J.* **49**, 570a.
60. Varga, S., Csermely, P., and Martonosi, A. (1985). *Eur. J. Biochem.*, **148**, 119–126.
61. Varga, S., Csermely, P., Müllner, N., and Martonosi, A. (1986). *Biophys. J.* **49**, 571a.
62. Varga, S. and Martonosi, A. manuscript in preparation.
63. Zampighi, G., Kyte, J., and Freytag, W. (1984). *J. Cell Biol.*, **98**, 1851–1864.

CHAPTER 16

CALCIUM ATPaseS: CONTRIBUTION OF MOLECULAR GENETICS TO OUR UNDERSTANDING OF STRUCTURE AND FUNCTION

DAVID H. MacLENNAN
CHRISTOPHER J. BRANDL
BOZENA KORCZAK
Banting and Best Department of Medical Research
C. H. Best Institute, University of Toronto
Toronto, Ontario, Canada

N. MICHAEL GREEN
National Institute for Medical Research
The Ridgeway, Mill Hill
London, United Kingdom

As new technologies emerge, their use broadens our knowledge in unpredictable ways. The Ca^{2+}–ATPase of sarcoplasmic reticulum has been investigated by use of many technologies since its discovery as a major ATPase activity in muscle homogenates in 1948 (23). As the technology for membrane fractionation became widely available, this enzyme was found to exist in microsomal fractions (10,15). It was shown to be a Ca^{2+}-dependent activity and, in analogy with the then recently discovered (Na^{+} + K^{+})–ATPase to transport Ca^{2+} against a concentration gradient and to store Ca^{2+} within luminal spaces (10,15). The availability of a purified membrane fraction offered enzymologists the opportunity to investigate the kinetics and dissect the partial reactions of ATP-dependent Ca^{2+} transport (8,19). The isolation of the ATPase as an active, detergent-soluble enzyme (26) provided material for refinement of kinetic analyses and opened the door to studies of reconstitution (37), biosynthesis (27,39), and structure (28) of this enzyme.

The most recent advance, the cloning of DNA encoding Ca^{2+}–ATPase has opened up new potential for investigation. In this chapter we review some of the advances that have already been made under the headings of "Molecular Genetics," "Structure of Ca^{2+}–ATPase," and "Mechanism of Action."

1. MOLECULAR GENETICS

We have cloned and characterized full-length cDNAs encoding two forms of Ca^{2+}–ATPase (5,29). These cDNAs have 76% sequence identity in their coding regions and code for proteins with 84% amino acid sequence identity. They display different codon usage and, as a result, have few common sites for restriction endonuclease digestion (Fig. 1). The 5′ untranslated regions of both clones are highly (G + C)-rich, but neither the 5′ nor 3′ untranslated regions bear any significant homology with each other.

One of the cDNAs clearly encodes the rabbit fast-twitch skeletal muscle Ca^{2+}–ATPase since partial amino acid sequences for this enzyme (2) have all been found in the complete amino acid sequence deduced from cDNA sequence (5). The other clone appears to encode a slow-twitch muscle form of the enzyme. There is evidence that the slow-twitch muscle clone may also encode all, or almost all, of the cardiac form of the enzyme. This evidence is derived from partial amino acid sequence identity, Northern and Southern blotting analysis and the demonstration of common epitopes.

The conclusion that the second clone encodes the slow-twitch muscle form of the enzyme is based on Northern blotting of mRNAs from rabbit fast-twitch, slow-twitch (soleus), and cardiac muscle (5). The use of homologous coding region probes from the two ATPase clones allowed us to identify two distinct transcripts in neonatal skeletal muscle mRNA (the source of our cDNA library) and in the three classes of muscle from year-old rabbits (5). The larger transcript (about 4 kilobases) was found predominantly in slow-twitch and cardiac muscle mRNA, while the smaller transcript (about 3.6 kilobases) was found predominantly in fast-twitch muscle mRNA. The

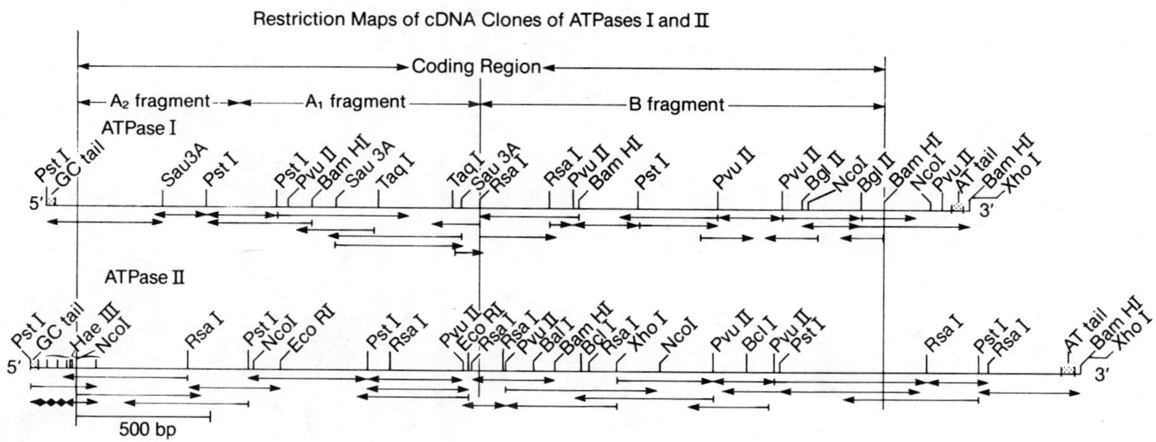

Figure 1. Restriction endonuclease map and sequencing strategy for cloned cDNA encoding two Ca^{2+}–ATPases from neonatal rabbit skeletal muscle sarcoplasmic reticulum. The cDNA library from which the clones were isolated was constructed in the vector pcD-X (34) with the generous assistance of Dr. H. Okayama. The restriction map displays six-base restriction sites and the four-base sites that were used in dideoxy sequencing according to the method of Sanger et al. (42). The coding regions are divided into three segments, A_2, A_1, and B, which correspond to the A and B fragments generated by brief tryptic digestion of the ATPase protein (see ref. 52). Top: ATPase I is the neonatal form of the fast-twitch Ca^{2+}–ATPase. Bottom: ATPase II is the neonatal form of the slow-twitch, cardiac form of the enzyme.

specific fast-twitch 3′ untranslated probe hybridized only with the smaller transcript in fast-twitch muscle mRNA, but the specific slow-twitch 3′ untranslated probe bound to the larger transcription in both slow-twitch and cardiac mRNAs.

Clearly, further information is required about the relationship between slow-twitch and cardiac forms of the enzymes, and we have begun to investigate sequence identity between these two proteins. A short amino acid sequence readily obtained from Ca^{2+}–ATPases is a tryptic fragment that extends from a major tryptic cleavage site (T_1) at residue 505–506 to residue 523 in the slow-twitch muscle sequence or to residue 524 in the fast-twitch muscle sequence. This fragment contains a Lys at residue 514 (or 515) which binds fluorescein isothiocyanate (FITC), thereby inhibiting ATP binding to the enzyme. The sequence of this fragment from the canine cardiac Ca^{2+}–ATPase was shown by Briggs et al. (7) to be identical to the sequence that we have deduced for our slow-muscle clone from rabbit skeletal muscle, suggesting at least partial sequence identity for cardiac and slow-twitch muscle ATPases.

Phospholamban, a phosphoprotein that regulates the activity of cardiac Ca^{2+}–ATPase (47), has been localized in immunohistochemical studies (21) in cardiac and slow-twitch fibers, but not in fast-twitch fibers. This would suggest a functional relationship between cardiac and slow-twitch ATPases. In other immunohistochemical studies Jorgensen and Campbell (20) used a monoclonal antibody against canine cardiac ATPase to show that the same epitope was present in ATPases in canine cardiac and slow-twitch fibers but was absent from canine fast-twitch fibers. Thus, the slow-twitch and cardiac ATPases have common transcript sequences, partial amino acid sequence identity, common epitopes, and, possibly, the ability to interact with a common regulatory protein, phospholamban.

We have used the technique of Southern blotting of rabbit genomic DNA and hybridization with nucleotide probes from the coding regions of our cDNA clones to demonstrate that only two closely cross-hybridizing genes for Ca^{2+}–ATPases exist in rabbit (5). (We can, however, detect five additional weakly cross-hybridizing genes in the same animal and these may represent other forms of ion-transporting enzymes). We have isolated one of these genes (24) and shown that it encodes the fast-twitch Ca^{2+}–ATPase. The other gene, by default, must encode both slow-twitch and cardiac forms of the enzyme.

If only two genes are available for encoding fast-twitch, slow-twitch and cardiac ATPases, can we expect additional diversity in these enzymes? Possibilities for diversity are to be found in the phenomenon of alternative splicing in which different exons encoded in a single hnRNA can be spliced to yield two or more mRNAs that share some exons and differ in others. There are many examples domonstrating that muscle has the potential to carry out alternative splicing (6,33). We have found that the fast skeletal ATPase transcript is alternatively spliced during the developmental transition of muscle from a neonatal to a mature form. Allen et al. (2) reported that the sequence of mature Ca^{2+}–ATPase ended with Gly. The amino acid sequence that we deduced from the cDNA sequence of the fast-twitch ATPase ended with

–Asp–Pro–Glu–Asp–Glu–Arg–Arg–Lys (5). Insight into the reason for this discrepancy was obtained when we isolated a 17-Kb fragment of genomic DNA that coded for all but the first 13 *N*-terminal residues of the fast-twitch ATPase (24). In analysis of this clone we found that an exon–intron boundary occurs at the point where the two sequences diverge. Between this point and the 3′ end of the gene we find one 42 bp exon encoding gly-stop and a downstream exon which codes for the novel, highly charged carboxyl terminal sequence plus the 3′ untranslated sequence found in our clone.

We expect that 3′ alternative splicing will be found to be developmentally regulated, with the exon for the highly charged *C*-terminal sequence predominating in neonatal transcripts and an exon for the less polar sequence predominating in adult transcripts. We do not know why this shift in splicing modes might occur or whether it has any functional significance. At this stage, the only clear deduction that we can make is that the highly charged carboxyl terminus must be located in the cytoplasm since it would be extremely unlikely that such a sequence could traverse the phospholipid bilayer once its synthesis were completed on cytoplasmic ribosomes. This tentative structural deduction leads logically to a fuller analysis of structure.

2. STRUCTURE OF Ca^{2+}–ATPase

In earlier reviews of the structure of the ATPase (28) we have made use of several topographical features to plot the interaction of the polypeptide chain with the hydrophobic bilayer of the membrane. We showed that the ATPase was synthesized without an *N*-terminal signal sequence, suggesting that the amino terminus would remain in the cytoplasm (39). We have confirmed the lack of an *N*-terminal signal sequence in our analyses of cloned ATPase cDNA by showing that the initiator Met in the deduced sequence (5) is the same as the initiator Met that we found in the mature ATPase sequence (39). We have also confirmed our deduction that the amino terminus of the ATPase is cytoplasmic by demonstrating the binding to Cys-12 of several ligands that are membrane impermeant (38). These findings, together with our speculation that the carboxyl terminus is cytoplasmic, leads us to the prediction that an even number of transmembrane passages must occur between the amino and carboxyl termini.

Other important cytoplasmic markers (see Fig. 2) are the Arg–Ala bonds at residues 198–199 (site T_2) and at 505–506 (site T_1), where rapid tryptic cleavage occurs, yielding linear fragments of M_r 22,000, 33,000, and 55,000 daltons (31,46,52). Since each of these tryptic fragments contains transmembrane segments and each begins and ends in the cytoplasm, we can, again, assume an even number of transmembrane passages for each segment.

Active sites can also be associated with specific, cytoplasmically localized amino acid residues. Aspartic acid-351 is the site at which the enzyme is phosphorylated in a

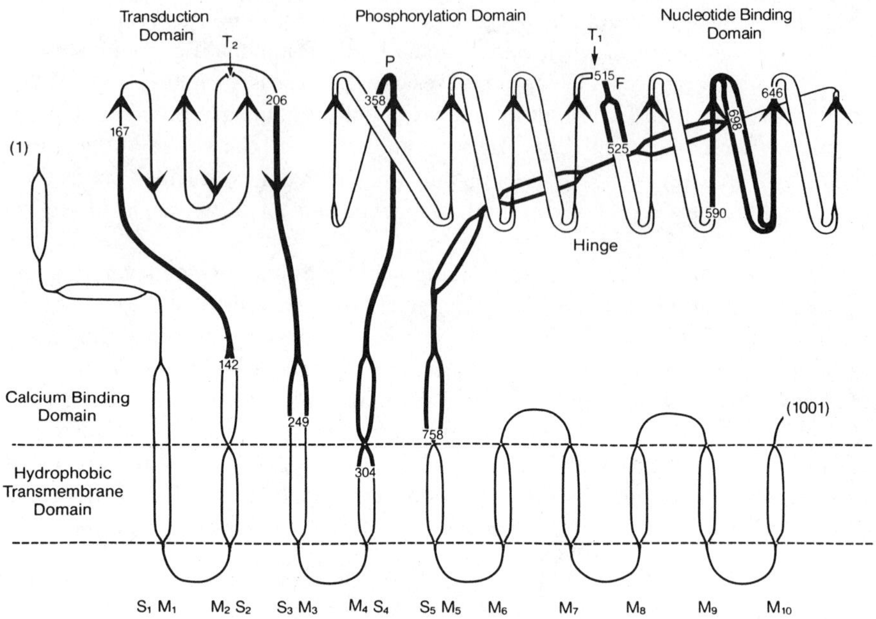

Figure 2. Model of Ca^{2+}–ATPase based on predicted secondary structure and on molecular shape (29). The molecule has been partly unfolded to give a planar layout and the scale of the membrane and stalk helices is larger than that of the helices in the globular domains. The main extramembranous domains are the nucleotide binding domain (parallel β sheet) hinged to the phosphorylation domain (parallel β sheet) and the transduction domain (antiparallel β sheet). These domains are linked to the membrane by two pairs of helices that, together with S_1 from the amino terminus, form the Ca^{2+} binding stalk. Ten helices (M_1–M_{10}), in an arrangement yet to be determined, form the transmembrane channel. The topology of the connections between helices and β strands is arbitrary since we have not attempted to predict the exact folding pattern. The three domains are likely to interact in a trigonal fashion, consistent with the electron micrographs of regular arrays (50). The heavy lines in the diagram show the segments of the sequence that show homology with other ion-transporting ATPases.

partial step of the reaction cycle (1). Lysine-515 is the site at which FITC binds to the enzyme, inhibiting ATP binding (32). Thus the adenine portion of ATP is bound to the enzyme at a site far distant in the linear chain from the site at which it phosphorylates the enzyme.

Sites of Ca^{2+} binding have not been precisely defined. The intact enzyme will induce an increase in Ca^{2+} conductance when incorporated into a planar phospholipid bilayer (44). This activity, in modified form, was traced to an *N*-terminal CnBr fragment of about 13,000 daltons (residues 2–126) (30) Dicyclohexylcarbodiimide (DCCD) binds to the 22,000-dalton *N*-terminal fragment, inhibiting ATPase activity (36). Since DCCD binding is observed only in the presence of the Ca^{2+} chelator EGTA, it is assumed that the two ligands compete for the same site and that Ca^{2+} binds in the *N*-terminal portion of the molecule.

We began our analyses of the secondary and tertiary structure of the ATPase by searching for potential transmembrane segments. Although we found five regions with low polarity indices over about 21 amino acid residues in the *N*-terminal quarter of the molecule (29), we knew that an even number of passages was mandatory. Accordingly, we predicted that four of these regions would form hydrophobic hairpin loops, and we integrated the fifth into a very plausible globular cytoplasmic domain made up of seven β strands (29). The central half of the molecule was relatively polar, but the *C*-terminal quarter was very nonpolar. We were aided in our prediction of where the transmembrane segments were located by at least three considerations: (1) we knew that the number of passages should be even; (2) we knew that Cys-876 and Cys-888 were likely to be involved in disulfide bond formation (14) and were, therefore, unlikely to be found in the cytoplasm, where the reducing potential is high; and (3) we had made site-directed polyclonal antibodies against several segments of the ATPase (J. Talbot and E. Zubrzycka, unpublished observations). Reaction of these antibodies with the ATPase in intact sarcoplasmic reticulum guided us in our topographical modelling. On the basis of these findings we have deduced that there are six transmembrane passages in the carboxyl terminus (5,29). These transmembrane regions are indicated as M_1–M_{10} in Fig. 2. Eight of these passages, M_1/M_2, M_3/M_4, M_5/M_6, and M_9/M_{10}, would form hydrophobic hairpin loops, while M_7/M_8 would contain a longer, luminal loop of about 30 amino acids that would include the Cys residues that are probably involved in disulfide bond formation. By analogy with the structure of transmembrane segments in rhodopsin (17), we predict that each of these transmembrane passages will be α helical.

Further analysis of the secondary structure was carried out by use of the predictive alogrithms of Garnier et al. (11), and these were refined by using the supersecondary predictions of Taylor and Thornton (51), which resolve some of the ambiguities in priority of strand and helix by comparing patterns of hydrophobicity and supersecondary structure with known structures. From these secondary structure analyses, several domains emerged, and we have used rather subjective interpretation of these domains to predict the tertiary structure of the enzyme (Fig. 2).

Perhaps the most interesting domain structure to emerge was that created by segments of the chain running into and out of transmembrane segments M_1 through M_5. In each case these segments formed α helices about 21 residues (2.8–3.1 nm) long, either continuous with the transmembrane helices or, in some case, separated by residues that would form bend or coil structures. We referred to this group of helices as *stalk sectors* S_1–S_5. Many of these sectors were highly charged, particularly at the membrane interface, but those that were charged formed amphipathic helices with charges aligned on one face of the helical structure. Negative staining of the enzyme (12) has shown that its cytoplasmic projections are composed of a narrow stalk connecting membranous domains to a headpiece, perhaps composed of several globular domains. We deduced that the five α helices projecting into the cytoplasm from transmembrane segments M_1–M_5 would constitute a narrow stalk structure about 3.0 nm long. We also proposed that the high density of negatively charged residues lying at the interface with the hydrophobic sector would constitute components of the Ca^{2+} binding sites in the amino terminus of the molecule.

The headpiece appeared to be made up of three additional cytoplasmic domains. The first of these, some 107 residues long, lay between S_2 and S_3 and contained the trypsin-sensitive site, T_2. We deduced that this segment was made up of β strands that would fold into a very plausible antiparallel β sandwich domain. The function of this domain is not clear, although we have referred to it as the transduction domain (29).

The segment lying between S_4 and S_5 is about 410 residues long. We predicted its structure to be a series of nine alternating β strands and α helices (29). General folding patterns for alternating α–β structures have been shown to form parallel β sheets in the crystal structures of several proteins (40). Of particular interest was the folding pattern for kinases such as phosphoglycerate kinase and hexokinase, which fold into two domains (3). These domains typically interact by movement about a double hinge constituted from a connecting bridge in the center of the molecule and an α-helical tail at the end of the second domain that crosses over from the second domain to rejoin the first domain. Such a structure was felt to be plausible for the central cytoplasmic ATPase sequence (residues 330–740). We know that the exposed tryptic site T_1 (residues 505 and 506) lies near the center of this sequence and that Lys-515 involved in ATP binding is only 10 residues downstream. Clearly, this exposed segment could constitute the first of two bridging structures that would connect two globular domains, each composed of parallel β sheets. At the end of the second domain we found a series of three α helices (residues 680–740) that could form the second hinge structure if they were to rejoin the first domain.

Our model for this large central cytoplasmic domain was that of a typical kinase structure. In our model, however, the kinase was bound through stalk segments to a transmembrane channel, and these stalk segments would be able to harness the energy released on enzyme phosphorylation and funnel it into gating and translocation steps required for transmembrane ion transport (whether membranous ion transport enzymes evolved by binding a kinase to a membrane channel or whether kinases evolved following their release from ancient membrane ion transport enzymes will eventually have to be dealt with by molecular evolutionists). In our model, ATP bound to one domain (the nucleotide binding domain) would phosphorylate an Asp residue on another domain (the phosphorylation domain). Although the analogy with soluble kinases is clear, the kinase function is different. Most kinases catalyze phosphorylation of small molecules or endogenous or exogenous phosphorylation of Thr, Ser, or Tyr residues on proteins. Transport enzymes, however, catalyze phosphorylation of only a single Asp residue within their own structures. The understanding that this is so might permit the design of experiments in which the central kinase domain can be cleaved from the membrane and analyzed as a soluble enzyme with a defined function.

Since the Na^+, K^+–ATPase (22,45) and the H^+–ATPase (43) have sequence homology with the Ca^{2+}–ATPase, it was of interest to see how the homologies related to our structural predictions. We found six regions of high amino acid sequence homology among these proteins (Fig. 3). Four of these regions, those

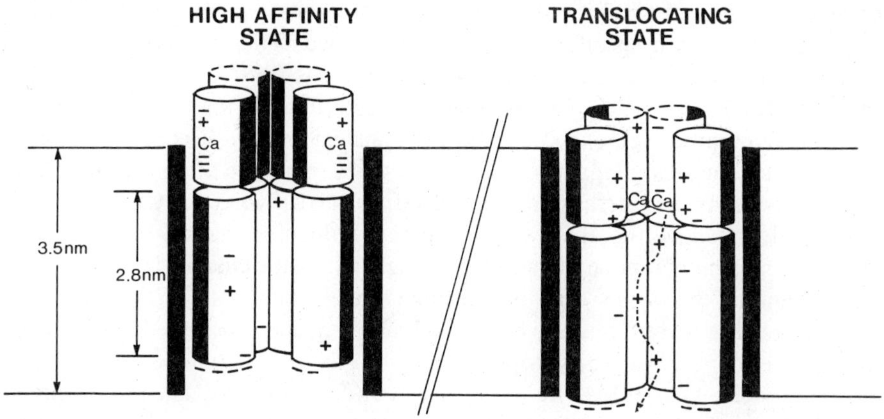

Figure 3. Model illustrating Ca^{2+} gating, perpendicular penetration of the ATPase into the membrane, and the translocation of Ca^{2+}. In the E_1 state the relatively short hydrophobic segments of the transmembrane sectors of the ATPase are suspended within a longer hydrophobic phospholipid bilayer. This configuration is stabilized by negative charges at either end that provide a counterbalance against movement in either direction. Here, Ca^{2+} is bound to the polar surface of amphipathic stalk sectors near the cytoplasmic membrane boundary. Hydrolysis of ATP in cytoplasmic domains leads to rotation of stalk sectors, gating of Ca^{2+} and exposure of hydrophobic surfaces of the stalk. The stalk moves passively into the membrane, and a combination of perpendicular and rotational movements in the predominantly hydrophobic, transmembrane sector aligns charges creating a transmembrane channel to which Ca^{2+} is now accessible; Ca^{2+} traverses the membrane by an ion-exchange process involving fixed, aligned charges.

extending from either side of the transduction domain into the tops of S_2 and S_3, and those extending from the phosphorylation and nucleotide binding domains deep into S_4 and S_5, were the structures most likely to funnel energy, released in the cytoplasmic domains, into the stalk and membranous segments that are directly involved in Ca^{2+} translocation. Regions of homology also included the phosphorylation site and sequences downstream of Lys-515 that may be involved in ATP binding. (There is little homology immediately upstream of Lys-515, even between the two Ca^{2+}–ATPases that we have analyzed.) The sixth region of homology lay in the middle of the nucleotide binding domain.

If our postulate is correct, specifically, that a kinase has been harnessed to provide the energy for ion translocation across a transmembrane channel, then these regions of homology among ATPases should be viewed in relation to their function in phosphorylation and energy transduction. The homologies that we have observed would suggest that the essential, conserved elements for the kinase are the phosphorylation site, the ATP binding site, and a third region that may well be involved in active-site formation. They would also suggest that the essential features of the structures that transmit energy from the cytoplasmic domains to the transmembrane domains must also be highly conserved. The corollary is that much of the precise sequence of the α–β and β structures of the cytoplasmic globular domains is not very significant, provided the overall structure is maintained. A different conclusion may hold for the stalk and transmembrane segments, however. It is in these regions that ion specificity is probably defined. Thus the lack of sequence homology in S_1 and S_2 stresses the importance of these regions for Ca^{2+} selectivity as opposed to Na^+ or K^+ selectivity. Similarly, the lack of homology in transmembrane regions may point to their specificity for the translocation of different cations (13).

The transmembrane domain was, in our analysis, made up of 10 transmembrane segments. Since five of these were attached to five stalks involved in Ca^{2+} binding, it was logical to believe that they would also be involved in channel formation. The additional transmembrane segments were thought to line the channel or be in a second tier surrounding the primary tier of transmembrane sequences.

An interesting feature of these transmembrane sequences was the high density of charges on either side of the hydrophobic sector. The M_1 sequence was preceded by the sequence –Glu–Gln–Phe–Glu–Asp– and succeeded by the sequence –Glu–Glu–Gly–Glu–Glu. The sequence also contained several charged residues within the predominantly hydrophobic regions. We postulated that there are nine positively charged residues (Arg, Lys, His) and seven negatively charged residues (Glu, Asp) within 10 transmembrane sequences. These polar residues were thought to be significant in the translocation of ions. A final interesting point was that many of the hydrophobic sequences were short. They extended for as little as 19 residues, which, in an α helix, might span only about 2.85 nm. Since the width of a phospholipid bilayer is believed to be about 3.5 nm, the hydrophobic stretches in the protein may not traverse the full width of the phospholipid bilayer.

3. MECHANISM OF ACTION

A major objective in all our studies has been to relate structural features of the ATPase molecule to the function of Ca^{2+} transport. Accordingly, recognition of important sites in the primary, secondary, and tertiary structure of the enzyme presented us with the new challenge of describing some of the mechanistic implications of the structural features that we had uncovered (5,13).

The major structural features of interest for the understanding of Ca^{2+} translocation were the Ca^{2+} binding sites in the amphipathic stalk helices. These would appear to lie at the boundary between the cytoplasm and the hydrophobic bilayer. These sites are of high affinity, but they were clearly not of the "*E–F* hand" configuration that has been associated with high-affinity Ca^{2+} binding to nontransport proteins (25). This appears reasonable since *E–F*-hand structures are stable and of constant affinity, whereas in the transport cycle Ca^{2+} is bound initially to sites of high affinity, but these sites disappear after the enzyme is phosphorylated. Until the enzyme is dephosphorylated, only low-affinity Ca^{2+} binding sites can be detected (53). These observations imply that the high-affinity Ca^{2+} binding sites are dependent on protein configuration and that an oscillation between high and low affinities can be achieved through alterations of configurational states.

The reaction cycle for Ca^{2+} translocation begins with binding of two Ca^{2+} and one ATP to high-affinity sites. This is followed by a rapid, Ca^{2+}-dependent phosphorylation of Asp-351 and a step usually described as occlusion (9,48), in which the rate of dissociation of bound Ca^{2+} is decreased. The central reaction is an energy-transducing step in which two Ca^{2+} are transferred from sites of high affinity to sites of low affinity in a process that is coupled to a decrease in the energy conserved in the aspartyl phosphate bond (49). The energy transduction step is followed by dissociation of Ca^{2+} from internal low-affinity sites, dephosphorylation, and regain of the E_1 configuration.

We assumed that the Ca^{2+} binding sites were made up of two or more chains that could rotate relative to each other (5). In one configuration the high-affinity sites would be formed; in another, energy-requiring step they would be disrupted. An intermediate configuration, brought about in the initial enzyme phosphorylation but requiring minimal energy input, would describe the occluded state. We further assumed that the sites were made up from the amphipathic stalk helices in which the high-affinity Ca^{2+} binding sites were exposed to the cytoplasm and their hydrophobic faces were sequestered in the interior of the cluster of helices (see Fig. 3). We predicted that, on utilization of the energy conserved in the aspartyl phosphate bond in globular cytoplasmic domains, the highly conserved sequences leading from these globular domains into the stalk sectors would rotate. On reaching the stalk sectors, these rotations would bring Ca^{2+} into the interior of the cluster of helices and, at the same time, lower its binding affinity. As a result, Ca^{2+} would be released into a cage that would not be accessible to the cytoplasm but would be accessible to a transmembrane channel.

The rotation of the amphipathic stalk helices would expose their hydrophobic faces to the cytoplasm. This would be energetically unfavorable and create a driving force for movement of the stalk sectors into the membrane. We then attempted to determine whether there was any reason to believe that perpendicular movements of the ATPase would occur in the reaction cycle. Blasie and his colleagues (4) have carried out detailed x-ray and neutron diffraction analysis of the relation of the mass of the ATPase molecule to the phospholipid bilayer during a single turnover induced by release of caged ATP. They have found that the enzyme becomes thinner and that about 8% of its mas moves into the membrane during the translocation cycle. Highsmith and Scales (18) and Peracchia et al. (35) have shown by freeze-fracture studies, and Hasselbach et al. (16) have shown by sectioning, that in two-dimensional, vanadate-induced crystals the ATPase molecules are more deeply buried in the bilayer than they are prior to crystallization. Since the vanadylated form of the enzyme is believed to correspond to the E_2P or translocating state of the enzyme, it is implied that movement of the enzyme perpendicular to the plane of the bilayer does occur during the reaction cycle.

In Fig. 3 we have provided a diagrammatic model of how Ca^{2+} might bind to the cytoplasmic surface of the stalk sectors and be translocated across a barrier in this sector by helix rotation induced by enzyme phosphorylation. The analogy of this model with that of a gated channel is obvious. Much of the energy of ATP hydrolysis could be utilized in rotation of stalk sectors which would be the essential component of this gating mechanism.

Figure 3 also presents our ideas concerning perpendicular movements of the channel-forming sectors and of the structure of the channel itself. The hydrophobic sector of the transmembrane helices would be relatively short, about 2.8 nm long, and highly charged on either end. By contrast, the hydrophobic region of the phospholipid bilayer might be about 3.5 nm wide. We have suggested (5) that the transmembrane helix might position itself in the center of the bilayer region with the charged sequences, providing a counterbalance against movement in either direction. Thus the symmetrical distribution of charges across the transmembrane helix might provide a plausible explanation for the incomplete passage of the transmembrane segment across the membrane that has been described in morphological studies (16,18,35).

Exposure of the hydrophobic faces during stalk sector rotation would extend the hydrophobic regions, previously buried or confined to the transmembrane sectors, into parts of the stalk sectors. The stalks might now be driven into the hydrophobic region of the phospholipid bilayer in a search for a more favorable environment. Neither this perpendicular movement nor its reversal would require direct energy input; the energy utilization would be concentrated on rotation of stalk helices and on its accompanying gating mechanism.

We do not believe that penetration of stalk sectors into the membrane would be a very important feature in the actual translocation of Ca^{2+} across the membrane since the same 2.8-nm sector of low polarity would still have to be traversed. We suspect that intramembranous realignments of transmembrane chains occurring during the perpendicular movements would either create an internal Ca^{2+} channel from previously densely packed sequences or, alternatively, align those charges on the interior of the channel that might be essential for efficient translocation of Ca^{2+} (5). As we pointed out earlier, the transmembrane segments contain about equal numbers of positive and negative charges. Therefore, it is plausible to think that they might align in a translocation channel to create a charge matrix with overall neutrality. Then Ca^{2+} could move across the hydrophobic sector by an ion-exchange mechanism involving sequential binding to and replacement from fixed negative charges found at different locations within the channel.

4. CONCLUSION

As we have demonstrated in this chapter, we have been able to gain new insights into the biogenesis, structure, and function of the Ca^{2+}–ATPase through the use of a remarkable technology, that of molecular cloning. Since the boundaries of this technology are rapidly expanding, it will be of great interest to observe the expansion of our knowledge of ion-transport enzymes that will occur as the technology is fully exploited.

ACKNOWLEDGMENTS

Original work described in this chapter was supported by grants to D. H. MacLennan from the Ontario Heart Foundation, The Muscular Dystrophy Association of Canada (MDAC), and the Medical Research Council of Canada (MRCC). C. J. Brandl was a predoctoral Fellow of the MRCC; B. Korczak was a postdoctoral fellow of the MDAC. The work of N. M. Green was supported by the Medical Research Council (UK).

REFERENCES

1. Allen, G, and Green, N. M. (1976). *FEBS Letts.*, **63**, 188–192.
2. Allen, G., Trinnaman, B. J., and Green, N. M. (1980). *Biocehm. J.*, **187**, 591–616.
3. Bennett, W. S. and Huber, R. (1984). *CRC Crit. Rev. Biochem.*, **15**, 291–384.
4. Blasie, J. K., Herbette, L. G., Pascolini, D., Skita, V., Pierce, D. H., and Scarpa, A. (1985). *Biophys. J.*, **48**, 9–18.
5. Brandl, C. J., Green, N. M., Korczak, B., and MacLennan, D. H. (1986). *Cell.*, **44**, 597–607.
6. Breitbart, R. E., Nguyen, H. T., Medford, R. M., Destree, A. T., Mahdavi, V., and Nadal-Ginard, B. (1985). *Cell*, **41**, 67–82.
7. Briggs, F. N., Cable, M. B., Geisow, and Green, N. M. (1986) *Biochem. Biophys. Res. Commun.*, **135**, 864–869.
8. de Meis, L. and Vianna, A. L. (1979). *Ann. Rev. Biochem.*, **48**, 275–292.
9. Dupont, Y. (1980). *Eur. J. Biochem.*, **109**, 231–238.
10. Ebashi, S. and Lipmann, F. (1962). *J. Cell Biol.*, **14**, 389–400.
11. Garnier, J., Osguthorpe, D. J., and Robson, B. (1978). *J. Molec. Biol.*, **120**, 97–120.
12. Greaser, M. L., Cassens, R. G., Hoekstra, W. G., and Briskey, E. J. (1969). *J. Cell Physiol.*, **74**, 37–50.
13. Green N. M., Taylor, W. R., Brandl, C. J., Korczak, B., and MacLennan, D. H. (in press). In *Calcium and the Cell*, CIBA Foundation Symposium No. 122, Pitman, London.
14. Green, N. M. and Toms, E. J. (1985). *Biochem. J.*, **231**, 425–429.
15. Hasselbach, W. and Makinose, M. (1961). *Biochem. Z.*, **333**, 518–528.
16. Hasselbach, W., Medda, P., Migala, A., and Agostini, B. (1983). *Z. Naturforsch.*, **38C**, 1015–1022.
17. Henderson, R. and Unwin, P. N. T. (1975). *Nature*, **257**, 28–32.
18. Highsmith, S. R. and Scales, D. (1984). *Z. Naturforsch.*, **39C**, 177–179.
19. Inesi, G. (1985). *Ann. Rev. Physiol.*, **47**, 573–601.
20. Jorgensen, A. O. and Campbell, K. P. (1986). *Biophys. J.*, **49**, 589a.
21. Jorgensen, A. O. and Jones, L. R. (1986). *J. Biol. Chem.*, **261**, 3775–3781.
22. Kawakami, K., Noguchi, S., Noda, M., Takahashi, H., Ohta, T., Kawamura, M., Nojima, H., Nagano, K., Hirose, T., Inayama, S., Hayashida, H., Miyata, T., and Numa, S. (1985). *Nature*, **316**, 733–736.
23. Kielley, W. W. and Meyerhof, O. (1948). *J. Biol. Chem.*, **176**, 591–601.
24. Korczak, B., Zarain-Herzberg, A., Brandl, C. J., Ingles, C. J., Green, N. M., and MacLennan, D. H. (in press). *Proc. Natl. Acad. Sci.* (USA).
25. Kretsinger, R. H. (1976). *Ann. Rev. Biochem.*, **45**, 239–266.
26. MacLennan, D. H. (1970). *J. Biol. Chem.*, 245, 4508–4518.

27. MacLennan, D. H. and Holland, P. C. (1976). *Ann. Rev. Biophys. Bioeng.*, **4**, 377–404.

28. MacLennan, D. H. and Reithmeier, R. A. F. (1982). In Martonosi, A. N., Ed., *Membranes and Transport*, Vol. 1, Plenum, New York, pp. 567–571.

29. MacLennan, D. H., Brandl, C. J., Korczak, B., and Green, N. M. (1985). *Nature*, **316**, 696–700.

30. MacLennan, D. H., Reithmeier, R. A. F., Shoshan, V. Campbell, K. P., LeBel, D., Herrmann, T. R., and Shamoo, A. E. (1980). *Ann. N. Y. Acad. Sci.*, **358**, 138–148.

31. Migala, A., Agostini, B., and Hasselbach, W. (1973) *Z. Naturforsch.*, **28**, 178–182.

32. Mitchinson, C., Wilderspin, A. F., Trinnaman, B. J., and Green, N. M. (1982). *FEBS Lett.*, **146**, 87–92.

33. Nabeshima, Y., Fujii-Kuriyama, Y., Muramatsu, M., and Ogata, K. (1984). *Nature*, **308**, 333–338.

34. Okayama, H. and Berg, P. (1983). *Molec. Cell Biol.*, **3**, 280–289.

35. Peracchia, C., Dux, L., and Martonosi, A. N. (1984). *J. Muscle Res. Cell Motil.*, **5**, 431–442.

36. Pick, U. and Racker, E. (1979). *Biochemistry*, **18**, 108–113.

37. Racker, E. (1972). *J. Biol Chem.*, **247**, 8198–8200.

38. Reithmeier, R. A. F. and MacLennan, D. H. (1981). *J. Biol. Chem.*, **256**, 5957–5960.

39. Reithmeier, R. A. F., de Leon, S., and MacLennan, D. H. (1980). *J. Biol. Chem.*, **255**, 11839–11846.

40. Richardson, J. S. (1981). *Adv. Protein Struct.*, **34**, 167–339.

41. Rizzolo, L. J. and Tanford, C. (1978). *Biochemistry*, **17**, 4044–4048.

42. Sanger, F., Nicklen, S., and Coulson, A. R. (1977). *Proc. Natl. Acad. Sci.* (USA), **74**, 5463–5467.

43. Serrano, R., Kielland-Brandt, M. C., and Fink, G. R. (in press) *Nature*.

44. Shamoo, A. E. and MacLennan, D. H. (1974). *Proc. Natl. Acad. Sci.* (USA), **71**, 3522–3526.

45. Shull, G., Schwartz, A., and Lingrel, G. (1985). *Nature*, **316**, 691–695.

46. Stewart, P. S. and MacLennan, D. H. (1974). *J. Biol. Chem.*, **249**, 985–993.

47. Tada M. and Katz, A. M. (1982). *Ann. Rev. Physiol*, **44**, 401–423.

48. Takisawa, H. and Makinose, M. (1983). *J. Biol. Chem.*, **258**, 2986–2992.

49. Tanford, C. (1984). *FEBS Lett.*, **166**, 1–7.

50. Taylor, K., Dux, L., and Martonosi, A. N. (1984). *J. Molec. Biol.*, **174**, 193–204.

51. Taylor, W. R. and Thornton, J. M. (1984). *J. Molec. Biol.*, **173**, 487–514.

52. Thorley-Lawson, D. A. and Green, N. M. (1973). *Eur. J. Biochem.*, **40**, 403–413.

53. Yamada, S. and Tonomura, Y. (1972). *J. Biochem.* (Tokyo), **72**, 417–425.

54. Zubrzycka-Gaarn, E., MacDonald, G., Phillips, L., Jorgensen, A. O., and MacLennan, D. H. (1984). *J. Bioenergetics, Biomembranes*, **16**, 441–464.

CHAPTER 17

ISOLATION AND CHARACTERIZATION OF A cDNA FOR THE CATALYTIC SUBUNIT OF THE (Na^+ + K^+)–ATPase

GARY E. SHULL
JERRY B. LINGREL
Department of Microbiology and Molecular Genetics
University of Cincinnati, College of Medicine
Cincinnati, Ohio

UNIVERSITY COLLEGE CARDIFF LIBRARY

The $(Na^+ + K^+)$–ATPase enzyme, discovered by Skou in 1957 (40), is responsible for establishing the sodium and potassium gradients across the plasma membrane of animal cells. It is composed of two subunits, a larger catalytic subunit termed α, with a molecular weight of 110,000, and a smaller glycoprotein subunit termed β, with a molecular weight of about 55,000 (8). During its catalytic cycle the enzyme hydrolyzes one molecule of ATP while transporting three Na^+ out of the cell and two K^+ into the cell. The ion gradients it produces are essential for the maintenance of cell volume (27) and the transport of a variety of organic and inorganic molecules (44). The $(Na^+ + K^+)$–ATPase is also responsible for establishing and maintaining membrane potentials, which are of particular importance in electrically excitable tissues such as those of the brain and heart (15).

Despite the central role of $(Na^+ + K^+)$–ATPase in animal metabolism and physiology, very little of its primary sequence has been determined (4,7,18,31). To obtain this information, we isolated a cDNA clone for the catalytic subunit of sheep kidney $(Na^+ + K^+)$–ATPase (39). In this chapter we present the procedures used to construct and screen the cDNA library and a discussion of the structural and functional domains of the protein.

1. CONSTRUCTION AND SCREENING OF cDNA LIBRARY

Over the past several years the procedures for constructing and screening cDNA libraries have improved steadily, and many of the steps that were difficult to control have been replaced by relatively straightforward procedures. These advances in technology, and the abundance of high-quality, commercially available products, have simplified the task of cDNA cloning to the point where these procedures are now accessible to any laboratory dealing with protein structure and function. Nowhere is this more evident than in the study of membrane-associated proteins, which is currently witnessing an explosion in the availability of complete primary sequence data for various receptors (42,43), channels (32), and transport proteins (17,23,28,39). However, it is often very difficult for investigators who have had no prior experience with cDNA cloning to select the most appropriate procedures for the particular application they have in mind. In cloning the catalytic subunit of $(Na^+ + K^+)$–ATPase we have utilized procedures that are currently the most suitable for isolating cDNAs of proteins for which some amino acid sequence data are available. In addition, we have devised a simple and very efficient cDNA size selection procedure that allows the cloning of cDNAs within a particular size range, thereby greatly reducing the number of colonies that must be screened in order to obtain a full-length copy of the cDNA of interest. Here we describe the procedures we used in the construction and oligonucleotide screening of a sheep kidney cDNA library.

1.1 Isolation of Messenger RNA

One of the most critical steps in cDNA cloning is the isolation of intact messenger RNA. This is particularly true when the protein of interest is encoded by a large mRNA, as in the case of $(Na^+ + K^+)$–ATPase, because the larger mRNAs are more easily degraded. In order to minimize such degradation, sheep kidney was removed and frozen in liquid nitrogen immediately after the animal was sacrificed. Tissue can be stored in this manner for extended periods of time with no evidence of mRNA degradation. Total cellular RNA was extracted basically according to the procedure of Chirgwin et al. (6) with minor modifications. The original procedure involved homogenizing tissue in guanidinium monothiocyanate solution containing 1 *M* β-mercaptoethanol followed by several steps in which a 7 *M* guanidine–HCl solution was used. Although guanidinium monothiocyanate should be used for tissues that have a high nuclease content, for most tissues the guanidine–HCl solution is quite satisfactory for the initial homogenization and is much easier to work with. Ten gram of frozen tissue was broken into small chunks and homogenized with a Polytron in the guanidine–HCl solution as described elsewhere (6). After three precipitations from guanidine–HCl the RNA was dissolved in 20 ml of 20 m*M* Tris, pH 7.6, 1 m*M* EDTA, 0.5% sarkosyl containing 100 μg Proteinase K/ml and incubated at 37°C for 1 hr. The sample was extracted one time with a 1:1 mixture of phenol and chloroform and one time with chloroform and then precipitated by the addition of 1/10 volume of 2 *N* potassium acetate, pH 5.0, and 2 volumes of ethanol. Poly A^+ RNA was obtained by two rounds of affinity chromatography on oligo(dT) cellulose basically as described by Maniatis et al. (29). However, we included 1% SDS and 5 m*M* EDTA in all buffers, as this appears to aid in the recovery of high-molecular-weight mRNA. Samples were precipitated from ethanol four times to remove all traces of SDS. The mRNA was dissolved in 0.1 *N* sodium acetate, pH 7.0, during these washes. After the last precipitation the mRNA was dried under vacuum, dissolved in water, and stored at −70°C.

1.2 Preparation of cDNA

The procedures used to prepare the sheep kidney cDNA are illustrated in Fig. 1. First- and second-strand cDNA syntheses were performed as described by Gubler and Hoffman (14). In the first-strand reaction 7.5 μg of mRNA was reverse transcribed in a reaction volume of 50 μl to yield 1.8 μg of single-stranded cDNA. A trace of (^{32}P)dCTP was included in order to determine the yield and analyze the length distribution of the cDNA. Only 1/15,000 dCTP molecules were labeled, so degradation of cDNA due to incorporation and later decay of isotope was negligible. After a 30-min incubation at 43°C, the reaction was stopped by the addition of 4 μl of 0.5 *M* EDTA, pH 8.0, and 3 μl of 10% SDS. A 2-μl aliquot was removed and used to determine the specific activity. The sample was extracted twice with a 1:1 mixture of

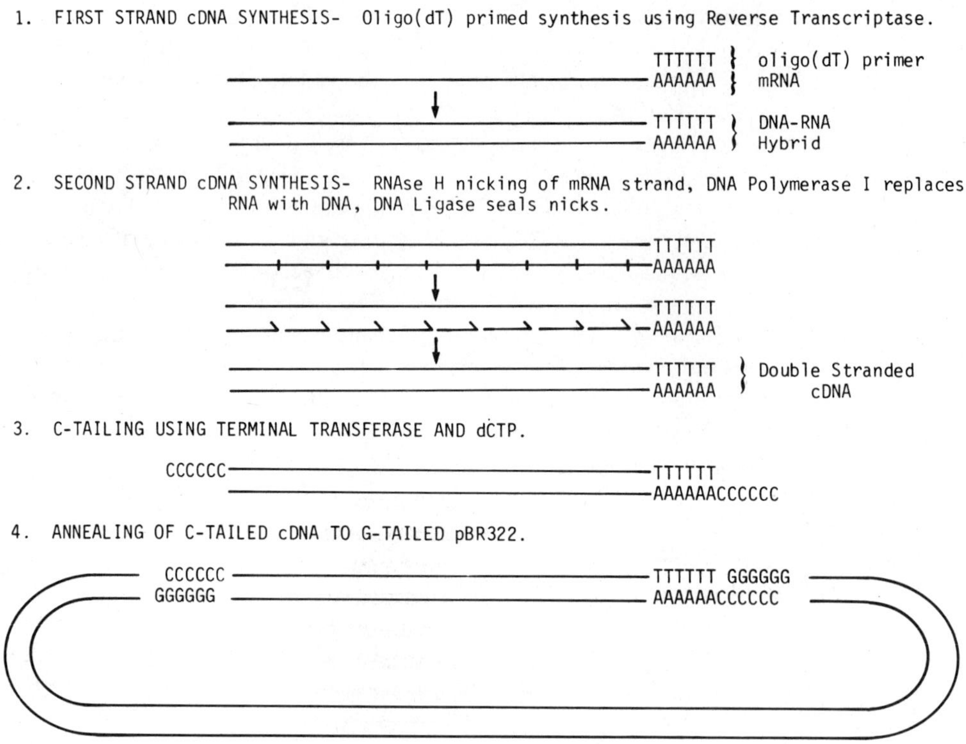

Figure 1. Schematic illustration of the procedures used to prepare the sheep kidney cDNA. First- and second-strand syntheses were performed basically as described by Gubler and Hoffman (14), and the C-tailing and annealing procedures were similar to those described by Peacock et al. (34) with minor modifications (see text).

phenol and chloroform, ammonium acetate (10 *M* stock) was added to a concentration of 2 *M*, and the nucleic acids precipitated by addition of 2 volumes of ethanol. (See ref. 33 for a detailed description of the extraction and precipitation procedure, which allows efficient removal of unincorporated nucleotides.) After centrifugation for 10 min in a microfuge the sample was dissolved in 10 m*M* Tris, 1 m*M* EDTA, pH 7.6 (TE), and precipitated a second time by the addition of ammonium acetate and ethanol. The pellet was rinsed with 70% ethanol containing 30 m*M* NaCl, dried, and dissolved in 100 μl of TE. A 2-μl aliquot was removed and used to determine the cDNA recovery and size.

The size distribution of the first strand was determined (Fig. 2) by analysis of the products on a denaturing gel as described by Maniatis et al. (29). It is essential to analyze the length of the first-strand cDNA because there is considerable variation in the efficiency of commercially available reverse transcriptase preparations, and in our experience they vary from lot to lot even when obtained from the same company.

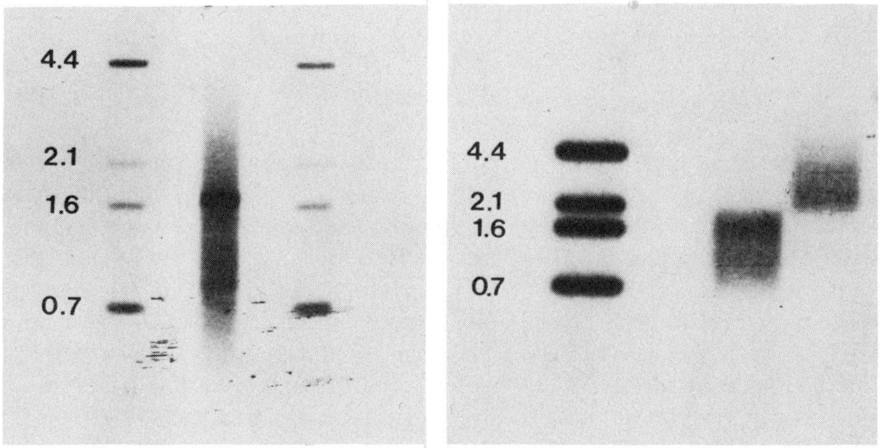

Figure 2. The cDNA size analysis. The products of the first-strand cDNA synthesis (left panel) were separated on an alkaline agarose gel (29), and the size distribution of the cDNA (center lane) was determined by autoradiography. The size markers are end-labeled restriction fragments of pBR322. The right panel shows the size distribution of the two double-stranded cDNA fractions isolated by capture on DEAE paper. The larger fraction was used to prepare the library.

The second-strand synthesis was performed with 1.5 μg of single-strand cDNA (3 μg of DNA–RNA hybrid) in a reaction volume of 300 μl; RNAse H is included in the reaction mix in order to produce nicks in the RNA of the DNA–RNA hybrid (Fig. 1). Deoxyribonucleic acid polymerase I initiates synthesis at each of these nicks, utilizing the RNA as a primer, and DNA ligase is included to join the newly synthesized fragments of DNA. At the end of the incubation period 30 μl of 10% SDS was added, and the sample was extracted three times with phenol–chloroform and precipitated once with ammonium acetate and ethanol.

Following the second-strand synthesis the double-stranded cDNA was size-fractionated (described in detail later), and the 2100–4500-base pair fraction was tailed with deoxycytidine (34). The tailing reaction was performed at 37°C in a volume of 100 μl containing 200 ng of cDNA, 5 μ*M* dCTP, 20 units of terminal transferase, and the buffer supplied with the enzyme (Bethesda Research Laboratories). The reaction mixture was incubated for 5 min, then stopped by the addition of 5 μl of 5% SDS, 100 m*M* EDTA, and extracted three times with phenol–chloroform. The sample was precipitated with ammonium acetate and ethanol twice, rinsed with 70% ethanol containing 30 m*M* NaCl, dried, and dissolved in water at 4 ng/μl.

The final step in the cDNA preparation procedure is the annealing of the C-tailed cDNA to G-tailed Pst I cut pBR322; G-tailed vector that gives very low background is commercially available from a number of sources. Both the cDNA and vector are dissolved in annealing buffer (10 m*M* Tris, pH 7.5, 1 m*M* EDTA, 150 m*M* NaCl) at 2 ng/μl, mixed at a 1:1 mole ratio, and incubated for 2 hr at 58°C (34). The annealed cDNA can be used immediately or frozen until needed.

1.3 cDNA Size-Fractionation Procedure

The need for a size-fractionation procedure is readily apparent when one examines the size distribution of the first-strand cDNA shown in Fig. 2. The intensity of the exposure reflects the mass distribution of the cDNA. The mass average is below 1.5 kilobases, much less than the 3.7-kilobase full-length (Na^+ + K^+)–ATPase cDNA. The number average of the cDNA distribution is even less favorable for the cloning of a large cDNA. For example, the intensity is approximately the same at 0.7 kilobase as at 2.1 kilobases, but this means that there are about three times as many 700-base molecules as there are 2100–base molecules. In the absence of a stringent size selection step, the library would consist primarily of short transcripts. Although it is possible to size-fractionate mRNA prior to cDNA synthesis, there is little advantage in doing this because the library would still contain many short transcripts of the longer mRNAs. Furthermore, mRNA degradation may occur during the procedure. A more satisfactory solution is to size-fractionate the cDNA.

The size-fractionation procedure we used involves the capture of the appropriate cDNA fraction on DEAE paper. First, 3 μg of double-stranded cDNA was dissolved in 40 μl of TE, electrophoresis dye was added, and the sample was loaded on a 0.8% agarose gel containing ethidium bromide. The markers placed in adjacent lanes were restriction fragments of pBR322. An electric current was applied, and the sample was examined periodically with a low-intensity hand-held UV light. After sufficient migration had taken place, the sample lane was cut horizontally with a scalpel at 700- and 2100-base pairs and strips of NA45 paper (Schleicher & Schuell) were inserted. Electrophoresis was then continued until the 700–2100- and 2100–4500-base pair fractions had been captured on the paper. The paper strips were removed and placed in an Eppendorf tube, and the cDNA was eluted at 68°C in 200-μl 50 m*M* arginine (free base), 1 *M* NaCl for 4 hr. Sodium dodecyl sulfate (10 μl) was added to each eluate, which was then extracted four times with phenol–chloroform and precipitated by the addition of 2 volumes of ethanol. After centrifugation in a microfuge the pellets were dissolved in 100 μl of TE and precipitated a second time by addition of ammonium acetate and ethanol. The pellets were rinsed with 70% ethanol containing 30 m*M* NaCl, dried, and dissolved in 25 μl of TE. A 1-μl aliguot was used to determine the radioactivity, and the concentration was then adjusted to 10 ng/μl. The size distribution of the double-stranded cDNA fractions is shown in Fig. 2. The library was prepared by using the 2100–4500-base pair fraction.

1.4 Plating of Library and Preparation of Filters

Escherichia coli RRI cells were transformed with 60-ng cDNA (20 ng of insert annealed to 40 ng of vector) and spread on 15 agar plates (150 mm) containing 10 μg tetracycline/ml. Colonies (approximately 3500/plate) were allowed to grow for 20 hr at 37°C and then transferred to nitrocellulose filters. This is a departure from the usual procedure of spreading the cells directly on a nitrocellulose filter that has been placed on an agar plate (16). We find that the plating efficiency is much better if the cells are spread on agar. Furthermore, after the colonies have been transferred to the nitrocellulose master filter, the agar plate can be placed at 37°C for 8–10 hr to regrow the colonies and then stored at 4°C until needed for retrieval of colonies that give positive signals. To transfer the colonies, a dry, sterile nitrocellulose filter was placed on the agar plate. As soon as it had wetted, the filter was keyed and removed from the plate. These master filters were used to prepare two replicas on Zeta Bind membrane (AMF Cuno) that had been wetted on fresh agar plates containing tetracycline. After replication the nitrocellulose master filters were processed according to the procedure of Hanahan and Meselson (16) and placed at −80°C for long-term storage. Colonies on the replica filters were grown for 10 hr at 37°C on plates containing tetracycline and then transferred to plates containing chloramphenicol (100 μg/ml) and incubated for 18 hr. To process the filters, a row of trays was set up with one sheet of Whatman paper on each and soaked in the appropriate solution. The filters were placed successively for 5 min each on (1) 0.5 *M* NaOH, 0.2% Triton X-100; (2) 0.5 *M* NaOH, 1.5 *M* NaCl; (3) 0.5 *M* Tris, 1.5 *M* NaCl, pH 7.5; and (4) 0.5 *M* Tris, 1.5 *M* NaCl, pH 7.5. They were then submerged in a dish of 0.5 *M* Tris, 1.5 *M* NaCl, pH 7.5, for 5 min. To remove bacterial debris, the filters were soaked in 10 m*M* Tris, pH 7.5, 100 m*M* NaCl, 0.5% SDS, 100 μg/ml Proteinase K for 30 min, followed by washing in 0.5 *M* Tris, pH 7.5, 1.5 *M* NaCl, with frequent changes of solution, for 30–60 min. A final wash was performed in 2× SSC (1× SSC is 0.15 *M* NaCl/0.015 *M* sodium citrate, pH 7.0). Filters were then placed between sheets of Whatman paper and baked for 2 hr at 80°C in a vacuum oven.

1.5 Screening with Oligonucleotide Probes

Approximately 10% of the amino acid sequence of the catalytic subunit of sheep kidney ($Na^+ + K^+$)ATPase was determined previously by Collins et al. (7), and two of the tryptic peptides (T_{14} and T_{16}) they described were suitable for deriving oligonucleotide hybridization probes; T_{14} has been identified (24,10) as the tryptic peptide that binds fluorescein isothiocyanate (FITC) and is thought to form part of the ATP binding site (10). The oligonucleotides were synthesized as a mixture containing all possible anticodon combinations for their amino acid sequences (Fig. 3). The probe derived from T_{14} is a 14 mer, which is a mix of 32 oligonucleotides, and the probe derived from T_{16} is a 20 mer, which is a mix of 256 oligonucleotides. The probes were 5′-end labeled, using T_4 polynucleotide kinase and (γ^{32}P) ATP to a specific activity of approximately 3×10^9 counts per minute (cpm)/μg.

The filters were prepared for hybridization by wetting in 6× SSC and then prehybridized overnight at 68°C in 6× SSC, 1× Denhardt's solution (29), 0.1% SDS, 100 μg salmon sperm DNA/ml. Filters were then hybridized in 60 ml of the same solution containing the 14-mer probe at 10^7 cpm/ml. We used the procedure of Boyd et al. (3) in which the temperature is dropped slowly from 68 to 30°C during the

```
TRYPTIC PEPTIDE 14:  -Val-Met-Lys-Gly-Ala-Pro-Glu-

        CODONS  5'   GUU AUG AAG GGU GCU CCU GAA   3'
                       C       A   C   C   C   G
                       A           A   A   A
                       G           G   G   G

        PROBE   3'       TAC TTC CCA CGA GG        5'
                               T   G   G
                                   T   T
                                   C   C

TRYPTIC PEPTIDE 16:  -Val-Phe-Gln-Ala-Asn-Gln-Asp

        CODONS  5'   GUU UUU CAA GCU AAU CAA GAU   3'
                       C   C   G   C   C   G   C
                       A           A
                       G           G

        PROBE   3'   CAA AAA GTT CGA TTA GTT CT    5'
                       G   G   C   G   G   C
                       T           T
                       C           C
```

Figure 3. Oligonucleotide probes were derived from amino acid sequences of two tryptic peptides of the sheep kidney enzyme (7). The probes are complimentary to the messenger RNA sequences encoding the peptides.

hybridization. This procedure appears to be much more effective than hybridization at a single fixed temperature, possibly because it allows a higher proportion of perfect hybrids to form between the probe and the target sequence. The filters were washed for 1 hr in 6× SSC, 0.1% SDS at 30°C with repeated changes of solution, placed between sheets of Saran wrap mounted on Whatman paper, and autoradiographed overnight. The Zeta Bind filters were not allowed to dry because this would cause the probe to bind irreversibly. However, excess moisture was removed by blotting briefly with Whatman paper before placement on the Saran wrap. After the first round of autoradiography the filters were washed at 35°C for 1 hr, followed by autoradiography, and then washed a third time at 40°C, followed by autoradiography. After each exposure colonies that gave signals on both replicas were marked.

Before probing with the second oligonucleotide the filters were stripped by placing them in 0.2 *N* NaOH, 47°C, for 30 min with continuous shaking followed by 30 min in 0.5 *M* Tris, pH 7.5, 1.5 *M* NaCl at 47°C. They were prehybridized and then hybridized with the 20-mer oligonucleotide at 3×10^7 cpm/ml. Washing and autoradiography were as described above, except that an additional wash was performed at 45°C. The advantage of this wash procedure is that it eliminates any guesswork regarding the annealing temperature of the probe to the target sequence and allows the examination of signals and background before each stepwise increase in stringency.

1.6 Positive Identification of (Na^+ + K^+)ATPase cDNAs

Hybridization of the 14-mer probe to the 50,000-colony library resulted in 155 positive signals on both replicas, whereas the 20-mer probe gave 62 positive signals.

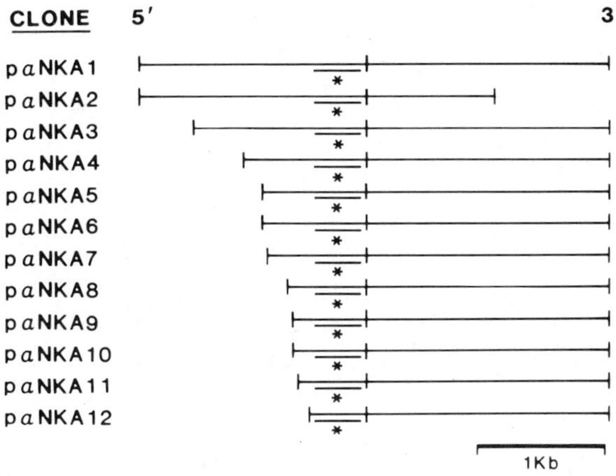

Figure 4. Schematic diagram of the inserts of 12 cDNAs for the sheep kidney ($Na^+ + K^+$)–ATPase α subunit that were isolated by screening with two oligonucleotide probes derived from known tryptic peptides of the sheep kidney enzyme (see Fig. 3). The vertical bars represent the single internal Pst I site and the Pst I sites at the 5′ and 3′ ends that were constructed during the cloning procedure. The region indicated by the asterisk is a 440-base pair Bam HI fragment that gave signals with both oligonucleotide probes during blot hydridization analysis.

Twelve colonies gave signals with both oligonucleotides, and these clones were isolated and analyzed by restriction mapping and blot hybridization analysis. As shown in Fig. 4, the size of the cDNA inserts ranged from 2.3 to 3.65 kilobases. Of the 12 cDNAs, 11 had a 1.8-kilobase Pst I fragment and a variable Pst I fragment ranging in size from 0.44 to 1.8 kilobases. From the initial mapping, it appeared that the 1.85-kilobase fragment would correspond to the 3′ end of the message, whereas the variable fragment would correspond to the 5′ end of the message. This was confirmed in the sequencing studies described in the following paragraphs. One of the clones, pαNKA2, had a 1.8-kilobase 5′ fragment and a 0.95-kilobase 3′ fragment, resulting from loss of a portion of the 3′ end during the cloning procedure. In each case a 0.44-kilobase Bam HI restriction fragment gave strong signals with both oligonucleotides when analyzed by blot hybridization. Mapping with additional restriction enzymes revealed no differences other than the length of the 5′ end as a result of premature termination by reverse transcriptase during first-strand cDNA synthesis.

To confirm the identity of these clones, we first determined the nucleotide sequence of pαNKA1 in the region containing homology to the hybridization probes. The nucleotide sequences coding for T_{14} and T_{16} were identified as well as those for tryptic peptides T_3, T_5, and T_8 (7). In addition, the codons for the sequence Cys–Ser–Asp–Lys, predicted by Bastide et al. (1) to occur at the phosphorylation site, were found at the expected location upstream from T_{14}, the FITC peptide. All 16 of the known tryptic peptides of the sheep kidney enzyme have been identified in the full sequence (39).

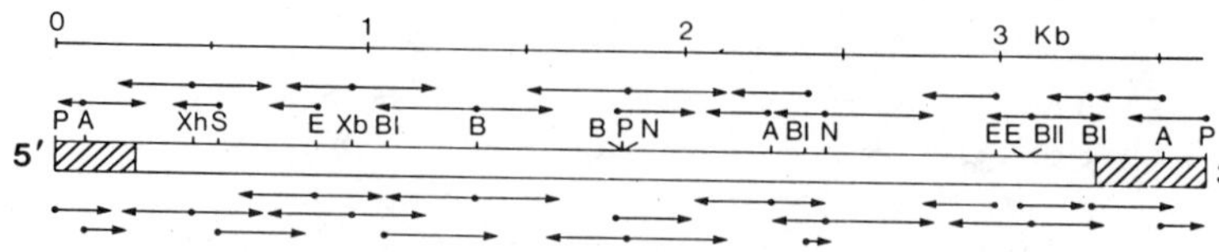

Figure 5. Restriction map and sequencing strategy of clone pαNKA1. The arrows indicate the direction and extent of sequencing. The lower arrows represent the coding strand and the upper arrows represent the noncoding strand. The restriction sites shown are those used in the sequencing studies (P, Pst I; A, Ava I; Xh, Xho I; S, Sal I; E, Eco RI; Xb, Xba I; BI, Bgl I; B, Bam HI; N, Nco I; BII, Bgl II). The hatched areas represent the 5′ and 3′ untranslated sequences. Deoxyribonucleic acid sequencing was performed by the procedure of Maxam and Gilbert (30).

2. CHARACTERIZATION OF (Na^+ + K^+)–ATPase cDNA

The restriction map and sequencing strategy of pαNKA1 are shown in Fig. 5; the complete nucleotide sequence and the deduced amino acid sequence are shown in Fig. 6. The nucleotide sequence includes an additional 10 bases at the 5′ end that were determined from pαNKA2. Although this cDNA was missing approximately 900 bases at its 3′ end, it extended slightly beyond the 5′ terminus of pαNKA1. The full sequence consists of 264 bases of 5′ untranslated sequence, a 3063-base open reading frame, and 337 bases of 3′ untranslated sequence, excluding the poly(A) tract. We confirmed the location of the poly(A) tract by sequencing the 3′ end of pαNKA3, which contained 14 A residues in this position rather than the 3 A residues found in pαNKA1. The canonical polyadenylation signal (35) is located in the expected region, 23 bases upstream from the poly(A) tract. The 5′ untranslated region is intensely (G + C)-rich (72%). This is one of many similarities with Ca^{2+}–ATPase, which also has a (G + C)-rich 5′ untranslated sequence (28). The function of these (G + C)-rich untranslated sequences is not known, but there is evidence that they may be a common feature of "housekeeping" genes (2,47).

The translation initiation site is located at residue 265. Our reasoning in designating this as the initiation site is as follows: (1) the ATG codon occurs in the sequence CCGCTATGG, which closely matches the consensus sequence CC(A,G)CCATGG, which is favorable for initiation of translation in eucaryotes (25); (2) no other ATG triplet occurs in the (G + C)-rich sequence preceding this site; and (3) the ATG codon is located five codons upstream from the sequence encoding

Figure 6. Nucleotide sequence and deduced amino acid sequence of clone pαNKA1. The DNA sequence includes 10 nucleotides at the 5′ end that were determined from clone pαNKA2. The amino acids are numbered starting from the *N* terminus of the mature protein (7) and are preceded by five amino acids that are apparently removed by posttranslational processing. Eight major hydrophobic regions (H_1–H_8), which are likely to traverse the membrane, are underlined. A portion of the phosphorylation site region (P) and of the FITC peptide (A), which is thought to form part of the ATP binding site, are indicated by brackets.

```
GCG GCG GGG TCT GGG GCT CAG AGC AGC GGC GGG AGG AGG CGG ACA CGT GGC AGC AGC AGC AGT AGC AGC CCC GGC GGC GGC GGC GGC GGC GGC AGA AGG CAT CGG CCC GAG CCG CCT GCC GTC TTC CCT CCT TCC CGC CTT GCG GCA GCC CTA GCT CCT TCC TCT TCG CTC CCC CGG CCG  180

                                                                                  -5              -1  1                               10                                          20
                                                                                  Met Gly Lys Gly Val Gly Arg Asp Lys Tyr Glu Pro Ala Ala Val Ser Glu His Gly Asp Lys Lys Lys Ala Lys Lys Glu Arg Asp Met Asp Glu
CGC TTG CTT GGC CGG GAG CCG CTC TCT TTT CTC TCT CAT TCT CTG AAC AGA AGG ACC CGG CGC GGG GCG CAG ACC ACC GCC GCT ATG GGG AAG GGG GTT GGA CGT GAT AAA TAT GAA CCG GCT GCT GTT TCA GAG CAT GGC GAC AAA AAA AAG GCC AAG AAA GAG AGG GAT ATG GAC GAG  360

30                                      40                                      50                                      60                                      70                                      80
Leu Lys Lys Glu Val Ser Met Asp Asp His Lys Leu Ser Leu Asp Glu Leu His Arg Lys Tyr Gly Thr Asp Leu Asn Arg Gly Leu Thr Thr Ala Arg Ala Ala Glu Ile Leu Ala Arg Asp Gly Pro Asn Ala Leu Thr Pro Pro Pro Thr Thr Pro Glu Trp Val Lys Phe Cys Arg
CTG AAG AAA GAA GTT TCT ATG GAT GAT CAT AAA CTT AGC CTT GAT GAA CTT CAT CGC AAA TAT GGA ACA GAC TTG AAC CGA GGC TTA ACA ACT GCT CGA GCT GCT GAG ATC CTA GCC AGA GAT GGA CCC AAT GCC CTC ACA CCC CCT CCC ACA ACC CCT GAA TGG GTC AAG TTC TGT CGA  540

90                                      100                                     110                                     120                                     130                                     140
Gln Leu Phe Gly Gly Phe Ser Met Leu Leu Trp Ile Gly Ala Val Leu Cys Phe Leu Ala Tyr Gly Ile Gln Ala Ala Thr Glu Glu Glu Pro Gln Asn Asp Asn Leu Tyr Leu Gly Val Val Leu Ser Ala Val Val Ile Ile Thr Gly Cys Phe Ser Tyr Tyr Gln Glu Ala Lys Ser
CAG CTG TTT GGG GGG TTC TCA ATG TTA CTG TGG ATT GGA GCA GTT CTT TGC TTC CTG GCC TAT GGC ATC CAA GCT GCT ACA GAA GAG GAA CCT CAA AAT GAT AAT CTG TAC CTT GGT GTG GTG CTG TCT GCC GTC GTC ATC ATA ACC GGG TGT TTC TCC TAC TAT CAA GAA GCT AAA AGC  720
------------------------- H1 -------------------------                                                                         ------------ H2 ------------

150                                     160                                     170                                     180                                     190                                     200
Ser Lys Ile Met Glu Ser Phe Lys Asn Met Val Pro Gln Gln Ala Leu Val Ile Arg Asn Gly Glu Lys Met Ser Ile Asn Ala Glu Glu Val Val Val Gly Asp Leu Val Glu Val Lys Gly Gly Asp Arg Ile Pro Ala Asp Leu Arg Ile Ile Ser Ala Asn Gly Cys Lys Val Asp
TCC AAG ATC ATG GAG TCC TTC AAA AAC ATG GTT CCT CAG CAA GCT CTC GTG ATT CGA AAT GGT GAA AAA ATG AGC ATA AAT GCG GAA GAA GTT GTC GTC GGA GAC CTG GTG GAA GTG AAA GGA GGG GAT CGA ATT CCT GCT GAT CTC AGG ATC ATT TCC GCG AAT GGC TGC AAG GTG GAT  900

210                                     220                                     230                                     240                                     250                                     260
Asn Ser Ser Leu Thr Gly Glu Ser Glu Pro Gln Thr Arg Ser Pro Asp Phe Thr Asn Glu Asn Pro Leu Glu Thr Arg Asn Ile Ala Phe Phe Ser Thr Asn Cys Val Glu Gly Thr Ala Arg Gly Ile Val Val Tyr Thr Gly Asp Arg Thr Val Met Gly Arg Ile Ala Thr Leu Ala
AAC TCC TCA CTC ACG GGA GAA TCA GAG CCA CAG ACC AGG TCT CCG GAT TTC ACC AAT GAA AAT CCT CTA GAG ACA CGG AAC ATT GCC TTT TTT TCA ACC AAC TGT GTT GAA GGC ACT GCA CGT GGC ATT GTT GTG TAC ACC GGG GAT CGA ACC GTG ATG GGC AGA ATT GCC ACC CTG GCT  1080

270                                     280                                     290                                     300                                     310                                     320
Ser Gly Leu Glu Gly Gly Gln Thr Pro Ile Ala Ala Glu Ile Glu His Phe Ile His Ile Ile Thr Gly Val Ala Val Phe Leu Gly Val Ser Phe Phe Ile Leu Ser Leu Ile Leu Glu Tyr Thr Trp Leu Glu Ala Val Ile Phe Leu Ile Gly Ile Ile Val Ala Asn Val Pro Glu
TCT GGG CTG GAA GGG GGC CAG ACT CCC ATC GCT GCG GAA ATT GAA CAT TTT ATC CAC ATC ATC ACG GGT GTG GCT GTG TTC CTG GGC GTG TCT TTC TTC ATC CTT TCT CTG ATC CTT GAG TAC ACC TGG TTG GAG GCT GTC ATC TTC CTC ATC GGC ATC ATT GTA GCC AAT GTG CCA GAG  1260
                                                                        ------------------------- H3 -------------------------

330                                     340                                     350                                     360                                     370                                     380
Gly Leu Leu Ala Thr Val Thr Val Cys Leu Thr Leu Thr Ala Lys Arg Met Ala Arg Lys Asn Cys Leu Val Lys Asn Leu Glu Ala Val Glu Thr Leu Gly Ser Thr Ser Thr Ile Cys Ser Asp Lys Thr Gly Thr Leu Thr Gln Asn Arg Met Thr Val Ala His Met Trp Phe Asp
GGT TTG CTG GCC ACC GTC ACG GTG TGC CTG ACC CTG ACT GCC AAG CGC ATG GCC AGG AAG AAC TGC TTA GTG AAG AAC CTG GAG GCT GTG GAG ACC TTG GGA TCC ACG TCC ACC ATC TGC TCA GAC AAA ACT GGA ACT CTG ACC CAG AAC CGA ATG ACA GTG GCC CAC ATG TGG TTC GAC  1440
-- H4 --------------------------                                                                           |--- P ---|

390                                     400                                     410                                     420                                     430                                     440
Asn Gln Ile His Glu Ala Asp Thr Thr Glu Asn Gln Ser Gly Val Ser Phe Asp Lys Thr Ser Ala Thr Trp Leu Ala Leu Ser Arg Ile Ala Gly Leu Cys Asn Arg Ala Val Phe Gln Ala Asn Gln Asp Asn Leu Pro Ile Leu Lys Arg Ala Val Ala Gly Asp Ala Ser Glu Ser
AAC CAA ATC CAT GAG GCG GAC ACG ACA GAG AAT CAG AGC GGT GTT TCA TTT GAC AAG ACT TCG GCC ACC TGG CTC GCT CTG TCC AGA ATT GCA GGT CTT TGT AAC AGG GCC GTG TTT CAG GCT AAC CAG GAT AAC CTG CCT ATC CTG AAG CGG GCT GTA GCA GGT GAT GCA TCA GAG TCT  1620

450                                     460                                     470                                     480                                     490                                     500
Ala Leu Leu Lys Cys Ile Glu Val Cys Cys Gly Ser Val Lys Glu Met Arg Glu Arg Tyr Ala Lys Ile Val Glu Ile Pro Phe Asn Ser Thr Asn Lys Tyr Gln Leu Ser Ile His Lys Asn Ala Asn Ala Gly Glu Pro Arg His Leu Leu Val Met Lys Gly Ala Pro Glu Arg Ile
GCG CTT CTG AAG TGC ATT GAG GTG TGC TGC GGT TCT GTG AAG GAG ATG AGA GAG AGA TAT GCC AAG ATC GTC GAG ATC CCG TTC AAC TCC ACC AAC AAG TAC CAG TTA TCC ATC CAC AAG AAT GCC AAT GCA GGT GAG CCC CGG CAC CTA CTG GTG ATG AAA GGT GCT CCG GAG AGG ATC  1800
                                                                                                                                   |----- A -----|

510                                     520                                     530                                     540                                     550                                     560
Leu Asp Arg Cys Ser Ser Ile Leu Ile His Gly Lys Glu Gln Pro Leu Asp Glu Glu Leu Lys Asp Ala Phe Gln Asn Ala Tyr Leu Glu Leu Gly Gly Leu Gly Glu Arg Val Leu Gly Phe Cys His Leu Met Leu Pro Asp Glu Gln Phe Pro Glu Gly Phe Gln Phe Asp Thr Asp
CTG GAT CGC TGC AGC TCC ATC CTC ATC CAT GGC AAG GAG CAG CCC CTG GAC GAG GAG CTG AAG GAC GCC TTT CAG AAC GCC TAC CTG GAG CTG GGT GGC CTG GGA GAA CGT GTG CTG GGT TTC TGC CAC TTG ATG CTG CCA GAT GAG CAG TTT CCT GAA GGC TTC CAG TTT GAC ACC GAC  1980

570                                     580                                     590                                     600                                     610                                     620
Asp Val Asn Phe Pro Val Asp Asn Leu Cys Phe Val Gly Leu Ile Ser Met Ile Asp Pro Pro Arg Ala Ala Val Pro Asp Ala Val Gly Lys Cys Arg Ser Ala Gly Ile Lys Val Ile Met Val Thr Gly Asp His Pro Ile Thr Ala Lys Ala Ile Ala Lys Gly Val Gly Ile Ile
GAT GTG AAC TTC CCT GTT GAT AAT CTC TGC TTC GTG GGG CTC ATC TCC ATG ATT GAC CCA CCA CGG GCT GCT GTC CCA GAT GCT GTG GGC AAA TGC CGA AGT GCT GGA ATC AAG GTC ATC ATG GTC ACC GGA GAC CAT CCC ATC ACT GCC AAA GCC ATT GCC AAA GGC GTG GGC ATC ATC  2160

630                                     640                                     650                                     660                                     670                                     680
Ser Glu Gly Asn Glu Thr Val Glu Asp Ile Ala Ala Arg Leu Asn Ile Pro Val Ser Gln Val Asn Pro Arg Asp Ala Arg Ala Cys Val Val His Gly Ser Asp Leu Lys Asp Met Thr Pro Glu Gln Leu Asp Asp Ile Leu Lys Tyr His Thr Glu Ile Val Phe Ala Arg Thr Ser
TCA GAA GGG AAT GAG ACT GTA GAA GAC ATT GCT GCC CGC CTC AAC ATC CCA GTG AGC CAG GTG AAT CCC AGG GAT GCC AGG GCC TGC GTG GTG CAT GGA AGT GAT CTG AAG GAC ATG ACT CCC GAG CAA CTG GAT GAC ATT TTG AAG TAC CAC ACG GAG ATT GTG TTC GCC AGG ACC TCT  2340

690                                     700                                     710                                     720                                     730                                     740
Pro Gln Gln Lys Leu Ile Ile Val Glu Gly Cys Gln Arg Gln Gly Ala Ile Val Ala Val Thr Gly Asp Gly Val Asn Asp Ser Pro Ala Leu Lys Lys Ala Asp Ile Gly Val Ala Met Gly Ile Ala Gly Ser Asp Val Ser Lys Gln Ala Ala Asp Met Ile Leu Leu Asp Asp Asn
CCG CAG CAG AAG CTC ATC ATT GTG GAA GGC TGC CAG AGA CAG GGT GCC ATT GTG GCT GTA ACT GGA GAC GGT GTC AAT GAC TCC CCG GCT TTG AAG AAG GCA GAC ATT GGG GTT GCC ATG GGG ATT GCA GGC TCA GAC GTG TCT AAG CAA GCT GCT GAC ATG ATC CTC CTA GAT GAC AAC  2520

750                                     760                                     770                                     780                                     790                                     800
Phe Ala Ser Ile Val Thr Gly Val Glu Glu Gly Arg Leu Ile Phe Asp Asn Leu Lys Lys Ser Ile Ala Tyr Thr Leu Thr Ser Asn Ile Pro Glu Ile Thr Pro Phe Leu Ile Phe Ile Ile Ala Asn Ile Pro Leu Pro Leu Gly Thr Val Thr Ile Leu Cys Ile Asp Leu Gly Thr
TTT GCT TCA ATT GTG ACT GGA GTA GAG GAA GGT CGT CTG ATC TTT GAT AAC TTG AAG AAA TCC ATT GCT TAT ACC CTC ACC AGT AAC ATT CCT GAG ATT ACC CCC TTC CTG ATA TTT ATT ATT GCA AAC ATT CCA CTC CCC CTG GGG ACC GTC ACC ATC CTC TGC ATA GAC TTG GGA ACA  2700
                                                                                                            ------------------------- H5 -------------------------

810                                     820                                     830                                     840                                     850                                     860
Asp Met Val Pro Ala Ile Ser Leu Ala Tyr Glu Gln Ala Glu Ser Asp Ile Met Lys Arg Gln Pro Arg Asn Pro Gln Thr Asp Lys Leu Val Asn Glu Arg Leu Ile Ser Met Ala Tyr Gly Gln Ile Gly Met Ile Gln Ala Leu Gly Gly Phe Phe Thr Tyr Phe Val Ile Met Ala
GAC ATG GTC CCT GCT ATC TCC CTG GCT TAT GAG CAA GCT GAG AGC GAC ATC ATG AAG AGA CAG CCC AGA AAC CCC CAG ACA GAC AAA CTC GTG AAC GAG CGG CTG ATC AGC ATG GCC TAC GGA CAG ATT GGT ATG ATC CAG GCC CTG GGG GGC TTC TTC ACT TAC TTT GTG ATC ATG GCT  2880
                                                                                                            ------------------------- H6 -------------------------

870                                     880                                     890                                     900                                     910                                     920
Glu Asn Gly Phe Leu Pro Asn His Leu Leu Gly Ile Arg Val Thr Trp Asp Asp Arg Trp Ile Asn Asp Val Glu Asp Ser Tyr Gly Gln Gln Trp Thr Tyr Glu Gln Arg Lys Ile Val Glu Phe Thr Cys His Thr Ala Phe Phe Val Ser Ile Val Val Val Gln Trp Ala Asp Leu
GAG AAC GGC TTC CTC CCT AAT CAC CTG CTG GGC ATC CGA GTG ACC TGG GAT GAC CGT TGG ATC AAC GAC GTG GAG GAC AGC TAT GGG CAG CAG TGG ACC TAT GAA CAG AGG AAG ATC GTG GAA TTC ACC TGC CAC ACG GCC TTC TTT GTC AGC ATC GTG GTG GTG CAG TGG GCT GAC TTG  3060
                                                                                                                                   ------------- H7 -------------

930                                     940                                     950                                     960                                     970                                     980
Val Ile Cys Lys Thr Arg Arg Asn Ser Val Phe Gln Gln Gly Met Lys Asn Lys Ile Leu Ile Phe Gly Leu Phe Glu Glu Thr Ala Leu Ala Ala Phe Leu Ser Tyr Cys Pro Gly Met Gly Val Ala Leu Arg Met Tyr Pro Leu Lys Pro Thr Trp Trp Phe Cys Ala Phe Pro Tyr
GTC ATC TGC AAG ACC AGG AGG AAT TCC GTC TTC CAG CAG GGG ATG AAG AAC AAG ATC TTA ATA TTT GGC CTC TTC GAA GAG ACA GCC CTT GCT GCT TTC CTT TCC TAC TGC CCT GGA ATG GGT GTT GCC CTG AGG ATG TAT CCC CTC AAA CCT ACC TGG TGG TTC TGT GCC TTC CCC TAC  3240
-------                                                             ------------------------- H8 -------------------------

990                                     1000                                    1010          1016
Ser Leu Leu Ile Phe Val Tyr Asp Glu Val Arg Lys Leu Ile Ile Arg Arg Arg Pro Gly Gly Trp Val Glu Lys Glu Thr Tyr Tyr END
TCG CTT CTC ATC TTC GTG TAT GAC GAA GTC CGG AAG CTC ATC ATC AGG CGA CGC CCT GGC GGC TGG GTG GAG AAG GAG ACC TAC TAC TAG CCC CTC CTG CAC GCC GCG GAG CAT CGC GCC GCA AGC TCT GCA CCC ACC ACC CAC CCC CTC TCT GTG TAC TTC AGT CTT GGA ATT CAG AAT  3420

TCC ACC CTG GTA GGA AAG CAC CGA AGC ATG TGG GGA AGC GAG ACG TCC CGG AAT GAA GCA TGT AGC TCT ATG GGG GGG GTG GGG GTG GGG GGG AGG GTT GCC CGA GAA ACA TCT GTC TGT GGA AAC AAC CGT GGG GAA GGT TTT TAT GTG CCT TTT TGT TTT TGT AAA ATG GAA AAC TCG  3600

GAA TGT ATG AAA GAA ATA CAT TTT ATA TTC GGA TTT TTA CAA ATA AAG ATG GCT ATT ATA ATG GAA  A                                                                   3667
```

the known *N*-terminal amino acid sequence of the mature protein (7). We conclude that the primary translation product lacks an *N*-terminal hydrophobic signal peptide. It is interesting that the sequence Met–Gly–Lys–Gly–Val, which is removed to yield the *N*-terminus of the mature protein, is very similar to the deduced *N*-terminus of $(Na^+ + K^+)$–ATPase of *Torpedo californica* electroplax, Met–Gly–Lys–Gly–Ala (23). It is not known whether this sequence is retained in the mature Torpedo enzyme. Although the presence of this sequence in the primary translation product could be coincidental, it might serve some function prior to posttranslational processing. The first in-frame stop codon is located at residue 3327 and is preceded by codons for the sequence Lys–Glu–Thr–Tyr–Tyr, which is very similar to the *C*–terminal sequence of the dog kidney enzyme, Lys–Ser–Thr–Tyr–Tyr (5). The mature protein contains 1016 amino acids and has a molecular weight of 112,177.

3. IDENTIFICATION OF STRUCTURAL AND FUNCTIONAL DOMAINS

3.1 Transmembrane Organization

A major question regarding the structure of $(Na^+ + K^+)$ATPase is the organization of the protein in the membrane. A definitive answer to this question will require additional experimental work, most likely involving the use of antibodies directed to synthetic peptides corresponding to various regions of the peptide chain. However, with the information now available it is possible to develop a reasonable hypothetical model for the path of the polypeptide through the membrane.

There is a great deal of information in the literature regarding the general location of various sites with respect to the membrane. For example, the phosphorylation site (1) and the ATP binding site (10) are both located on the cytoplasmic side of the membrane. In addition to these functional domains, there are a number of conformation-dependent proteolysis sites that have been localized on the cytoplasmic side (see ref. 22 for a review). When potential transmembrane domains are identified, these sites can be used to determine the orientation of the protein in the membrane.

The location and extent of the major hydrophobic domains were determined by using the hydropathy program of Kyte and Doolittle (26). This program allows the identification of regions of the polypeptide chain that have sufficient length and hydrophobicity to represent transmembrane regions. The hydropathy plot for the deduced amino acid sequence is shown in the top panel of Fig. 7. The bottom panel is an expanded hydropathy plot of a region that contains the phosphorylation site, the putative cardiac glycoside binding site, and a region that may be a component of the energy transduction system. These functional domains are discussed in detail in the following paragraphs.

The protein has eight major hydrophobic domains H_1–H_8, which range in size from 17 to 29 amino acids (Figs. 6 and 7), and it seems likely that each of these

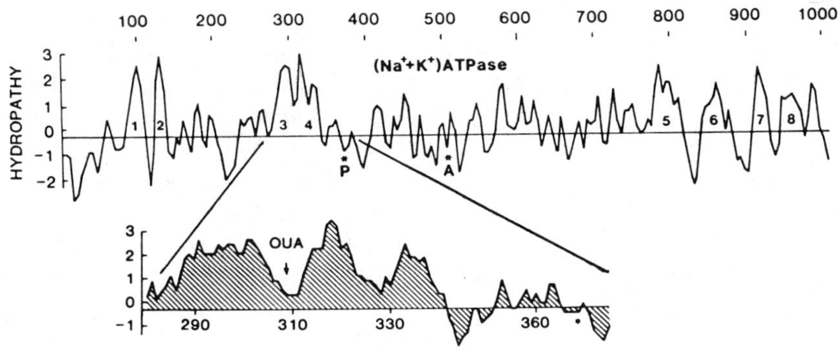

Figure 7. Hydropathy analysis (26) of the $(Na^+ + K^+)$–ATPase catalytic subunit. The hydropathy plot for the entire protein is shown in the top panel; the bottom panel is an expanded hydroplot of the region including the phosphorylation site (indicated by the asterisk), hydrophobic domains H_3 and H_4, and the putative ouabain binding site. Hydrophobic regions are plotted above the X axis and hydrophilic regions below it.

segments traverses the membrane. The hydropathy values for the four hydrophobic domains in the N-terminal half of the protein are well above the hydropathy average calculated for a number of known transmembrane sequences (26). The N-terminus of the protein and the region between H_2 and H_3 are located on the cytoplasmic side of the membrane (21). As the phosphorylation site, located at Asp-369 on the C-terminal side of H_4, is also in the cytoplasm, it is likely that the first four major hydrophobic domains each pass through the membrane one time. The junction between H_1 and H_2 and that between H_3 and H_4 would be located on the extracellular side of the membrane. This assignment is in agreement with that suggested for the Ca^{2+}–ATPase of rabbit muscle (28) and the $(Na^+ + K^+)$–ATPase of *Torpedo californica* electroplax (23).

On the basis of their hydropathy values, the four major hydrophobic domains in the C-terminal region of the protein (H_5–H_8) would each appear to span the membrane once. If this is the case, the H_5–H_6 and H_7–H_8 junctions would both be located on the extracellular side and the C terminus would be located in the cytoplasm. However, there is no experimental evidence concerning the location of the sequences joining these domains, nor do we know which side of the membrane the C terminus is on. Furthermore, an examination of these hydrophobic domains in the Ca^{2+}–ATPase (28) and the $(Na^+ + K^+)$–ATPase of *T. californica* (23) has led to quite different conclusions regarding the number of transmembrane domains. Given the extensive homology between these proteins and the striking similarity in their hydropathy plots (23,39), it seems very unlikely that they differ in the number of transmembrane domains. The H_5 domain, like H_4, is particularly long, and it has been suggested that the corresponding hydrophobic domain of Ca^{2+}–ATPase, which is even more extensive than that of $(Na^+ + K^+)$–ATPase, may pass twice through the membrane (28). If this is the case, $(Na^+ + K^+)$–ATPase must be arranged in a similar manner as this domain is flanked on both sides by regions of amino acid homology between the two proteins (39). An alternative explanation for the length of both H_4 and H_5 is that the cytoplasmic portion connected to the transmembrane sequence is buried within the protein. Several investigators have suggested that both $(Na^+ + K^+)$–ATPase (45) and Ca^{2+}–ATPase (28) are composed of a globular cytoplasmic head group connected to the transmembrane domains by a stalk region. If the transmembrane and stalk regions resemble a shaft composed of eight helices, then two of these helices, H_4 and H_5, could reside primarily inside this shaft, thus accounting for their additional length.

3.2 Amino Acid Homology with Other Cation Transport ATPases

Full-sequence data are now available for a number of cation transport ATPases and partial sequence is available for several more. An evolutionary relationship between the K^+–ATPase KdpB protein of *E. coli* and the rabbit muscle Ca^{2+}–ATPase was clearly demonstrated by Hesse et al. (17) and a comparison of the $(Na^+ + K^+)$–ATPase and the Ca^{2+}–ATPase demonstrated a striking degree of amino acid homology in scattered blocks throughout the peptide chains (39). There is also considerable homology between the FITC peptide of the $(H^+ + K^+)$ATPase and those of the $(Na^+ + K^+)$–ATPase and the Ca^{2+}–ATPase (9), and the phosphorylation site of the corn root H^+–ATPase and the $(H^+ + K^+)$–ATPase appear to have the Cys–Ser–Asp–Lys sequence found in Ca^{2+}–ATPase and $(Na^+ + K^+)$–ATPase (46). Comparisons of the amino acid sequences of these proteins should be useful in identifying conserved domains that are involved in the general function of active cation transport. Those regions involved in ion specific functions would be expected to be quite different.

When the full sequence of the $(Na^+ + K^+)$–ATPase (39), the Ca^{2+}–ATPase (28), and the K^+–ATPase KdpB protein of *E. coli* were compared, four major regions of homology, spanning 40 amino acids or more, were identified. The amino acid homology is presented in Fig. 8, and the locations of these regions of homology are illustrated in Fig. 10. Although the FITC peptide is quite homologous among the eucaryotic transport ATPases (9), the homology with the corresponding region of the procaryotic K^+–ATPase is relatively low (17) and is not included as a major homologous region.

Although the function of these conserved regions is not clear, it is apparent that there have been strong evolutionary constraints on their amino acid sequences. A possible explanation is that some of these sequences may provide the contact points between various functional domains. An additional constraint might be provided by the different conformations of the protein. For example, if two adjacent domains had different conformations in the E_1 and E_2 states, then any amino acid substitution at the interface of these domains would have to be compatible in both enzyme conformations.

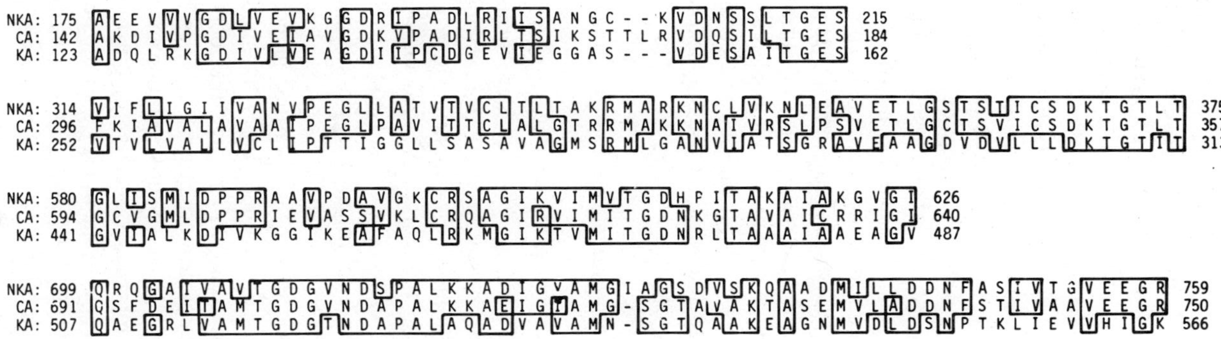

Figure 8. Major homologous regions of cation-transport ATPases. Homology comparisons were performed using the full sequence of the (Na^+ + K^+)–ATPase (NKA) from sheep kidney (39), the Ca^{2+}–ATPase (CA) from neonatal rabbit skeletal muscle (28), and the K^+–ATPase KdpB protein (KA) from *E. coli* (17). Identical amino acid residues are boxed. The locations of these four regions of homology are illustrated in Fig. 10.

MacLennan et al. (28) have suggested that the cytoplasmic regions of the Ca^{2+}–ATPase may consist of three major functional domains that they term the *transduction domain*, the *phosphorylation domain*, and the *nucleotide binding domain*. The first homologous region, which lies between H_2 and H_3, is within the proposed transduction domain. The second conserved region includes the phosphorylation site and much of H_4. The degree of homology in the H_4 sequence is particularly interesting as it is the only major hydrophobic domain exhibiting homology among all three proteins. A likely possibility, discussed in the next section, is that this sequence serves to transmit conformational changes into the transmembrane regions of the protein. The third conserved region, located approximately 75 amino acids beyond the FITC peptide, lies within the proposed nucleotide binding domain. In the fourth major homologous region, which lies on the *N*-terminal side of H_5, there is greater than 60% homology between the two eucaryotic ATPases and 36% of the residues are identical in all three proteins. This region corresponds to the α-helical subdomain described by MacLennan et al. (28). These investigators made the interesting suggestion that this subdomain could serve as a hinge between the phosphorylation domain and the nucleotide binding domain. The high degree of homology among both procaryotic and eucaryotic ATPases is consistent with this suggestion, as such a function would appear to impose a severe evolutionary constraint.

3.3 Energy Transduction System

During its catalytic cycle, (Na^+ + K^+)–ATPase alternates between two conformations. The conformational transition between these forms is extensive, and there is evidence for the involvement of at least 80 amino acids that undergo an α-helix→β-sheet transition (13). The energy for this transition is provided by the transfer of a high-energy phosphate bond from ATP to the aspartate residue at position 369 (36,39). Thus Asp-369 is the central residue involved in the conversion of the chemical energy of ATP to the mechanical energy of the polypeptide chain. It seems clear that various regions of the molecule are strongly interconnected as the binding of various ligands such as Na^+ and K^+, cardiac glycosides and ATP all influence the conformational state of the enzyme (19). A detailed understanding of the mechanism of active cation transport will require the identification of those regions whose function is to transmit conformational changes from the catalytic center to the domains directly involved in the selective binding and vectorial transport of Na^+ and K^+ through the membrane. Collectively, the regions involved in transmitting conformational changes form the energy transduction system.

Given the magnitude of the conformational changes occurring during the catalytic cycle, the energy transduction system is likely to be extensive and appears to involve cytoplasmic, transmembrane, and extracellular sequences. The accessibility of a number of well-studied proteolysis sites, located on the cytoplasmic side of the membrane and distantly separated in the primary sequence, are dependent on the conformation of the enzyme (22). The involvement of extracellular and transmembrane sequences is indicated by the fact that ouabain, which binds on the extracellular side of the membrane, blocks the conformational changes by stabilizing the phosphorylated intermediate (19).

The region extending from Arg-262 to Arg-423 (see bottom panel of Fig. 7) includes the phosphorylation site, the transmembrane domains H_3 and H_4, and a binding site for a ouabain photoaffinity derivative (21). The boundaries of this region, Arg-262 and Arg-423, appear to correspond to the well-known conformation-dependent trypsin sites 3 and 1, respectively (22,39). Phenylalanine-284, the first residue of H_3, is a likely possibility for the conformation-dependent chymotrypsin site near bond 3 (22). As described previously, the region extending from the phosphorylation site into H_4 is conserved among procaryotic and eucaryotic transport ATPases. This suggests that this region may function as a component of the energy transduction system that transmits conformational changes across the membrane.

3.4 Cardiac Glycoside Binding Site

The location of the cardiac glycoside binding site on the catalytic subunit is of considerable interest as it appears to be the pharmacological receptor site for the digitalis class of drugs and may also be the binding site for putative endogenous molecules that regulate the activity of the pump (38). Investigations with a variety of ouabain derivatives, including a number of photoaffinity derivatives that covalently bind to the catalytic subunit, have provided some information concerning the location and characteristics of this site. Fortes (11) conducted studies using anthroylouabain that indicated that the binding site is hydrophobic and shielded from water. Also, energy transfer from a tryptophan residue was detected, suggesting its presence near the binding site. Goeldner et al. (12) developed a ouabain photoaffinity derivative that was activated by energy transfer from an excited tryptophan residue. This derivative bound only to the 41-kilodalton *N*-terminal tryptic fragment of the protein. Binding of an alkylating ouabain derivative to this region has also been demonstrated (37). Other ouabain photoaffinity derivatives have been shown to bind covalently to both the *N*- and *C*-terminal regions of the protein. However, it is likely that labeling at the *C*-terminal site is due to binding of the sugar moiety of ouabain rather than the steroidlike portion (12).

Jørgensen et al. (21) have shown that the binding site in the *N*-terminal half of the protein is localized to the H_3–H_4 region. The only portion of this region of the molecule that is likely to be located on the extracellular side is the H_3–H_4 junction (Fig. 8). This region appears to have the characteristics predicted for the ouabain binding site as it is relatively hydrophobic, contains a Trp residue, and is well situated with respect to the phosphorylation site and the proposed energy transduction pathway. Furthermore, it is not homologous with the corresponding site of Ca^{2+}–ATPase, which is insensitive to ouabain.

3.5 Ion Specificity

The specificity of (Na + K^+)–ATPase for the transport of Na^+ and K^+ is remarkable when one considers that other cation pumps, such as Ca^{2+}–ATPase, are very similar in their overall organization (28,39) yet have a very different ion specificity. A detailed knowledge of the structure and location of the regions involved in cation discrimination, binding, and transport are essential for an understanding of the mechanisms of cation transport.

One might expect that a functional domain that discriminates between various cations would include a number of both negatively and positively charged residues as this would allow the attraction and repulsion of their charges to be utilized in the process of ion selection. This region of the enzyme should be highly conserved among species widely separated in evolution yet very different from the corresponding region of cation transport ATPases having a different ion specificity. The hydrophilic *N*-terminal domain of (Na^+ + K^+)–ATPase fits these expectations. It is intensely hydrophilic, with a high proportion of charged residues, and is not homologous with the *N*-terminal region of Ca^{2+}–ATPase. In fact, when compared with Ca^{2+}–ATPase, (Na^+ + K^+)–ATPase appears to have approximately 35 additional residues in this region (39). Despite the lack of homology with Ca^{2+}–ATPase, it is highly homologous with the *N*-terminal sequences of mammalian, avian, and crustacean (Na^+ + K^+)–ATPases (Fig. 9). Particularly striking is the conservation of lysine residues between the sheep and brine shrimp enzymes, which diverged in evolution over 400 million years ago. The *Torpedo californica* enzyme (23) is also quite homologous in this region (data not shown), although there appear to be several insertions and deletions of amino acids between Glu-12 and Asp-23. However, the five lysine residues in this region are retained.

There is evidence that the *N*-terminal region may be involved in the cation-dependent equilibrium between the E_1 and E_2 conformations (reviewed in ref. 22). Tryptic bond 2, which has been estimated to be approximately 20 amino acids

Duck Salt Gland[a]	NH_2	Gly	Arg	Asn	Lys	Tyr	Glu	Thr	Thr	Ala	(?)	Ser	Glu
Dog Kidney[b]	NH_2	Gly	Arg	Asn	Lys	Tyr	Glu	Pro	Ala	Ala	(?)	Ser	Glu
Sheep Kidney	NH_2	Gly	Arg	Asp	Lys	Tyr	Glu	Pro	Ala	Ala	Val	Ser	Glu

Sheep Kidney	His	Gly	Asp	Lys	Lys	Lys	Ala	Lys	Lys	Glu	Arg	Asp	Met	Asp
Brine Shrimp[c]		NH_2	Ala	Lys	Lys	Lys	Gln	Lys	Lys	Gly	Lys	Asp	Leu	Asn

Sheep Kidney	Glu	Leu	Lys	Lys	Glu	Val	Ser	Met	Asp	Asp	His	Lys	Leu	Ser
Brine Shrimp	Glu	Leu	Lys	Lys	Glu	Leu	Asp	Ile	Asp	Phe	His	Lys	Ile	Pro

Figure 9. Homology comparison of the *N*-terminal sequences of the (Na^+ + K^+)–ATPase from sheep kidney (39), duck salt gland (18), dog kidney (4), and brine shrimp (31). Identical residues are boxed.

from the *N* terminus, is cleaved rapidly when the enzyme is in the E_1Na form and yields an *N*-terminal Ala (5,22). Lysine 18 is the probable location of this site. Following cleavage of bond 2, the equilibrium is shifted toward the E_1 form (20). It has been suggested that this cleavage removes a His residue that is important in the E_1–E_2 equilibrium (41); the only histidine in this sequence is at position 14. Interestingly, the enzymes of *Torpedo californica* and *Artemia salina* (brine shrimp), both of which are exposed to a salt environment very different from that of mammals, are missing His-14. This evidence suggests the possibility that the *N*-terminal hydrophilic domain may play a role in cation discrimination, binding, and access to the transmembrane domains through which transport occurs.

4. SUMMARY

Two general cloning and screening procedures are commonly used to isolate cDNAs for proteins that have been purified to homogeneity: (1) insert the cDNA into an expression vector and screen with antibodies to the purified protein (48) and (2) insert the cDNA into a plasmid vector and screen with oligonucleotides derived from fragments of the amino acid sequence. Although it may be more difficult to obtain amino acid sequence data than raise antibodies, there are significant advantages in the oligonucleotide screening procedure. The procedures used to prepare a cDNA library in a plasmid vector (Fig. 1) and screen with oligonucleotides (Section 1.5) are much easier than those used to prepare and screen an expression library, and it is not necessary to subclone the insert before beginning the sequence analysis. An additional advantage is that the amino acid sequence data are useful in confirming the identity of the protein.

The task of cloning the α subunit was greatly simplified by the use of the cDNA size-fractionation procedure described in Section 1.3. If this procedure had not been included, it would have been necessary to screen a library at least 10 times the size of

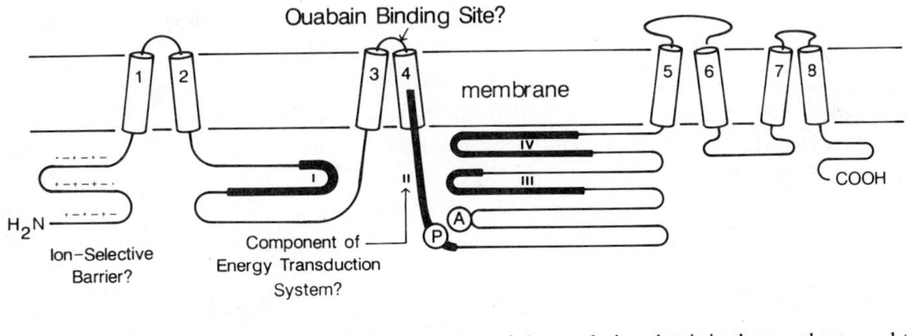

Figure 10. Hypothetical model for the arrangement of the catalytic subunit in the membrane and the location of various functional domains, including the phosphorylation site (P) and the ATP binding site (A). The four major regions of homology (I–IV) with other cation transport ATPases (see Fig. 8) are represented by the heavy lines. See text for additional explanations.

the one that yielded pαNKA1. The size-fractionation method described may be of considerable value in the isolation of cDNAs for ion channel proteins, many of which would be very large.

Analysis of the deduced amino acid sequence of the α subunit has allowed the tentative identification of a number of structural and functional domains of the protein and the development of a model for the organization of the protein in the membrane (Fig. 10). On the basis of hydropathy analysis, the polypeptide chain appears to traverse the membrane eight times. A comparison of the amino acid sequence of eucaryotic and procaryotic ATPases reveals four major homologous regions (I–IV in Fig. 10). The second region, which includes the phosphorylation site and much of H_4, may be a component of the energy transduction system. The H_3–H_4 junction, which is located on the *N*-terminal side of region II, is the probable location of the cardiac glycoside binding site and could be involved in the binding of endogenous regulators of the pump. The hydrophilic *N*-terminal domain may play a role in cation selectivity.

ACKNOWLEDGMENT

This work was supported by NIH grants HL 28573 and HL 22619.

REFERENCES

1. Bastide, F., Meissner, G., Fleischer, S., and Post, R. L. (1973). *J. Biol. Chem.*, **248**, 8385–8391.
2. Bird, A., Taggart, M., Frommer, M., Miller, O. J., and Macleod, D. (1985). *Cell*, **40**, 91–99.
3. Boyd, D., Jain, S. K., Crampton, J., Barrett, K. J., and Drysdale, J. (1984). *Proc. Natl. Acad. Sci.* (USA), **81**, 4751–4755.
4. Cantley, L. C. (1981). *Curr. Top. Bioenerg.*, **11**, 201–237.
5. Castro, J. and Farley, R. A. (1979). *J. Biol. Chem.*, **254**, 2221–2228.
6. Chirgwin, J. M., Przybyla, A. E., MacDonald, R. J., and Rutter, W. J. (1979). *Biochemistry*, **18**, 5294–5299.
7. Collins, J. H., Zot, A. S., Ball, W. J., Jr., Lane, L. K., and Schwartz, A. (1982). *Biochim. Biophys. Acta*, **742**, 358–365.
8. Craig, W. S. and Kyte, J. (1980). *J. Biol. Chem.*, **255**, 6262–6269.
9. Farley, R. A. and Faller, L. D. (1985). *J. Biol. Chem.*, **260**, 3899–3901.
10. Farley, R. A., Tran, C. M., Carilli, C. T., Hawke, D., and Shively, J. E. (1984). *J. Biol. Chem.*, **259**, 9532–9535.
11. Fortes, P. A. G. (1977). *Biochemistry*, **16**, 531–540.
12. Goeldner, M. P., Hirth, C. G., Rossi, B., Ponzio, G., and Lazdunski, M. (1983). *Biochemistry*, **22**, 4685–4690.
13. Gresalfi, T. J. and Wallace, B. A. (1984). *J. Biol. Chem.*, **259**, 2622–2628.
14. Gubler, U. and Hoffman, B. J. (1983). *Gene*, **25**, 263–269.
15. Guyton, A. C. (1981). In *Textbook of Medical Physiology*, Saunders, London, pp. 104–108.
16. Hanahan, D. and Meselson, M. (1980). *Gene*, **10**, 63–67.

17. Hesse, J. E., Wieczorek, L., Altendorf, K., Reicin, A. S., Dorus, E., and Epstein, W. (1984). *Proc. Natl. Acad. Sci.* (USA), **81**, 4746–4750.

18. Hopkins, B. E., Wagner, H., Jr., and Smith, T. W. (1976). *J. Biol. Chem.*, **251**, 4365–4371.

19. Jørgensen, P. L. (1982). In Martonosi, A. N., Ed., *Membranes and Transport*, Vol. I, Plenum, New York, pp. 537–546.

20. Jørgensen, P. L. and Karlish, S. J. D. (1980). *Biochim. Biophys. Acta*, **597**, 305–317.

21. Jørgensen, P. L., Karlish, S. J. D., and Gitler, C. (1982). *J. Biol. Chem.*, **257**, 7435–7442.

22. Jørgensen, P. L., Skriver, E., Hebert, H., and Maunsbach, A. B. (1982). *Ann. N. Y. Acad. Sci.*, **402**, 207–225.

23. Kawakami, K., et al. (1985). *Nature*, **316**, 733–736.

24. Kirley, T. L., Wallick, E. T., and Lane, L. K. (1984). *Biochem. Biophys. Res. Commun.*, **125**, 767–773.

25. Kozak, M. (1984). *Nucleic Acids Res.*, **12**, 857–872.

26. Kyte, J. and Doolittle, R. F. (1982). *J. Molec. Biol.*, **157**, 105–132.

27. Macknight, A. D. C. and Leaf, A. (1977). *Physiol. Rev.*, **57**, 510–573.

28. MacLennan, D. H., Brandl, C. J., Korczak, B., and Green, N. M. (1985). *Nature*, **316**, 696–700.

29. Maniatis, T., Fritsch, E. F., and Sambrook, J. (1982). In *Molecular Cloning: A Laboratory Manual*, Cold Spring Harbor Laboratory, Cold Spring Harbor, NY.

30. Maxam, A. M. and Gilbert, W. (1980). *Methods Enzymol.*, **65**, 499–560.

31. Morohashi, M. and Kawamura, M. (1984). *J. Biol. Chem.*, **259**, 14928–14934.

32. Noda, M., et al. (1984). *Nature*, **312**, 121–127.

33. Okayama, H. and Berg, P. (1982). *Molec. Cell. Biol.*, **2**, 161–170.

34. Peacock, S. L., McIver, C. M., and Monahan, J. J. (1981). *Biochim. Biophys. Acta*, **655**, 243–250.

35. Proudfoot, N. J. and Brownlee, G. G. (1976). *Nature*, **263**, 211–214.

36. Repke, K. R. H. (1982). *Ann. N. Y. Acad. Sci.*, **402**, 272–286.

37. Rossi, B., Ponzio, G., Lazdunski, M., Goeldner, M., and Hirth, C. (1983). *Curr. Top. Membrane Transp.*, **19**, 271–276.

38. Schwartz, A., Whitmer, K., Grupp, G., Grupp, I., Adams, R. J., and Lee, S. W. (1982). *Ann. N. Y. Acad. Sci.*, **402**, 253–271.

39. Shull, G. E., Schwartz, A., and Lingrel, J. B. (1985). *Nature*, **316**, 691–695.

40. Skou, J. C. (1957). *Biochim. Biophys. Acta*, **23**, 394–401.

41. Skou, J. C. and Esman, M. (1980). *Biochim. Biophys. Acta*, **601**, 386–402.

42. Takai, T. et al. (1985). *Nature*, **315**, 761–764.

43. Ullrich, A., et. al. (1985). *Nature*, **313**, 756–761.

44. Ullrich, K. J. (1979). *Ann. Rev. Physiol.*, **41**, 181–195.

45. Vogel, F., Meyer, H. W., Grosse, R., and Repke, K. R. H. (1977). *Biochim. Biophys. Acta*, **470**, 497–502.

46. Walderhaug, M. O., Post, R. L., Saccomani, G., Leonard, R. T., and Briskin, D. P. (1985). *J. Biol. Chem.*, **260**, 3852–3859.

47. Wolf, S. F. and Migeon, B. R. (1985). *Nature*, **314**, 467–469.

48. Young, R. A. and Davis, R. W. (1983). *Proc. Natl. Acad. Sci.* (USA), **80**, 1194–1198.

INDEX

UNIVERSITY COLLEGE CARDIFF LIBRARY